Springer Series in Vision Research

Series Editors

For further volumes:
http://www.springer.com/series/10633

About the Series

The Springer Series in Vision Research is a comprehensive update and overview of cutting edge vision research, exploring, in depth, current breakthroughs at a conceptual level. It details the whole visual system, from molecular processes to anatomy, physiology, and behavior; and covers both invertebrate and vertebrate organisms from terrestrial and aquatic habitats. Each book in the Series is aimed at all individuals with interests in vision including advanced graduate students, post-doctoral researchers, established vision scientists and clinical investigators. The series editors are N. Justin Marshall, Queensland Brain Institute, The University of Queensland, Australia and Shaun P. Collin, Neuroecology Group within the School of Animal Biology and the Oceans Institute at the University of Western Australia.

Gianluca Tosini • P. Michael Iuvone
Douglas G. McMahon • Shaun P. Collin
Editors

The Retina and Circadian Rhythms

Springer

Editors
Gianluca Tosini
Morehouse School of Medicine
Atlanta, GA, USA

P. Michael Iuvone
Emory University
Atlanta, GA, USA

Douglas G. McMahon
Vanderbilt University
Nashville, TN, USA

Shaun P. Collin
The University of Western Australia
Crawley, WA, Australia

ISBN 978-1-4614-9612-0 ISBN 978-1-4614-9613-7 (eBook)
DOI 10.1007/978-1-4614-9613-7
Springer New York Heidelberg Dordrecht London

Library of Congress Control Number: 2013957057

Printed on acid-free paper

Springer is part of Springer Science+Business Media (www.springer.com)

Contents

Contributors

Annette E. Allen, B.Sc., Ph.D. Faculty of Life Sciences, University of Manchester, Manchester, UK

Gene D. Block, Ph.D. Laboratory of Circadian and Sleep Medicine, Department of Psychiatry and Biobehavioral Sciences, David Geffen School of Medicine, University of California, Los Angeles, CA, USA

Morven A. Cameron, B.Sc., Ph.D. School of Medicine, University of Western Sydney, Campbelltown, NSW, Australia

M.A. Chrenek, B.Sc. Department of Ophthalmology, Emory University, Atlanta, GA, USA

Shaun P. Collin, B.Sc., M.Sc., Ph.D. The University of Western Australia, Animal Biology and UWA Oceans Institute, Crawley, WA, Australia

Christopher S. Colwell, Ph.D. Laboratory of Circadian and Sleep Medicine, Department of Psychiatry and Biobehavioral Sciences, David Geffen School of Medicine, University of California, Los Angeles, CA, USA

Hester C. van Diepen, M.Sc. Laboratory for Neurophysiology, Department of Molecular Cell Biology, Leiden University Medical Center, Leiden, The Netherlands

John E. Dowling, Ph.D. Department of Molecular and Cellular Biology, Harvard University, Cambridge, MA, USA

Farida Emran, Ph.D. Centre for Research in Neuroscience, Montreal General Hospital, Montreal, QC, Canada

Silvia C. Finnemann, Ph.D. Department of Biological Sciences, Center for Cancer, Genetic Diseases and Gene Regulation, Fordham University, Bronx, NY, USA

Russell G. Foster, B.Sc., Ph.D., F.R.S. Nuffield Laboratory of Ophthalmology, Nuffield Department of Clinical Neurosciences, Levels 5-6 West Wing, University of Oxford, John Radcliffe Hospital, Headley Way, Oxford, UK

Erik D. Herzog, Ph.D. Department of Biology, Washington University in St. Louis, St. Louis, MO, USA

P. Michael Iuvone, Ph.D. Departments of Ophthalmology and Pharmacology, Emory University, Atlanta, GA, USA

Robert J. Lucas, B.Sc., Ph.D. Faculty of Life Sciences, University of Manchester, Manchester, UK

Douglas G. McMahon, Ph.D. Department of Biological Sciences, Vanderbilt University, Nashville, TN, USA

Johanna H. Meijer, Ph.D. Laboratory for Neurophysiology, Leiden University Medical Center, Leiden, The Netherlands

Daniel T. Organisciak, Ph.D. Petticrew Research Laboratory, Department of Biochemistry and Molecular Biology, Boonshoft School of Medicine, Wright State University, Dayton, OH, USA

Vincenzo Parisi, M.D. GB Bietti Foundation for Ophthalmology, Via Livenza 3, Rome, Italy

Christopher L. Passaglia, Ph.D. Department of Chemical & Biomedical Engineering, University of Southern Florida, Tampa, FL, USA

M.L. Patterson, B.Sc., Ph.D. Department of Biological Sciences, University of Alberta, CW405 Biological Sciences Bldg, Edmonton, AB, Canada

S.N. Peirson, B.Sc., Ph.D. Nuffield Laboratory of Ophthalmology, Nuffield Department of Clinical Neurosciences, Levels 5-6 West Wing, University of Oxford, John Radcliffe Hospital, Headley Way, Oxford, UK

Linda Ruggiero, Ph.D. Department of Biological Sciences, Center for Cancer, Genetic Diseases and Gene Regulation, Fordham University, Bronx, NY, USA

Enrica Strettoi, Ph.D. Neuroscience Institute, Italian National Research Council, Pisa, Italy

Gianluca Tosini, Ph.D. Department of Pharmacology, Morehouse School of Medicine, Atlanta, GA, USA

P. Wong, Ph.D. Department of Ophthalmology, Emory University, Atlanta, GA, USA

A.C. Ziesel, B.Sc. Department of Ophthalmology, Emory University, Atlanta, GA, USA

Chapter 1
Introduction

Gianluca Tosini, Douglas G. McMahon, and P. Michael Iuvone

Daily rhythms are a ubiquitous feature of living systems. Generally, these rhythms are not just passive consequences of cyclic fluctuations in the environment, but instead originate within the organism. In mammals, including humans, the master pacemaker controlling 24-h rhythms is localized in the suprachiasmatic nuclei (SCN) of the hypothalamus. This circadian clock is responsible for the temporal organization of a wide variety of functions, ranging from sleep and food intake to physiological measures such as body temperature, heart rate, and hormone release. Moreover, accumulating evidence suggests that dysfunction of the circadian rhythms due to genetic mutations or environmental factors (i.e., jet lag or shift work) contributes to the development of many pathologies, including sleep disorders, mood and affective disorders such as major depression, bipolar disorder, and schizophrenia, as well as the risk of metabolic and cardiovascular disorders.

The retina plays a critical role in the organization of the circadian system by synchronizing the brain's central clock with the external day through transduction of the daily light/dark cycle. However, the substantial variation in luminance imposed on the retina between day and night also poses a challenge to its role as a sensory tissue—how is it possible to faithfully encode the enormous dynamic range of luminance that can exceed ten orders of magnitude? In this regard, the retina has

G. Tosini, Ph.D. (✉)
Morehouse School of Medicine, 720 Westview Drive SW,
Atlanta, GA 30310, USA
e-mail: gtosini@msm.edu

D.G. McMahon, Ph.D.
Department of Biological Sciences, Vanderbilt University, VU Station B, Box 1634,
Nashville, TN 37235, USA
e-mail: douglas.g.mcmahon@vanderbilt.edu

P.M. Iuvone, Ph.D.
Departments of Ophthalmology and Pharmacology, Emory University, Atlanta, GA, USA

G. Tosini et al. (eds.), *The Retina and Circadian Rhythms*, Springer Series in Vision Research 1, DOI 10.1007/978-1-4614-9613-7_1,

evolved not only sophisticated mechanisms of light and dark adaptation, but its own local circadian clock to allow anticipation of the regular cycle of the solar day.

Indeed, the retinal circadian clock was the first extra-SCN circadian oscillator to be discovered in mammals, and several studies have now demonstrated that many of the physiological, cellular, and molecular rhythms that are present within the retina are under the control of a retinal circadian clock or, more likely, a network of hierarchically organized circadian clocks that are present within the retina, where the network of circadian clocks regulates several physiologically relevant functions.

In this book, we summarize the knowledge accumulated over the last 30 years about the organization of the retinal circadian clock in many different species, concentrating on the roles that this circadian system plays in retinal function. The first chapter will provide the reader with a general description of the mammalian retina and its circuitry. The second chapter will describe how light information is perceived by the classical and nonclassical photoreceptors and is used to entrain the circadian clock within the retina and in the brain. Chapters 3–5 will be dedicated to the description of the circadian organization of the mammalian retina, the contribution of melatonin and dopamine to the regulation of ocular circadian rhythms and on the regulation of circadian rhythmicity in the retinal pigment epithelium. Chapters 6 and 7 will focus on how the retinal circadian clock modulates visual processing and the contribution of the retinal circadian clock to light-induced retinal damage. Finally, Chaps. 8–10 will describe the circadian organization in the zebrafish retina and in the eyes of two species of invertebrates (e.g., *Bulla* and *Limulus*).

In conclusion, the main purpose of this book is to promote and encourage research in retinal circadian rhythms by reaching out to vision scientists and interested biologists so that they will consider this an important system in the design of their research studies.

Chapter 2
Fundamental Retinal Circuitry for Circadian Rhythms

Enrica Strettoi and Vincenzo Parisi

Abstract A remarkable piece of tissue, the retina is a true outpost of the brain, peripheral only for its location on the back of the eye. Downstream of the photoreceptors, the specialized cells which transduce light energy into electric signals then conveyed to the brain by the optic nerve, approximately 60 types of neurons belonging to five classes are arranged in a sophisticated architecture and provide the substrate for extracting information pertinent to contrast, position in space, intensity, chromatic content, and movement. Light reaching photoreceptors and other photosensitive retinal neurons is also coded as temporal information pertinent to the alternation of night and day and to seasonal changes. This information is transmitted to a central clock located in the brain, which tunes biological rhythms to environmental light–dark cycles. Thus, a single sensory organ, the retina, informs the brain of light changes functional to vision, as well as to variations of light occurring in time, providing the core information for the existence of circadian rhythms. Correspondingly, this chapter summarizes fundamental features of retinal organization providing an overview of the main principles according to which the mammalian retina is built and operates as an organ of the visual system. The focus is, however, on retinal neuronal types and circuits forming the substrate for the establishment and function of circadian rhythms. Indications are given for appreciating the elaborate architecture of the whole retinal neurome and the likely existence of retinal channels deputed to code features of the visual scene of so far unsuspected complexity.

E. Strettoi, Ph.D. (✉)
Department of Pharmacology, Neuroscience Institute, Italian National Research Council (CNR), Area della Ricerca CNR, 56124 Pisa, Italy
e-mail: enrica.strettoi@in.cnr.it

V. Parisi, M.D.
GB Bietti Foundation for Ophthalmology, Via Livenza 3, 00198 Rome, Italy
e-mail: vmparisi@gmail.com

G. Tosini et al. (eds.), *The Retina and Circadian Rhythms*, Springer Series in Vision Research 1, DOI 10.1007/978-1-4614-9613-7_2,

Keywords Rods and cones • Bipolar cells • Horizontal cells • Amacrine cells • Ganglion cells • ON and OFF channels • Rod pathway • Cone pathway • Parallel processing • Gap junctions • Dopaminergic amacrines • Melanopsin

2.1 Introduction

The retina is essentially a piece of the brain with a peripheral location. For its embryological origin (the neuroepithelium of the optic cup, an evagination of the diencephalon) and structural organization, comprising neurons and glia, the retina is part of the CNS. Functionally, it is much more complex than a plain relay station or than the simple photographic camera to which it is often compared. Rather, photoreceptors, the specialized sensors, transmit light-initiated signals to a cohort of about 60 types of neurons, arranged in an exquisitely ordered architecture, altogether extracting from light information pertinent to intensity, chromatic content, contrast, movement, and position. This bulk of data is then conveyed to the visual centers of the brain by the optic nerve using a complex (and partially undeciphered) code. Noticeably, the light reaching photoreceptors and other photosensitive retinal neurons is also coded as temporal information pertinent to the alternation of night and day and to seasonal changes. This information is transmitted to a central clock located in the brain that tunes various, and even distant, biological rhythms (body temperature, heart rate, fertility, etc.) to environmental light–dark cycles.

This book deals principally with circadian rhythms, a direct consequence of the ability of the retina to inform the brain about changes of light in time [1]; hence, this chapter will focus on retinal neuronal types directly involved in such a functional task. Also, it will illustrate fundamental principles of organization typical of the retina of mammals only and provide a necessarily simplified information, which is instrumental to the following chapters. Accordingly, most of the references quoted consist of review articles meant to address the reader toward fundamental literature, condensing decades of studies of many laboratories in the field of retinal organization.

2.2 Classes and Types of Retinal Neurons

The mammalian retina is only about 200 μm thick: light has to traverse it completely and with minimal distortions to reach the photosensitive elements placed on the opposite side of the leaflet. Within this minimum width, stereotyped networks of neurons (functional units) are repeated in an orderly fashion to ensure total coverage of the retinal surface and, correspondingly, adequate sampling of the outside world (Fig. 2.1).

Retinal neurons belong to five classes only: photoreceptors (rods and cones); horizontal cells (occurring in two main types); bipolar cells (further divided into rod

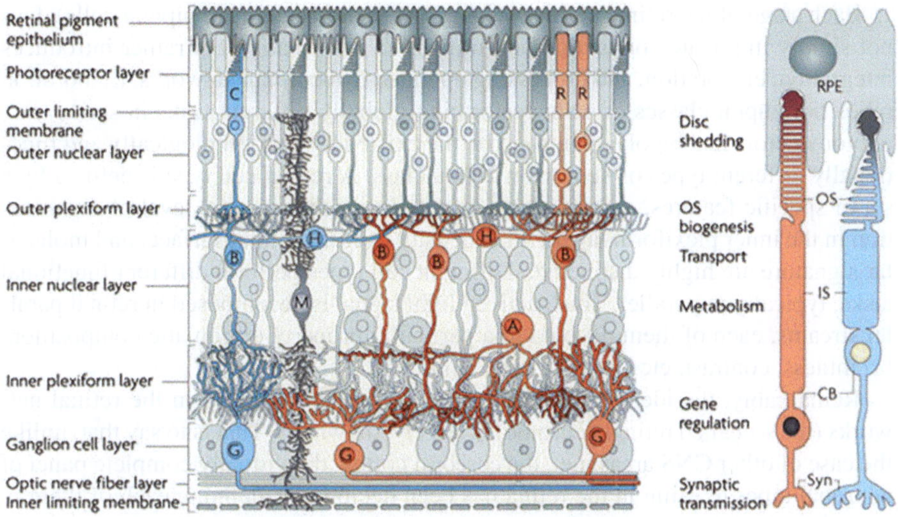

Fig. 2.1 Retinal organization. *Left side*: Rod (R) and cone (C) photoreceptors have cell bodies in the outer nuclear layer. Rod and cone inner segments (IS), rich in mitochondria and other organelles, and outer segments (OS) lay beyond the outer limiting membrane. Outer segments penetrate the microvilli of the retinal pigment epithelium (RPE). Photoreceptor synaptic terminals reside in the outer plexiform layer where synapses with horizontal (H) and bipolar (B) cells are established. The inner nuclear layer contains the cell bodies of bipolar, horizontal, and amacrine (A) cells, as well as those of Müller glial (M) cells. In the inner plexiform layer, bipolar cells establish synapses with amacrine and ganglion (G) cells: these send their axons to the optic nerve that carries signals to the brain. Processes of Müller cells form the outer limiting membrane, while their end-feet form the inner limiting membrane. A representative cone pathway is shown in *blue* and a representative rod pathway is shown in *orange*. *Right side*: Illustrations of rod and cone morphologies that include subcellular locations of functions (Modified with permission from [52])

bipolars, existing as a single type, and cone bipolars, occurring in a dozen different types); amacrine cells (approximately 30 types); and, finally, ganglion cells, which are believed to occur in some 20 distinct varieties [2]. Different classes of retinal neurons subserve globally different functions according to a hierarchical arrangement: upon light absorption and subsequent excitation of rods and cones, the visual signal has to travel through a cascade of other neurons (first bipolar and horizontal cells, then amacrine cells), to reach the final retinal station, represented by ganglion cells, which, with their axons, form the optic nerve and transport the signal to the visual centers of the brain. Traditionally, photoreceptors and bipolar and ganglion cells are assigned to the vertical, retinofugal pathway, while horizontal and amacrine cells are considered as modulatory interneurons with a horizontal arrangement. Exceptions apply to this rule: amacrine cells occur in two broad categories, i.e., wide-field and small-field neurons. While the wide-field amacrine cells have a large lateral spread and perform what is called "lateral integration," some small-field amacrines are true components of the vertical retinal pathway, as explained later for the case of AII cells.

Each stage of the retinal pathway (i.e., from photoreceptors to bipolar cells) does not simply implicate conduction and transfer of information but rather introduces integration, elaboration, and processing operated by neuronal networks. Elaboration relies both upon classes of cells performing globally different tasks in series (i.e., photoreceptors and bipolar cells) and on the existence of morphologically and functionally different types of neurons within a class. A retinal cell type is defined by a set of specific features, among which morphology (shape), size, level of stratification in the inner plexiform layer (IPL), density over the retinal surface, and molecular signature are highly distinctive. Different cell types perform different functional tasks, typically in parallel. Thus, a given light signal is decomposed in retinal parallel streams, each of them devoted to a particular property (chromatic composition, brightness, contrast, etc.) of the visual stimulus.

Remarkably, the identity of the neuronal types participating in the retinal networks (the so-called retinal neurome) is largely known [2]. That is to say that, unlike the case of other CNS areas, i.e., the cerebral cortex, the virtually complete panel of neuronal types existing in the retina has been deciphered, defining each of them in terms of stringent morphological and topographic properties as well as according to broad functional characteristics. However, data about exact physiological roles of known morphological types are still missing. The disparity of available morphological and physiological information is especially evident for amacrine and ganglion cells, which occur in many types. Even the code used to deliver information toward the brain through the optic nerve is partially obscure and the variety of visual properties embedded in a scene and encoded in ganglion cell signals more complex than previously believed [3].

2.3 Photoreceptors: Cones and Rods

Most mammalian retinas possess two or three types of cones and one type of rods distinguishable by their slightly different size and shape but equally perfectly aligned as it can be appreciated in a retinal vertical section (Figs. 2.2 and 2.3).

Cones are larger than rods, display an elliptical structure, and have their cell bodies (containing the nucleus and a small rim of cytoplasm) aligned in a row immediately below the outer limiting membrane (OLM). Their outermost portion is composed of an inner segment, mostly comprised of endoplasmic reticulum, Golgi apparatus, and a rich complement of mitochondria, in addition to a tapered outer segment. This is a modified cilium consisting of regularly tiled and tightly packed membranes containing the visual pigments (cone opsins). Inner and outer segments of cones project into the subretinal space toward the retinal pigment epithelium. Cones are specialized for vision in bright light (photopic vision), have low sensitivity, and also carry chromatic information. Their response to light is rapid.

Rods are more slender than cones and have a cylindrical (rod-like) shape. They also have thin inner and outer segments located beyond the OLM and the tips of their outer segments are equally surrounded by the apical processes of the retinal

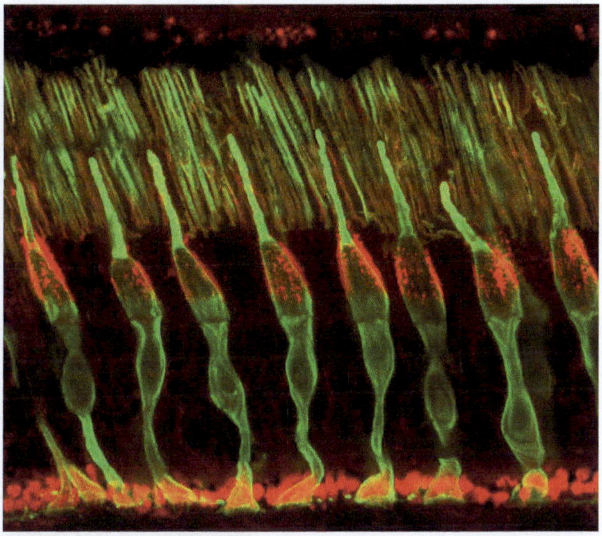

Fig. 2.2 Vertical section from a monkey retina stained with photoreceptor-specific antibodies revealing the detailed morphology of these cells. Cones (*green*, elongated cells) have *red*, punctate inner segments. Slender rods are visible in the background as *vertical, green lines*, corresponding to their outer segments. Synaptic terminals of rods appear as *bright red spherules* (Picture by Nicholas Cuenca, reproduced with permission from http://www.vision-research.eu/index.php?id=471&no_cache=1&sword_list[]=cuenca)

pigment epithelial cells. While cone cell bodies (mostly occupied by their nuclei) form a single row in the outer nuclear layer (ONL), rod nuclei are distributed in multiple layers, whose number changes in different species and is typically quite high (10–12 rows) in the small eye of a mouse. Rods have extremely high sensitivity and are specialized for vision at low light levels (scotopic vision); they actually detect single quanta of light, with an efficiency similar to the most sensitive of physical instruments. However, rods respond to light stimulation much more slowly than cones.

The morphological similarity of rods and cones is maintained at the ultrastructural level, which allows appreciation of the extremely specialized organization of these small neurons in spatial terms. Outer segments are comprised of stacks of regularly arranged membranes containing the highest density of proteins found in mammals. Besides photosensitive elements (rhodopsin for rods and cone opsins for cones), the various proteins constituting and controlling the phototransduction cascade are located here.

A noticeable difference between rods and cones is that in rods, the outer segments are separated from the cell membrane (they are called disks) (Fig. 2.4). In cones, they are rather invaginations of the plasma membrane itself. Inner segments of both rods and cones are rich in organelles and are very similar. Nuclei are different as the chromatin of rods is typically more condensed; this feature is often used to distinguish rods and cones by simple DNA staining methods. Another important

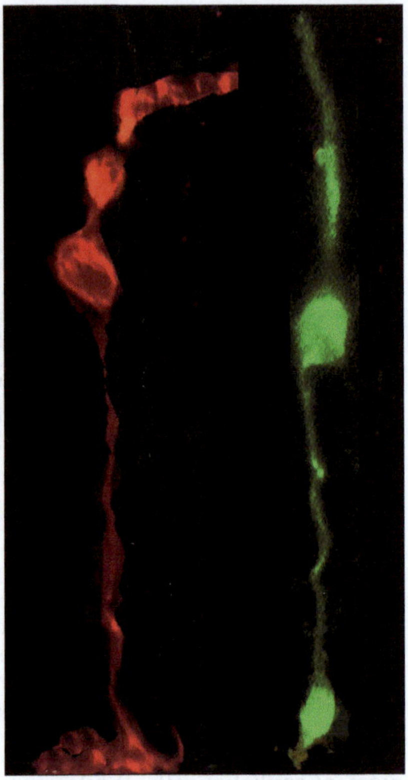

Fig. 2.3 Montage of photomicrographs of cone (*red*) and rod (*green*) photoreceptors from the mouse retina, stained by gun delivery of fluorescent nanoparticles [53]. The different size of their synaptic endings, at the bottom of the picture, is clearly visible

element of distinction is given by the size and shape of synaptic endings: these are simple ball-shaped terminals, called spherules, in rods, and larger, more complex, pyramid-shaped endings (pedicles) in cones (Fig. 2.5). In both cases, the main constituents of the ending are synaptic vesicles, filled with glutamate. However, there is dissimilarity in synaptic connectivity: a rod spherule typically establishes one or two synaptic connections with the dendrites of its dedicated second order neurons (rod bipolar and horizontal cells), while a cone makes numerous connections with second order neurons (cone bipolar and horizontal cells).

The cone ending is considered one of the most elaborate synaptic complexes of the whole CNS: in the human retina it can engage as many as 500 synaptic connections [4]. The reason of this diversity will be explained later.

Rods and cones are joined by tiny gap junctions made out of connexin 36. These contribute to the transfer of visual signals from the scotopic to the photopic channels and might be also implicated in what is known as the bystander effect. This is the complex of non-cell-autonomous events occurring in a cell type due to the effect of factors deriving from nearby cells. In the case of photoreceptors, this effect is believed to play a role when a mutation occurring in the rods, and consequently causing their death, leads to the secondary degeneration of nearby cones, such as happens in retinitis pigmentosa, a severe disease leading to near blindness [5, 6].

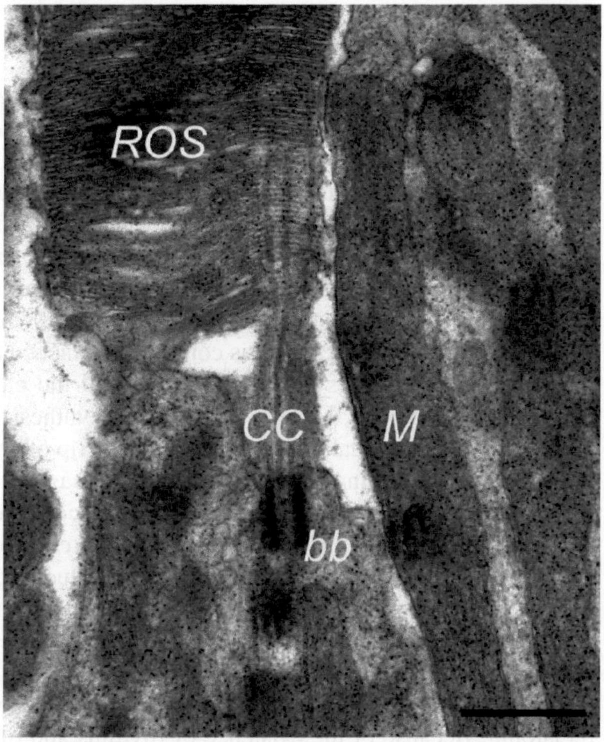

Fig. 2.4 Electron micrograph of photoreceptor outer and inner segments. The regularly stacked disks of the rod outer segment (ROS) and the inner segment, rich in mitochondria (M), are shown. *CC* connecting cilium, *bb* basal body, Bar is 0.5 μm

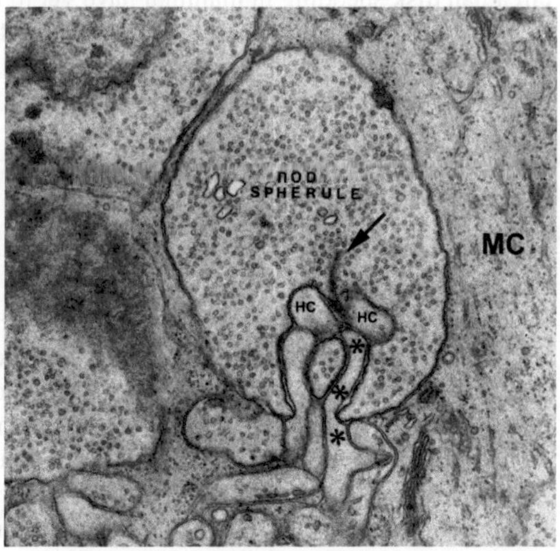

Fig. 2.5 Electron micrograph of the outer plexiform layer of the rabbit retina, showing the ultrastructure of a rod spherule, filled with synaptic vesicles. The presynaptic site is marked by a characteristic ribbon (arrow); postsynaptic processes are arranged in a triad, where the central element is the terminal dendrite of a rod bipolar cell (*asterisks*) and the two lateral elements belong to horizontal cells (HC). The large process of a Müller cell (MC) is also visible (Modified with permission from [54])

2.4 Outer Segment Renewal

Outer segments of rods and cones, containing the visual pigments and thus constituting the photosensitive part of the cell, are membranous protrusions evaginating from the base, near the photoreceptor cilium. Here, the opsin protein, synthesized in the inner segments and modified in the Golgi apparatus, becomes incorporated in the outer membrane by means of specialized areas and mechanisms of fusion [7]. The protein is synthesized by the photoreceptor itself, and the vitamin moiety of the photopigment is provided by the pigment epithelium. Binding of the two components takes place at the base of the outer segments. While the base is actively growing, the tip of the outer segment is constantly detached and eliminated: the pigment epithelium phagocytoses the apical portions of the outer segments that are being continuously renewed. Molecules of newly synthesized opsins can be followed while they move gradually from the base to the tip of an outer segment until the latter is "shed" [8]. Remnants of outer segments phagocytosed by the pigment epithelium are degraded by lysis. Shedding takes place regularly, so that a whole outer segment is completely renewed in a time interval of approximately 10 days. Moreover, the process of shedding is deeply influenced by external factors and by light in first instance. In amphibians, disk shedding in rods is activated by melatonin, which is produced by photoreceptors at night. Melatonin production is inhibited by light and dopamine. Conversely, dopamine (that is synthesized in neurons of the inner retina) is stimulated by light and inhibited by dark and the presence of melatonin. Because of the existence of rhythms and of opposite actions, rod outer segment disks are shed at light onset (in the morning), while cone outer segments are eliminated at the onset of darkness (at dusk) [9, 10]. The mechanisms regulating circadian disk shedding in mammal have yet to be clarified. For more details about disk shedding rhythms and their regulation by dopamine and melatonin, the reader is invited to refer to the Chap. "Role of Melatonin and Dopamine in the Regulation of Retinal Circadian Rhythms".

2.5 Horizontal Cells

Horizontal cells (HCs) are laterally interconnecting neurons with large cell bodies located in the outermost region of the inner nuclear layer (INL) and processes restricted to the outer plexiform layer (OPL). These neurons occupy a strategic position as they control signal processing in the outer retina and are themselves under neuromodulatory control from the retina and from the brain [11].

Most mammalian retinas have two types of horizontal cells, which can be named A- and B-types. Remarkably, the mouse retina has only one type of HC, morphologically similar to the B-type HC of other species. A-type HCs (Fig. 2.6) are large neurons with stout primary dendrites emerging directly from the cell body with a radial orientation giving rise to small, vertical terminals reaching the synaptic base of the photoreceptors. The B-type HC exhibits a smaller, bushy arborization with

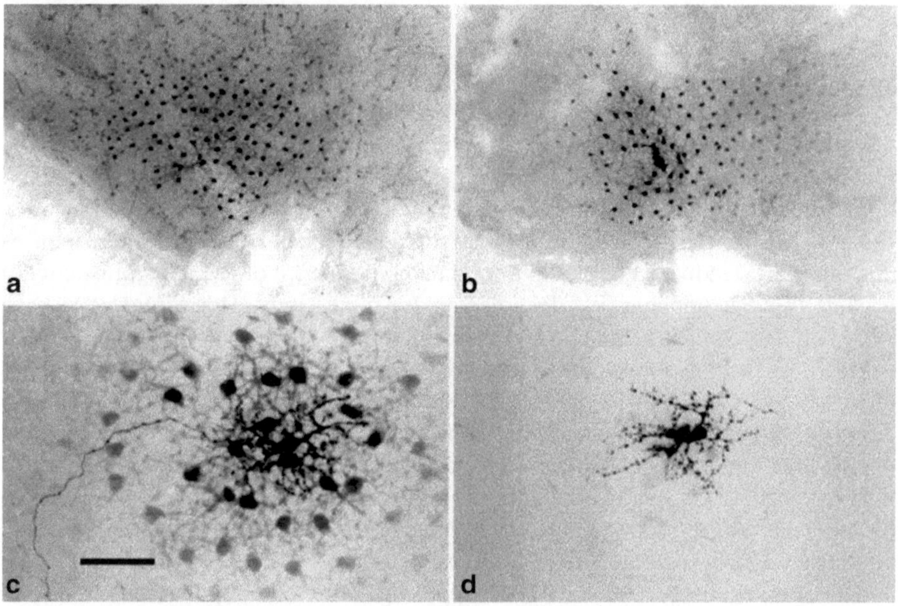

Fig. 2.6 Gap junctional coupling between horizontal cells in the mouse retina and regulation by Dopamine. (**a**) Neurobiotin tracer coupling in basal conditions. (**b**) Application of dopamine (100 μM) greatly reduces coupling. (**c**) Application of a pharmacological D_1 antagonist increases coupling. (**d**) Conversely, application of analogs of cAMP reduces the extent of coupling, as expected for a D_1 mechanism. Scale bar is 50 μm (Reproduced with permission from [55])

radial orientation, as well as a long, thin axon with horizontal course in the OPL. This atypical axon gives rise to a wide axonal ending, usually larger than the dendritic tree, with numerous terminals in the OPL. Because such axons are very long and thin, passive electrotonic spread of signals from one side of the cell to the other is negligible, and the two parts of the same cell behave independently. The dendrites of both A-type and B-type HCs are postsynaptic to cone pedicles, where they occupy the lateral elements of ribbon synapses. The axon terminals of the B-type HCs end in rod spherules, also constituting the lateral elements of ribbon synapses (Fig. 2.5).

Horizontal cells receive excitatory input directly from photoreceptors via chemical synapses mediated by glutamate acting on AMPA receptors and respond to light with graded hyperpolarization. They use GABA as a neurotransmitter but are considered unconventional GABAergic neurons for their unusual morphology and mode of transmitter release, which is not vesicular for the major part. An important feature of horizontal cells is that cells of the same type are connected by large gap junctions, the largest in the retina, through which they become excited by neighboring (homologous) neurons. For this reason, their receptive field is very wide, well beyond the extent of direct synaptic contact with photoreceptors.

The extent of HC coupling (as for other gap junction-connected neurons) can be visualized by intracellular injections of dyes that cross the gap junctions, thus revealing their morphological and functional syncytium across the whole OPL

(Fig. 2.6) [12]. Gap junctions might be constituted by connexin 57 or by connexin 50, where the diversity in their molecular composition contributes to the different properties of the various HC types.

A specific function of HCs is to send visual information back to cones by means of feedback connections established by dendrites penetrating cone pedicles. The functional effects of HC feedback onto cones are measurable even though conventional chemical synapses from HCs onto photoreceptors are observed only in a few species [13]. Responses of cones to light are antagonized from opposite responses due to the HC feedback effect. An important consequence of HC antagonistic feedback is "spatial opponency," documented for cones only. Horizontal cells have extremely large receptive fields, and their influences on photoreceptors and bipolar cells, whose receptive fields are much narrower, can be observed in the far periphery of these narrow-field cells. Wide light stimuli exert a depolarizing (antagonistic) effect on cones as an effect of horizontal cell feedback; such a depolarization opposes the direct hyperpolarization produced by the direct excitation of light falling onto cones. This spatial opponency is retained in the transmission from cones to bipolar cells to ganglion cells and contributes to the fundamental center-surround organization of the receptive fields of these neurons.

Horizontal cells also exert a direct, opposing action onto bipolar cells, reinforcing the effect of feedback onto cones. This feedforward action is particularly clear in OFF cone bipolars, which undergo delayed depolarization as an effect of wide-field stimuli affecting HCs. Thus, the "center-surround" spatial organization of bipolar cells is shaped due to the contribution of HCs.

The functional properties of horizontal cells can be modified as an effect of various substances released by retinal neurons as a function of changes in conditions of illumination. These substances (neuromodulators) have the important role of matching retinal physiology to ambient conditions. Dopamine is the best known neuromodulator and is released by dopaminergic amacrines (see below), which exerts multiple effects on horizontal cells acting on D1 and D2 receptors. The most relevant of these effects is a decrease in the extent of gap junction coupling and a consequent reduction of the receptive field size of HCs. This, in turn, is reflected in their changing outcome onto bipolar cells and onto cones through feedback action. In summary, horizontal cells contribute to the organization of spatially opponent receptive fields of bipolar cells and modulate the photoreceptor signal with different lighting conditions also adjusting what is known as the synaptic gain in the outer retina. Further, the horizontal cell itself is under the control of dopamine, which deeply affects its receptive field properties according to light conditions [11].

2.6 Bipolar Cells and ON and OFF Channels

In mammals, cones and rods converge upon separate sets of bipolar cells, correspondingly named cone and rod bipolars [2] (Fig. 2.7). All bipolar cells, like photoreceptors, use glutamate as a neurotransmitter and possess the adequate machinery for transport and release of this amino acid.

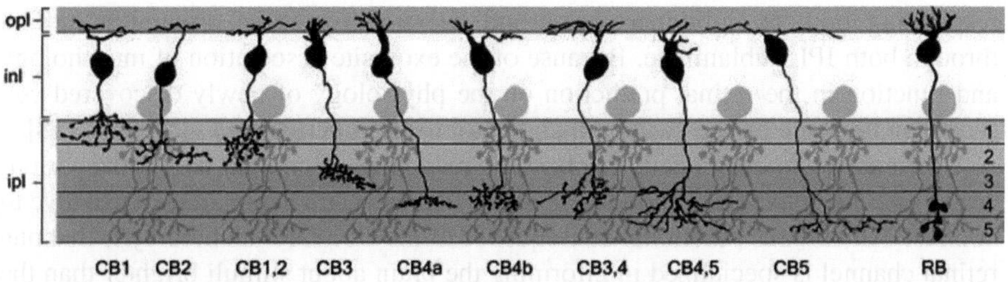

Fig. 2.7 Drawing of the types of mouse bipolar cells identified by means of gun delivery of fluorescent nanoparticles. One single type of rod bipolar cell (RBC) and 9 types of cone bipolar cells are illustrated (Reproduced with permission from [17])

Cone bipolar cells are now known to occur in a dozen different types [14, 15]. Each cone pedicle establishes multiple contacts in the OPL, making at least one synapse with each type of cone bipolar cell and all the bipolars spanning its territory. In the primate retina, cone pedicles have a highly complex structure and are engaged in up to 500 synaptic connections [4]. This arrangement splits signals generated in cone-mediated illumination conditions (the photopic range) into parallel streams, each extracting a specific parameter from the stimulus and, in turn, informing ganglion cells about this parameter.

The first known set of parallel channels generated at the cone-to-cone bipolar synapse is the dichotomous separation of the signal into ON and OFF channels. Actually, roughly one half of the existing types of cone bipolar cells belong to the so-called ON-center type, for they respond with a graded depolarization to light falling in the center of their receptive field. Since cones respond to light with graded hyperpolarizations, the synapse between cones and ON-center cone bipolars is called sign inverting. Inversion is achieved by the interaction of glutamate released by cones with mGluR6, a distinct type of metabotropic glutamate receptor, almost exclusively located in the retina on the dendritic tips of ON bipolar cells [16]. Although ON cone bipolars occur in different types (5–6 according to various classifications) [14, 17], with peculiar morphological and molecular signatures, all their terminal axonal arborizations remain confined to the innermost two thirds of the IPL, near the ganglion cell bodies. This is the so-called sublamina ON or, in anatomical terms, sublamina b. The other half of the cone bipolar cell population belongs instead to the OFF-center variety. These cells carry ionotropic glutamate receptors that maintain the cone-to-cone bipolar synapse sign conserving. Thus, as photoreceptors, these cells respond to light falling in the center of their receptive fields with graded hyperpolarizations; accordingly, their axonal arbors end in the outer part of the IPL, also called sublamina OFF or sublamina a. Fundamental rules of retinal organization are that (a) cone bipolars are presynaptic to ganglion cell dendrites in the corresponding (and spatially restricted) sublamina of the IPL and (b) cone bipolars establish sign-conserving synapses with ganglion cell dendrites. Ganglion cells also occur in ON and OFF functional types and their dendrites follow the rule of being stratified within the innermost and outermost portions of the IPL, respectively. This holds true for dendrites of amacrine cells as well; when cells with ON–OFF functional properties are found (i.e., amacrines and ganglion cells responding both at light increments and decrements), these are multistratified through both IPL sublaminae. Because of the exquisite association of morphology and function in the retina, prediction on the physiology of newly discovered cell types can be made on the basis of their anatomical properties and vice versa [18].

Noticeably, ON-center cells display OFF responses when light falls in the periphery of their receptive field and vice versa, so that each cell is maximally sensitive to contrast. An obvious functional consequence of the ON–OFF dichotomy is that one retinal channel is specialized in informing the brain about stimuli brighter than the background (the ON-center channel), while the second is better tuned for stimuli darker than the background (the OFF-center channel). This property contributes to the high sensitivity of the retina (and more generally of the visual system) to stimulus contrast, as first demonstrated by Kuffler from cat ganglion cells recordings [19].

ON and OFF functional differences among cone bipolar cells are established due to molecular diversity in glutamate receptors (i.e., metabotropic versus ionotropic). Glutamate receptors expressed by different types of cone bipolar cells vary in their inactivation kinetics, which might be rapid or slow. Corresponding cone bipolar types are transient or sustained, respectively [20, 21], and are tuned with phasic or tonic temporal properties of light stimuli. Also, certain types of cone bipolars are dedicated to process precise chromatic features of light signals [22]; in particular, an ancient retinal pathway has been described dedicated to blue light where a "blue cone bipolar" selectively contacts the short wavelength sensitive cones, thus ensuring that the chromatic information is not degraded while moving along the vertical retinal pathway [23]. In this particular channel, light is coded in terms of chromaticity, while other channels respond better to luminosity content. Additional elements of variety among bipolar cell types are found in the molecular composition of gap junctions in the IPL, which in turn determine their electrical properties and response to external control factors including calcium, pH, and dopamine [24, 25].

Our understanding of the functional abilities specific to each cone bipolar cell types is still limited. Bringing to light the full panel of molecular differences subserving specific functional prerogatives of these neurons represents one of the challenges of modern retinal research.

2.7 The Rod Pathway and the Piggyback Arrangement

Rod bipolar cells belong to a single morphological and functional type. Characterized by cells with cell bodies located in the outer part of the INL with a profuse dendritic arborization terminating in small tips, each of them receives a synaptic contact by a single rod terminal in the OPL. Although numerical variations in different species occur, the principle holds true that convergence in the rod pathway is high and that a large number of rods (i.e., from about 20 in a mouse to up to 80 in a rabbit) make connections with one postsynaptic bipolar cell. Thanks to this arrangement, summation of inputs at a postsynaptic level is achieved and sensitivity of the rod system

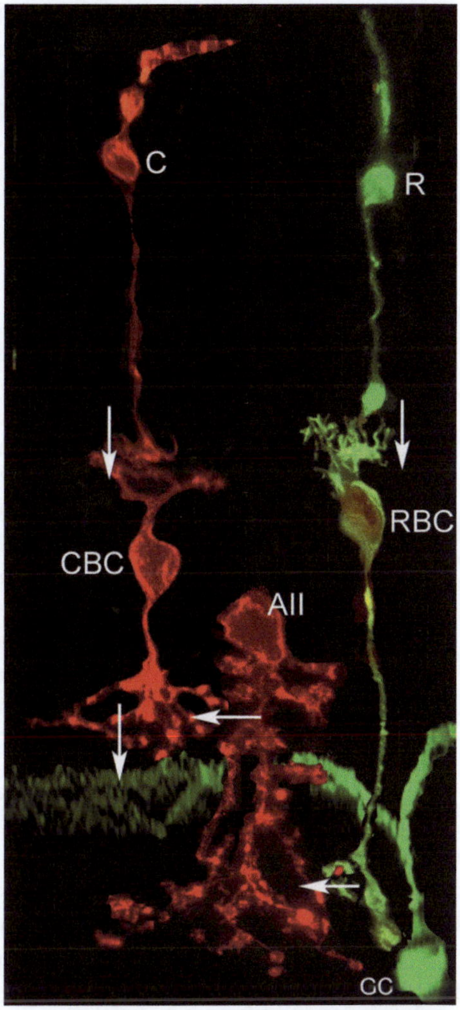

Fig. 2.8 Semi-schematic representation of the rod pathway. Images of neurons have been obtained from mouse retinal sections stained by delivery of lipophilic fluorescent dies, in the mouse retina. Rods (R) and cones (C) converge upon separate sets of bipolar cells. Rod bipolar (RB) cells are presynaptic s to AII amacrine cells, which, in turn, send the information to cone bipolar (CB) cells. Retinofugal connections with ganglion cells (GCs) are established by axonal endings of CBs making synapses onto GC dendrites (Reproduced with permission from [56])

increased. Extreme amplification in rods by the phototransduction cascade [26] and high convergence along all the steps of the rod pathway [27] result in enormous sensitivity of the retina to dim light: a change in cat ganglion cell electrophysiological response can be recorded even when only one of the thousands of rods within its receptive field captures a single photon [18]. In turn, each rod bipolar engages connections with all the rods within its reach. From the inner part of the rod bipolar cell body, a stout axon emerges, whose bulbous endings terminate in the deepest part of the IPL, near the cell bodies of ganglion cells. Here, synapses to and from third order neurons are established (Fig. 2.8). Functionally, rod bipolar cells have a center-surround organization of their receptive fields like cone bipolars. Their membrane potential undergoes a graded depolarization in response to stimulation of photoreceptors located in the central area of their receptive field, while a graded hyperpolarization is evoked by peripheral (annular) light stimulation. Hence, the synapses between rods and rod bipolars are sign inverting, exactly like ON-center cone bipolars, where inversion is achieved by the interaction of glutamate released by rods with mGluR6 receptors: rod bipolars are "ON-center" neurons showing a typical "OFF response" when the periphery of their circular receptive field is stimulated [4].

2.8 AII Amacrine Cells

The specialized chain of neurons carrying signals generated in rods across the retina to the ganglion cells is called the rod pathway and represents one of the most studied and best characterized networks of the mammalian retina. Here, a summary of this fundamental module of neuronal architecture is schematized in Fig. 2.8.

Axonal arborizations of rod bipolar cells occupy the deepest part of the ON sublamina of the IPL. Here, they establish multiple synaptic contacts with the dendrites of "dedicated" narrow-field amacrine cells called AII. Only exceptionally do the axonal arbors of rod bipolars make direct connections with ganglion cells. AII amacrines were first described in the cat [28] and are true hallmarks of the mammalian retina. These cells exhibit a typical bistratified morphology (Fig. 2.8). The outer dendritic arbor, restricted to the OFF sublamina of the IPL, consists of round, synaptic endings, known as lobular appendages, while the innermost arborization, spreading in the ON part of the layer, is comprised of thin, elongated dendrites, with tangentially oriented tips called "tufted processes." Such processes (a) receive multiple synapses from the axonal endings of rod bipolar cells, for which AII amacrines are the principal postsynaptic target; (b) engage homologous gap junctions with similar dendrites of nearby AIIs; and (c) form large, heterologous gap junctions with the axonal endings of most types of cone bipolar cells that ramify in the deepest layer of the IPL (ON cone bipolars) [29, 30]. AII amacrines respond to light with a graded depolarization similar (but more transient) to that generated in rod bipolar cells and, strictly speaking, they belong to the ON channel. The ON–OFF dichotomy generated in the cone pathway at the first cone-to-bipolar synaptic station becomes available to the rod pathway thanks to the bistratified morphology of AII amacrines. In fact, in the outermost part of the IPL, the lobular appendages of AII cells form sign-inverting, glycinergic synapses with the axonal arborizations of OFF cone bipolars, thus "feeding" the rod-generated signal in the OFF channel. Electrophysiological studies have shown that the heterologous gap junctions established in sublamina ON allow sign-conserving transfer of information between AII amacrines and cone bipolars terminating there. Transmission in the opposite direction occurs but it is less effective [31]. By means of this peculiar 5-neuron chain, the depolarizing response generated by light excitation of the central receptive field of a rod bipolar cell is transferred to AII amacrine cells, then split into the ON and the OFF retinal sublaminae and fed into the homologous channels of the cone pathway; output synapses of cone bipolar axonal arborizations onto ON and OFF ganglion cells dendrites in the corresponding sublaminae of the IPL are the gate through which the scotopic signal gains access to the retinal exit.

Although few direct synapses exist linking AII amacrine cells and dendrites of ganglion cells directly, physiology indicates that rarely ganglion cells are purely rod driven. Through AII amacrines, the bulk of the scotopic signal is injected directly into the cone pathway, in what has been named a "piggyback" array. AII amacrine cells, therefore, can be regarded as true elements of the "vertical" retinal pathway, even though usually cells of their class are considered as modulatory interneurons with a lateral arrangement.

More than simple conduits of signals, AII amacrine cells constitute nodal points of intersection of both the rod and cone systems and the ON and OFF channels. In the scotopic range, light excites rods, which, in turn, evokes a depolarization in rod bipolar cells, causing the release of glutamate from their axonal endings. Glutamate excites AII amacrines, causing simultaneously (1) the release of glycine onto OFF cone bipolar cells, translated into inhibition of OFF ganglion cells; (2) the depolarization of ON cone bipolar cells via gap junctions, coded as excitation of ON ganglion cells; and (3) the diffusion of the signal through the network of AIIs, through their homologous gap junctions. In summary, rod-generated signals produce inhibition of the OFF pathway and excitation of the ON pathway.

In photopic conditions, AII amacrine cells receive inputs from both ON (through gap junctions) and OFF cone bipolars (by means of chemical synapses), which also independently feed the corresponding ganglion cells. ON-center responses cause strong glycine release from AII lobular appendages, which, accordingly, inhibits OFF cone bipolar cells. Hence, in bright light the OFF channel is inhibited by the ON channel [24].

2.9 The Rod Pathway and Dopaminergic Amacrine Cells: The Power of Being Few

The piggyback arrangement does not implicate unidirectional information flow: first of all, cone bipolar cells make feedback synapses onto the lobular appendages of AII amacrines, thus creating the ground for transfer of data from the photopic to the scotopic channel. In addition, each AII cell receives a profuse innervation from dopaminergic amacrine (DA) cells. These are wide-field neurons, first described as A18 amacrines, occurring at low density (only 600 of them in a mouse retina, or 0.1 % of the total amacrine cell population in this species). However, these cells have very wide dendritic arbors with extensive ramifications ensuring efficient coverage of the retinal surface. Their dendrites are largely restricted in the outermost portion of the IPL (the sublamina 1), in close proximity to the amacrine cell bodies, and entangle a dense network of processes forming characteristic rings. These are essentially synaptic varicosities encircling the main dendrites of AII amacrines that therefore receive multiple contacts from dopaminergic endings [32]. In addition,

A18 amacrines give rise to long axon-like processes running in different strata of the IPL and forming a plexus in the OPL, for which they are also named interplexiform cells. A second type of dopaminergic cell (type 2) has been described using genetic labeling for catecholamines [33]. Electrophysiological evidence suggests that the two types respond to light in sustained and transient manners, respectively.

By means of their endings as well as their distal processes in the outer retina, dopaminergic amacrines can control their targets both through synaptic release of dopamine and via a paracrine discharge of the neurotransmitter [34]. Dopamine is a very important global regulator of retinal sensitivity to light and a powerful modulator of gap junction permeability. This transmitter controls multiple elements of the retinal circuitry; it alters the gap-junctional conductance between photoreceptors [35], horizontal cells, and amacrine cells; increases the responses of ionotropic glutamate receptors in bipolar cells; and ultimately affects the center-surround balance of ganglion cells. In scotopic and photopic conditions, and with the contribution of the control exerted by dopaminergic innervation, the AII amacrine cell operates as a switch from one input pathway to another, with high efficiency [36]. Because of a weaker coupling of gap junctions in photopic conditions, the dissipation of cone-initiated signals through the homologous network of AII amacrines is limited, so that this can reach ganglion cells more effectively. Dopaminergic control of gap junction permeability is, among others, an important component of the retinal circadian clock, as extensively described in the following chapters.

2.10 Possible Explanation and Advantages of the Piggyback Arrangement

Additional pathways exist in the retina through which light signals generated in rods can reach ganglion cells [37]. First of all, rods are linked by very small gap junctions, mainly established by telodendria, thin processes at the base of rod spherules and consisting of few connexons each. One of the advantages of these gap junctions is that noise arising in single rods (because of quantal fluctuation of neurotransmitter) is reduced by electrical coupling. Moreover, in some species rods are coupled electrically to cones, and in theory scotopic information can enter the cone system and bypass the rod bipolars. Actually, the functional effect of rod–cone gap junctions does not seem very high on ganglion cells, and their precise role is still debated. In several species, rods (albeit only a fraction) make direct connections with cone bipolar cells of the OFF pathway that are therefore called mixed bipolars. The effect on ganglion cell physiology varies among mammals.

Noticeably, all these pathways originate from rods but directly, or later along the retinal pathway, converge upon cone bipolars to access ganglion cells [2]. A possibility is that all the rod systems, and particularly the piggyback arrangement, originate evolutionarily from a common, ancestral pathway dedicated to photopic vision. It is known that rods and vision in dim light appear only after the evolution of the jawed vertebrates [38]. Hence, retinal network profiles were initially shaped by cones. Because of the creation of parallel channels built at the cone-to-bipolar synapses, it is possible to imagine an antique vertebrate retina comprised of cones, various types of cone bipolars, and cone-driven amacrine cells ultimately driving ganglion cells.

Thanks to the evolution of rhodopsin and its segregation into a new type of photoreceptor, the prototype of a rod, the retina became duplex. However, needless duplication of the whole inner circuitry was avoided by making preexisting pathways accessible to the rod system by means of rod bipolars and AII amacrine cells. High convergence of rods upon dedicated bipolars made the latter highly sensitive. AII amacrines ensured appropriate connectivity to the ON and OFF channels also thereby injecting the rod-generated information into the diversity (and processing abilities) of the various types of cone bipolars. The shared inner retinal circuitry was made more efficient by a dedicated network of dopaminergic amacrines controlling sensitivity by affecting gap junction permeability

A reflection of the fact that the original retinal circuitry was shaped by cones can be appreciated considering that, although rods are usually one order of magnitude more numerous than cones in most mammalian species, cone bipolars are instead much more numerous than rod bipolar cells [15], a fact only partially explained by the high convergence of the rod pathway mentioned above. Cone bipolars are numerous because they come in different types, each of them dedicated to a specific function and each cone has to contact them all to ensure parallel processing of the signal. However, the piggyback arrangement guarantees the access of the scotopic signal to the whole retinal processing originally evolved in the cone system.

2.11 Diverse and Complex Retinal Neurons: Amacrine and Ganglion Cells

The variety of cell types within a class increases from the outer to the inner retina and reaches its maximum for amacrines and ganglion cells [2]. Giving an account of such a complexity is well beyond the scope of this chapter, which is primarily focused upon those retinal neurons more directly involved in circadian rhythms. However, some general and important rules underlying the main properties of these cell classes should be mentioned.

Amacrine cells are by definition neurons without an axon, although some of them break this (quite atypical) rule. They occur in a large variety probably not yet completely exploited, and this, together with the relatively inaccessible position they occupy in the wiring diagram of the retina, makes their physiology still largely unexplored. As already mentioned, amacrine cells occur in two broad varieties, i.e., small-field and large-field cells. In general, small-field cells (like AII amacrines) have a radial arrangement in the IPL, might span across the ON and OFF sublaminae, and use glycine as primary neurotransmitter. Conversely, large-field amacrines (such as dopaminergic amacrines, or cholinergic, starburst, amacrine cells) have a radial, starlike morphology, with a tangential spread in the IPL and dendrites which are usually restricted in specific sublaminae. They use GABA as primary neurotransmitter.

Amacrines in general receive their synaptic inputs from bipolar cells, on which they might return feedback synapses; amacrine cells are also pre- and postsynaptic to each other and might be joined by gap junctions. Finally, they are presynaptic to ganglion cells. Connections involving amacrine cells can occur laterally as well as vertically across the ON and OFF sublaminae of the IPL. Since amacrine cells are inhibitory interneurons, the synaptic arrangements described above mediate feedback inhibition (i.e., from amacrines back to bipolar cells), lateral inhibition (i.e., between amacrine cells arranged horizontally), crossover inhibition (i.e., involving amacrines spanning radially across the two sublaminae of the IPL), and feedforward inhibition (i.e., from amacrines to other amacrines or to postsynaptic bipolar cells). While synapses established by bipolar cells in the IPL can be distinguished ultrastructurally for the presence of characteristic ribbons, amacrine cells make conventional chemical synapses by means of varicosities filled with synaptic vesicles. Such varicosities are quite similar for most amacrine cell types, making it hard to attribute a given contact to a specific parent amacrine cell. Hence, their circuitry in the IPL has yet to be explored. A connectome approach (i.e., the reconstruction of a complete map of the retinal neural connections), in which neuronal networks are traced by large-scale electron microscopy and extensive computation, is going to contribute sensibly to an advancement in this arduous task [39]. Functionally, amacrine cells refine the output of bipolar cells, performing computations that, ultimately, shape the physiological properties of ganglion cells and create task-specific types among them. Recognized examples of tasks to which the computational capabilities of amacrine cells are known to contribute are center and surround effects and directional selectivity [40] (i.e., the capability of certain ganglion cells to become excited by a stimulus moving in one direction and inhibited by a stimulus moving in the opposite direction). The list of newly discovered properties of amacrine cells continues to grow [41].

2.12 Ganglion Cells

Ganglion cells have been hard to classify so far, although many different types have been labeled by antibody staining, intracellular injections, genetic targeting, etc. (Fig. 2.9). Recent data suggest that the retina of common laboratory mammals has approximately 20 ganglion cell types [2]. This number reflects the fact that besides an input from the 12 types of bipolar cells described above, ganglion cells are shaped also by the contribution of amacrine cells, so that additional functional channels are created at the retinal exit [42]. Thus, moving from the outer to the inner retina, the number of visual channels expands, increasing the computational power

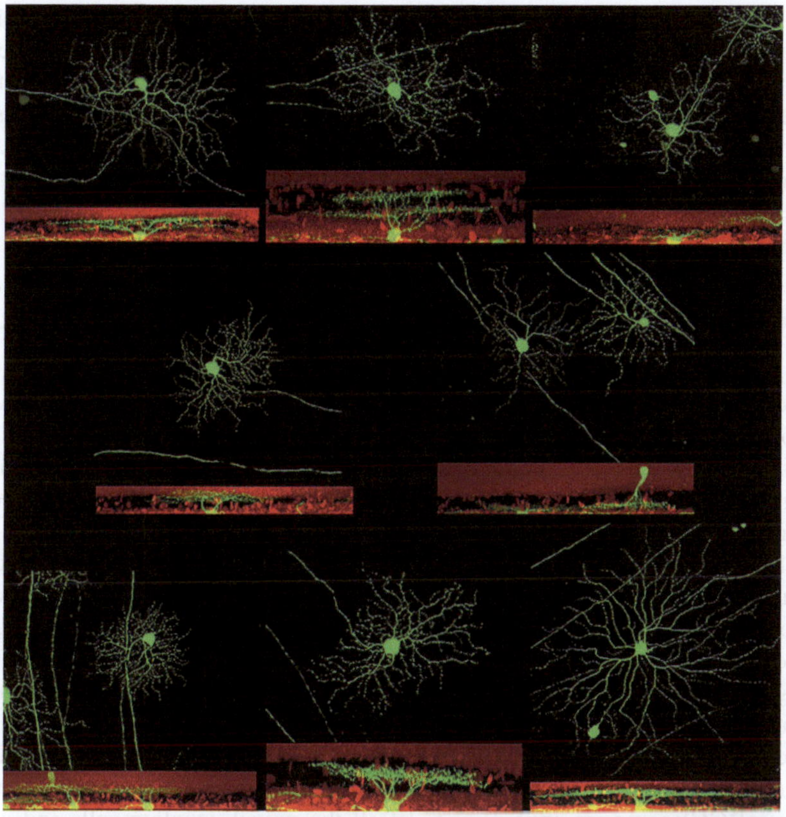

Fig. 2.9 Composite of different types of ganglion cells from the mouse retina. Transgenic expression of GFP in ganglion cells (shown in *green*) reveals their detailed morphology in retinal whole mounts. Computer rotations in the *vertical plane* show the different levels of stratification of their dendrites in the IPL, whose boundaries are marked by the nuclei of cells located in the inner nuclear and ganglion cells layers, labeled by a *red* DNA-binding molecule (see [47])

and feature selectivity operated on the signal. The visual scene is then presented to the brain by means of trains and patterns of action potentials traveling along the axons of ganglion cells constituting the optic nerve fibers and the optic nerve. As stated before, the functional properties of ganglion cells (best known from electrophysiological studies conducted in the rabbit retina) are actively explored and still partially obscure. Especially intriguing is the task of deciphering the code through which ganglion cells provide the results of the computation performed in the IPL to downstream brain areas through the optic nerve [3]. This code appears to have (previously unsuspected) multiple levels of complexity.

Morphological and physiological correlates of ganglion cell properties have allowed the identification of various types comprising (among others) ON-tonic and OFF-tonic cells; blue-ON and blue-OFF ganglion cells; an ON direction selective cell, which projects to the accessory optic system and forms the basis of optokinetic nystagmus; and ON–OFF directionally selective cells of various types. A recently discovered type, the most common in the retina of the mouse, was demonstrated to detect small moving objects down to the receptive field size of bipolar cells, but only on a featureless or stationary background. These cells may serve as "alarm neurons" for overhead predators [2]. A type of its own is represented by the intrinsically photosensitive (melanopsin) ganglion cells described below.

2.13 Melanopsin Ganglion Cells

A second role for the eye has been described besides sight: independently of vision of forms, built-in sensors paired to canonical photoreceptors have the function of measuring ambient light. These dedicated sensors are intrinsically light-sensitive ganglion cells, whose discovery represents a breakthrough in the field of retinal organization [43, 44].

These neurons (wide and sparse, see Fig. 2.10) contain a photopigment called melanopsin (with an absorption peak of the light at a ~480 nm) and are involved in various reflexive responses of the brain and body to the presence of (day)light, such as the regulation of circadian rhythms, pupillary reflex, and other nonvisual responses to light. Melanopsin ganglion cells project to various brain targets including the olivary pretectal nucleus (responsible for controlling the pupil of the eye), the lateral geniculate nucleus (LGN), and the suprachiasmatic nucleus of the hypothalamus (the master clock of circadian rhythms). Intrinsically photosensitive retinal ganglion cells (ipRGC) comprise only ~1–3 % of all the retinal ganglion cells and can be visualized selectively with melanopsin-specific antibodies (Fig. 2.10) [45].

Studies using profoundly blind humans lacking functional rods and cones showed that, similarly to other mammals, the human retina contains some type of non-rod, non-cone photoreceptor, whose identity was eventually found to be a ganglion cell. These studies were performed examining patients with rare diseases leading to rod and cone degeneration but preserving ganglion cells. Similarly to mice with inherited photoreceptor diseases leading to progressive death of rods and cones, these patients continued to exhibit circadian photoentrainment, circadian behavioral patterns, and pupil reactions, with peak spectral sensitivities compatible to that for melanopsin photopigment [46]. In both humans and mice, melanopsin ganglion cells appear particularly robust to secondary effects of degeneration affecting the survival of photoreceptors, such as retinitis pigmentosa [47, 48], as well as in pathologies primarily affecting ganglion cells, like mitochondrial optic neuropathies (i.e., Leber hereditary optic neuropathy and dominant optic atrophy). The latter selectively involve ganglion cells and cause major visual loss with a relatively preserved pupillary light reflex. Recent studies show melanopsin retinal ganglion cells are resistant to neurodegeneration caused by dysfunction of mitochondria as

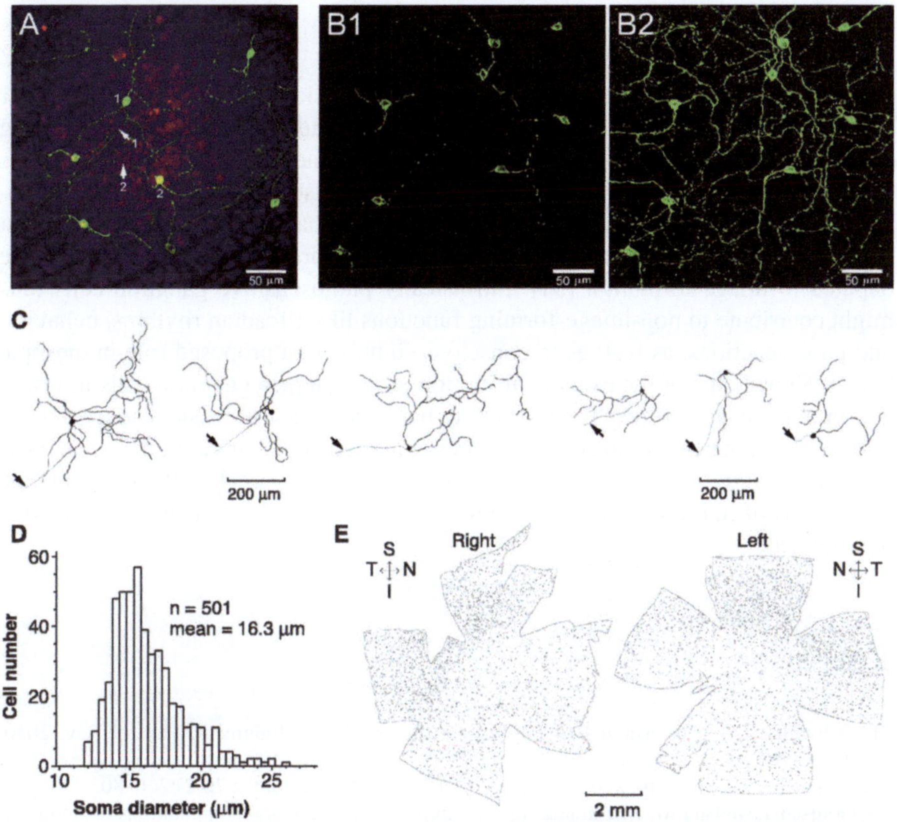

Fig. 2.10 Ganglion cells in the flat-mounted rat retina after melanopsin-antibody staining (*green* signal in **A**, **B1** and **B2**). (**C**) Drawings of examples of melanopsin GCs, where *arrows* indicate axons. (**D**) Soma-size distribution of melanopsin GCs. (**E**) Distribution of melanopsin-positive GCs on whole-mount retinas, showing the higher cell density in the superior (S) and temporal (T) quadrants (Reproduced with permission from [45])

shown by retention of non-image-forming functions in visually impaired patients [48]. Intrinsic resistance of melanopsin ganglion cells to degeneration might open new avenues for vision restoration based on these cells; their robustness to metabolic insults might be explored in search of intrinsic protective mechanisms that might be applied to glaucoma or similar disorders [49]. Indeed, the existence of intrinsically photosensitive ganglion cells has inspired experiments of vision restoration based on inner retinal intervention: transgenic technology can target the expression of melanopsin in inner retinal neurons (bipolar or ganglion cells) of individuals in which primary photoreceptors have been irretrievably lost by degenerative diseases [50].

2.14 Conclusions

In summary, there are two parallel pathways for vision: one arising in the outer retina and based upon rod and cone photoreceptors and a second channel detecting visual brightness arising from the inner retina. The outer and inner retinal channels are not entirely separate: rods and cones themselves also feed into intrinsically photosensitive ganglion cells, which are more complex than initially assumed, at least in some species, and send projections to multiple brain targets, including those deputed to image formation [51]. Intrinsically photosensitive ganglion cells thus might contribute to non-image-forming functions like circadian rhythms, behavior, and pupil reactions, as well as to conscious sight, with a proposed role in mesopic vision. Knowledge of the exact contribution of melanopsin ganglion cells in vision and in non-image-forming functions may have an impact on basic science and on human health in general: the consequences of altered circadian rhythms, when these cells are defective, are already of obvious relevance to clinical medicine, while the possibility of targeting these neurons for vision repair strategies is actively investigated.

References

1. Golombek DA, Rosenstein RE. Physiology of circadian entrainment. Physiol Rev. 2010; 90(3):1063–102.
2. Masland RH. The neuronal organization of the retina. Neuron. 2012;76(2):266–80.
3. Gollisch T, Meister M. Eye smarter than scientists believed: neural computations in circuits of the retina. Neuron. 2010;65(2):150–64.
4. Wassle H. Parallel processing in the mammalian retina. Nat Rev Neurosci. 2004; 5(10):747–57.
5. Punzo C, Kornacker K, Cepko CL. Stimulation of the insulin/mTOR pathway delays cone death in a mouse model of retinitis pigmentosa. Nat Neurosci. 2009;12(1):44–52.
6. Organisciak DT, Vaughan DK. Retinal light damage: mechanisms and protection. Prog Retin Eye Res. 2010;29(2):113–34.
7. Goldberg AF. Role of peripherin/rds in vertebrate photoreceptor architecture and inherited retinal degenerations. Int Rev Cytol. 2006;253:131–75.
8. Insinna C, Pathak N, Perkins B, Drummond I, Besharse JC. The homodimeric kinesin, Kif17, is essential for vertebrate photoreceptor sensory outer segment development. Dev Biol. 2008;316(1):160–70.
9. Huang H, Wang Z, Weng SJ, Sun XH, Yang XL. Neuromodulatory role of melatonin in retinal information processing. Prog Retin Eye Res. 2013;32:64–87.
10. Tosini G, Baba K, Hwang CK, Iuvone PM. Melatonin: an underappreciated player in retinal physiology and pathophysiology. Exp Eye Res. 2012;103:82–9.
11. Thoreson WB, Mangel SC. Lateral interactions in the outer retina. Prog Retin Eye Res. 2012;31(5):407–41.
12. Mills SL, Massey SC. A series of biotinylated tracers distinguishes three types of gap junction in retina. J Neurosci. 2000;20(22):8629–36.
13. Linberg KA, Fisher SK. Ultrastructural evidence that horizontal cell axon terminals are presynaptic in the human retina. J Comp Neurol. 1988;268(2):281–97.

14. Ghosh KK, Bujan S, Haverkamp S, Feigenspan A, Wassle H. Types of bipolar cells in the mouse retina. J Comp Neurol. 2004;469(1):70–82.
15. Strettoi E, Novelli E, Mazzoni F, Barone I, Damiani D. Complexity of retinal cone bipolar cells. Prog Retin Eye Res. 2010;29(4):272–83.
16. Snellman J, Kaur T, Shen Y, Nawy S. Regulation of ON bipolar cell activity. Prog Retin Eye Res. 2008;27(4):450–63.
17. Pignatelli V, Strettoi E. Bipolar cells of the mouse retina: a gene gun, morphological study. J Comp Neurol. 2004;476(3):254–66.
18. Field GD, Chichilnisky EJ. Information processing in the primate retina: circuitry and coding. Annu Rev Neurosci. 2007;30:1–30.
19. Kuffler SW. The single-cell approach in the visual system and the study of receptive fields. Invest Ophthalmol. 1973;12(11):794–813.
20. Awatramani GB, Slaughter MM. Origin of transient and sustained responses in ganglion cells of the retina. J Neurosci. 2000;20(18):7087–95.
21. DeVries SH. Bipolar cells use kainate and AMPA receptors to filter visual information into separate channels. Neuron. 2000;28(3):847–56.
22. Li W, DeVries SH. Bipolar cell pathways for color and luminance vision in a dichromatic mammalian retina. Nat Neurosci. 2006;9(5):669–75.
23. Haverkamp S, Wassle H, Duebel J, Kuner T, Augustine GJ, Feng G, Euler T. The primordial, blue-cone color system of the mouse retina. J Neurosci. 2005;25(22):5438–45.
24. Demb JB, Singer JH. Intrinsic properties and functional circuitry of the AII amacrine cell. Vis Neurosci. 2012;29(1):51–60.
25. Han Y, Massey SC. Electrical synapses in retinal ON cone bipolar cells: subtype-specific expression of connexins. Proc Natl Acad Sci U S A. 2005;102(37):13313–8.
26. Lagnado L. Signal amplification: let's turn down the lights. Curr Biol. 2002;12(6):R215–7.
27. Sterling P, Freed MA, Smith RG. Architecture of rod and cone circuits to the on-beta ganglion cell. J Neurosci. 1988;8(2):623–42.
28. Kolb H, Famiglietti EV. Rod and cone pathways in the inner plexiform layer of cat retina. Science. 1974;186(4158):47–9.
29. Strettoi E, Raviola E, Dacheux RF. Synaptic connections of the narrow-field, bistratified rod amacrine cell (AII) in the rabbit retina. J Comp Neurol. 1992;325(2):152–68.
30. Strettoi E, Dacheux RF, Raviola E. Cone bipolar cells as interneurons in the rod pathway of the rabbit retina. J Comp Neurol. 1994;347(1):139–49.
31. Veruki ML, Hartveit E. Electrical synapses mediate signal transmission in the rod pathway of the mammalian retina. J Neurosci. 2002;22(24):10558–66.
32. Gustincich S, Feigenspan A, Wu DK, Koopman LJ, Raviola E. Control of dopamine release in the retina: a transgenic approach to neural networks. Neuron. 1997;18(5):723–36.
33. Contini M, Lin B, Kobayashi K, Okano H, Masland RH, Raviola E. Synaptic input of ON-bipolar cells onto the dopaminergic neurons of the mouse retina. J Comp Neurol. 2010;518(11):2035–50.
34. Witkovsky P. Dopamine and retinal function. Doc Ophthalmol. 2004;108(1):17–40.
35. Ribelayga C, Cao Y, Mangel SC. The circadian clock in the retina controls rod-cone coupling. Neuron. 2008;59(5):790–801.
36. Bloomfield SA, Volgyi B. The diverse functional roles and regulation of neuronal gap junctions in the retina. Nat Rev Neurosci. 2009;10(7):495–506.
37. Tsukamoto Y, Morigiwa K, Ueda M, Sterling P. Microcircuits for night vision in mouse retina. J Neurosci. 2001;21(21):8616–23.
38. Bowmaker JK. Evolution of colour vision in vertebrates. Eye (Lond). 1998;12(Pt 3b):541–7.
39. Briggman KL, Helmstaedter M, Denk W. Wiring specificity in the direction-selectivity circuit of the retina. Nature. 2011;471(7337):183–8.
40. Masland RH. The many roles of starburst amacrine cells. Trends Neurosci. 2005;28(8):395–6.
41. Masland RH. The tasks of amacrine cells. Vis Neurosci. 2012;29(1):3–9.

42. Taylor WR, Smith RG. Trigger features and excitation in the retina. Curr Opin Neurobiol. 2011;21(5):672–8.
43. Rollag MD, Berson DM, Provencio I. Melanopsin, ganglion-cell photoreceptors, and mammalian photoentrainment. J Biol Rhythms. 2003;18(3):227–34.
44. Berson DM. Strange vision: ganglion cells as circadian photoreceptors. Trends Neurosci. 2003;26(6):314–20.
45. Hattar S, Liao HW, Takao M, Berson DM, Yau KW. Melanopsin-containing retinal ganglion cells: architecture, projections, and intrinsic photosensitivity. Science. 2002;295(5557): 1065–70.
46. Zaidi FH, Hull JT, Peirson SN, Wulff K, Aeschbach D, Gooley JJ, Brainard GC, Gregory-Evans K, Rizzo 3rd JF, Czeisler CA, Foster RG, Moseley MJ, Lockley SW. Short-wavelength light sensitivity of circadian, pupillary, and visual awareness in humans lacking an outer retina. Curr Biol. 2007;17(24):2122–8.
47. Damiani D, Novelli E, Mazzoni F, Strettoi E. Undersized dendritic arborizations in retinal ganglion cells of the rd1 mutant mouse: a paradigm of early onset photoreceptor degeneration. J Comp Neurol. 2012;520(7):1406–23.
48. La Morgia C, Ross-Cisneros FN, Sadun AA, Hannibal J, Munarini A, Mantovani V, Barboni P, Cantalupo G, Tozer KR, Sancisi E, Salomao SR, Moraes MN, Moraes-Filho MN, Heegaard S, Milea D, Kjer P, Montagna P, Carelli V. Melanopsin retinal ganglion cells are resistant to neurodegeneration in mitochondrial optic neuropathies. Brain. 2010;133(Pt 8):2426–38.
49. La Morgia C, Ross-Cisneros FN, Hannibal J, Montagna P, Sadun AA, Carelli V. Melanopsin-expressing retinal ganglion cells: implications for human diseases. Vision Res. 2011;51(2): 296–302.
50. Lin B, Koizumi A, Tanaka N, Panda S, Masland RH. Restoration of visual function in retinal degeneration mice by ectopic expression of melanopsin. Proc Natl Acad Sci U S A. 2008; 105(41):16009–14.
51. Dacey DM, Liao HW, Peterson BB, Robinson FR, Smith VC, Pokorny J, Yau KW, Gamlin PD. Melanopsin-expressing ganglion cells in primate retina signal colour and irradiance and project to the LGN. Nature. 2005;433(7027):749–54.
52. Swaroop A, Kim D, Forrest D. Transcriptional regulation of photoreceptor development and homeostasis in the mammalian retina. Nat Rev Neurosci. 2010;11(8):563–76.
53. Strettoi E, Pignatelli V. Bipolar cells of the mouse retina: a gene gun, morphological study. ARVO Meeting Abstracts. 2004;45(5):5365.
54. Strettoi E, Dacheux RF, Raviola E. Synaptic connections of rod bipolar cells in the inner plexiform layer of the rabbit retina. J Comp Neurol. 1990;295(3):449–66.
55. Weiler R, Pottek M, He S, Vaney DI. Modulation of coupling between retinal horizontal cells by retinoic acid and endogenous dopamine. Brain Res Brain Res Rev. 2000;32(1):121–9.
56. Strettoi E. Morphology of interneurons: amacrine cells. In: Besharse J, Dana R, editors. Encyclopedia of the eye, vol. 4. Amsterdam: Elsevier/Academic Press; 2010. p. 60–4.

Chapter 3
Circadian Photoreception: From Phototransduction to Behaviour

Hester C. van Diepen, Johanna H. Meijer, Stuart N. Peirson, and Russell G. Foster

Abstract Environmental light is detected by three classes of ocular photoreceptor, the rods and cones of the outer retina and photosensitive retinal ganglion cells (pRGCs) that utilise the photopigment melanopsin. In addition to their endogenous photosensitivity, pRGCs receive indirect inputs from the rods and cones. The primary role of the rods and cones is to collect light information for the construction of a visual image, whilst the pRGCs send direct projections to many regions of the brain to regulate nonimage-forming responses to light, including the entrainment of circadian rhythms. The master circadian pacemaker of mammals resides within the suprachiasmatic nuclei (SCN), and light information is transmitted from the pRGCs via their axonal projections which form the retinohypothalamic tract. Neurones within the SCN respond to retinal illumination with changes in electrical activity showing characteristic fast-transient components at the light transitions and a sustained response throughout light exposure. Until recently it was assumed that the fast-transient responses were driven by the rods and/or cones, via their inputs to the pRGCs, whilst the sustained responses to light were mediated by melanopsin. However, more recent studies show that both transient and sustained responses can be generated in the absence of melanopsin, suggesting a significant level of functional redundancy between photoreceptor classes. Although SCN electrical activity appears normal in melanopsin-deficient mice, these animals show attenuated light-induced phase-shifting response in behavioural activity. Collectively these findings

H.C. van Diepen, M.Sc. • J.H. Meijer, Ph.D. (✉)
Laboratory for Neurophysiology, Department of Molecular Cell Biology, Leiden University Medical Center, Einthovenweg 20, Mailbox S5-P, PO Box 9600, 2300 RC Leiden, The Netherlands
e-mail: H.C.van_Diepen@lumc.nl; J.H.Meijer@lumc.nl

S.N. Peirson, B.Sc., Ph.D. • R.G. Foster, B.Sc., Ph.D., F.R.S.
Nuffield Laboratory of Ophthalmology, Nuffield Department of Clinical Neurosciences, Levels 5-6 West Wing, University of Oxford, John Radcliffe Hospital, Headley Way, Oxford OX3 9DU, UK
e-mail: russell.foster@eye.ox.ac.uk

G. Tosini et al. (eds.), *The Retina and Circadian Rhythms*, Springer Series in Vision Research 1, DOI 10.1007/978-1-4614-9613-7_3,

raise intriguing questions about how different classes of retinal photoreceptor interact to encode nonimage-forming responses to light and suggest that recordings of multiunit activity at the level of the SCN do not entirely mirror the effects of light on circadian behaviour.

Keywords Circadian • Photoreceptors • Melanopsin • Suprachiasmatic nuclei (SCN) • Electrophysiology • Photoentrainment • pRGC • Retina • Retinohypothalamic tract (RHT) • Light

Abbreviations

AMPA	α-Amino-3-hydroxy-5-methyl-4-isoxazolepropionic acid
CREB	cAMP response element binding
GHT	Geniculohypothalamic tract
IPL	Inner plexiform layer
MAPK	Mitogen-activated protein kinase
MWS	Mid-wavelength sensitive
NMDA	Night-time application of *N*-methyl-D-aspartate
NO	Nitric oxide
Opn4	Melanopsin
PACAP	Pituitary adenylate cyclase-activating peptide
pRGC	Photosensitive retinal ganglion cell
rd/rd	Retinal degeneration
RGC	Retinal ganglion cell
RHT	Retinohypothalamic tract
SCN	Suprachiasmatic nuclei
TRP	Transient receptor potential
TTFL	Transcriptional-translational feedback loop
UVS	Ultraviolet sensitive
VA	Vertebrate ancient

3.1 Introduction

Circadian rhythms evolved as an adaptation to the predictable daily rotation of the Earth around its axis. In mammals, these rhythms are coordinated by the major circadian pacemaker in the suprachiasmatic nuclei (SCN), a bilateral structure that in rodents contains approximately 10,000 neurones in each hemisphere [1, 2]. The SCN are located in the anterior hypothalamus, immediately superior to the optic chiasm. In the absence of environmental time cues, the SCN generates an endogenous free-running rhythm of approximately 24 h. However, this "internal day" would be of no use unless it can be synchronised (entrained) with the environmental

24 h day. Once entrained, the electrical activity of the SCN is high during the day and low during the night, and this information is transmitted directly to many parts of the central nervous system and indirectly to the peripheral organs. The prevailing view is that the SCN function as the pacemaker in a hierarchically organised network, coordinating the rhythms of peripheral oscillators throughout the body.

In isolation, SCN neurones maintain their ability to generate ~24 h rhythms, demonstrating that the generation of this oscillation is a cell-autonomous property [3]. The rhythms themselves are generated by a transcriptional-translational feedback loop (TTFL). During the day, the transcriptional regulators CLOCK and BMAL1 dimerise and initiate transcription of the *Per1-2* and *Cry1-2* genes via an E-box enhancer. These PER and CRY proteins function as the negative limb of the molecular clock—PER and CRY dimerise and translocate to the nucleus, where they inhibit the transcription activation of CLOCK and BMAL1. Reduction of CLOCK/BMAL1 activity inhibits transcription of *Per* and *Cry* genes, resulting in a decrease in PER and CRY protein levels and a disinhibition of CLOCK and BMAL1. Collectively, these molecular processes form a feedback loop with a cycle of approximately 24 h. In addition, a number of other key clock genes are involved in this process, as well as a range of post-translational modifications, most notably phosphorylation. These processes determine the period of the TTFL, which gives rise to the period of SCN neuronal activity and ultimately overt behavioural rhythms [4]. Entrainment of the endogenous rhythm of the SCN with the external environmental cycle is a prerequisite for its adaptive function, and in mammals the primary time cue (zeitgeber) is light. Light information reaches the SCN exclusively via the eyes through a specialised monosynaptic pathway called the retinohypothalamic tract (RHT). Stimulation of the RHT results in a shift in the phase of the endogenous circadian oscillation of the SCN, thereby synchronising the internal clock with the external light–dark cycle.

In this chapter, we will discuss the role of different retinal photoreceptors in the response characteristics of SCN cells, as well as in photoentrainment. We will first describe how light is absorbed by the light-sensitive photopigments of the retina and then consider the signalling pathways whereby light information is integrated from different photoreceptors and then relayed to the SCN via the RHT. Finally, the role of photoreceptors in both behavioural photoentrainment and the responses of the SCN to light will be discussed.

3.2 Retinal Photoreceptors

In the eye, light is sensed by specialised photosensitive cells in the inner and outer layers of the retina. In the outer retina, light is detected by rod and cone photoreceptor cells, which contain the photopigment molecule. In all vertebrates, these photopigments consist of a vitamin A-based chromophore called 11-*cis*-retinaldehyde, which is covalently bound to an opsin protein. Upon activation by light, 11-*cis* retinal photoisomerises, resulting in the opsin protein undergoing a conformational change from the inactive state to the active state. This transition enables binding of the G-protein transducin and activates a signalling cascade that ultimately results in hyperpolarisation of the photoreceptor cell membrane and a reduction in glutamate release at the photoreceptor synapse [5]. But the rods and cones are not the only photoreceptors of the eye. In 1998 a new opsin family termed *vertebrate ancient* (VA) opsin was shown to form a functional photopigment and to be expressed within the teleost retina in a small number of retinal horizontal cells and ganglion cells, but not the rods and cones [6]. The demonstration that retinal neurones, other than the rods and cones, could detect light provided timely support for a growing body of behavioural data suggesting that the mammalian retina might also contain a non-rod, non-cone photoreceptors which entrain the SCN. In the early 1990s mice homozygous for retinal mutations (e.g. *retinal degeneration*, *rd/rd*) and with no apparent visual responses were studied to determine the impact of rod/cone loss on photoentrainment. Mice lacking all their rods and most cones (*rd/rd*) were able to regulate their circadian rhythms with the same sensitivity as fully sighted animals [7]. These, and a host of subsequent experiments on mice and human subjects with genetic defects of the eye [8–11], showed that the processing of light information by the circadian and classical visual systems must be different and raised the possibility that the eye might contain an additional non-rod, non-cone photoreceptor. These data were far from conclusive, however, because there remained the possibility that only small numbers of rods and/or cones are necessary for normal photoentrainment. As a result, mouse models were created in which all rods and cones were ablated (e.g. *rd/rd cl*). Remarkably, the loss of all types of known photoreceptor had little effect on photoentrainment, although loss of the eyes abolished this capacity completely [9]. Thus mammals, like teleost fish, had to possess another ocular photoreceptor.

Although it was clear that novel ocular photoreceptors drive photoentrainment, the identity of these cells remained unclear. Retinal ganglion cells (RGCs) had been implicated as these photoreceptors [12–14], but the final proof came from two independent approaches, one in rats the other in *rd/rd cl* mice. Both showed that the retina contains a small number of pRGCs. In rats, a subpopulation of retinal ganglion cells was labelled by retrograde dye injected into the SCN. The retina was removed and the electrical activity of individually labelled RGCs monitored by intracellular recording. These cells responded to bright light. This in itself was no surprise, because the rods and cones were present in the retina. However, light-evoked depolarisations persisted in the presence of a cocktail of drugs considered sufficient to block all retinal intercellular communication and even persisted in RGCs that were microdissected from the surrounding retinal tissue [15]. The second approach exploited the *rd/rd cl* mouse retina in combination with calcium (Ca^{2+}) imaging. This technique is capable of detecting small changes in the concentration of intracellular Ca^{2+} in large numbers of individual living cells across the entire retina. The use of the *rd/rd cl* mouse meant that pharmacological or surgical isolation of ganglion cells from the rod and cone photoreceptors was unnecessary. This study demonstrated that approximately 3 % of the neurones in the retinal ganglion cell layer responded to light, but after blocking gap junctions, the number of RGCs responding to light dropped to ~1 %. These results showed that there exists an extensive network of pRGCs that can be uncoupled by application of gap-junctional blockers. Furthermore, three types of light-evoked Ca^{2+} influx were observed in these neurones: a sustained, a transient and a repetitive response. Collectively, the studies in *rd/rd cl* mice identified a heterogeneous-coupled syncytium of pRGCs [16].

The photopigment of the pRGCs was defined in the first instance by action spectroscopy. This powerful approach rests upon the fact that a photopigment has a characteristic absorbance spectrum or profile, which describes the likelihood of photons being absorbed at different wavelengths. Thus, a description of the spectral sensitivity profile (action spectrum) of any light-dependent response will describe the absorbance spectrum of the photopigment upon which the response is based [17]. An action spectrum in *rd/rd cl* mice described a previously uncharacterised, opsin-/vitamin A-based photopigment with peak sensitivity in the "blue" region of the spectrum near 480 nm (opsin photopigment/OP^{479}) [18]. Since 2001, a series of action spectra, in mice [19], rats [15], non-human primates [20] and humans [21, 22] all demonstrated the existence of a single novel opsin photopigment with a λ_{max} of around 480 nm.

Although the photopigment of the pRGCs had been defined in a variety of mammals using action spectroscopy, the opsin gene remained unknown. It was first assumed that there would be a mammalian orthologue of teleost VA-opsin, but no such orthologues were found. Instead, another newly discovered opsin family, the melanopsins, soon emerged as the strongest candidate. The *melanopsin* gene family, also designated *Opn4*, was first identified in *Xenopus* photosensitive pigment cells (melanophores—hence the name melanopsin) [23], and then orthologues were isolated from other vertebrate classes including zebrafish [24] and many mammalian species including humans and mice [25], cats [26] and marsupials [27]. Melanopsin was immediately implicated as the photopigment, as it is expressed in pRGCs, and ablation of the melanopsin gene abolishes the pRGC responses to light [15, 16]. Furthermore, mice in which rods, cones and melanopsin have all been ablated fail to show circadian responses to light, arguing that these three classes of photoreceptor can fully account for all light detection within the eye [19].

The melanopsin phototransduction cascade differs from that of rods and cones and shares greater similarity to the signalling pathways used by invertebrate photoreceptors [23] and vertebrate sensory neurones that result in a depolarisation of membrane potential [28]. Similar to rods and cones, however, the absorption of a photon by 11-*cis* retinal bound to the melanopsin protein results in photoisomerisation and activation of a G-protein signalling cascade [29–32]. However, unlike rods and cones, this cascade uses a different intracellular G-protein signalling pathway, resulting in opening of transient receptor potential (TRP) cation channels, causing the membrane to depolarise [33]. In addition to their endogenous photosensitivity, pRGCs receive input from rod and cone photoreceptors in the outer retina [20, 34–38], and light information detected by rods, cones and melanopsin is integrated at the level of the pRGCs. In this sense, rods, cones and melanopsin can all transduce light information to the SCN via the RHT [39].

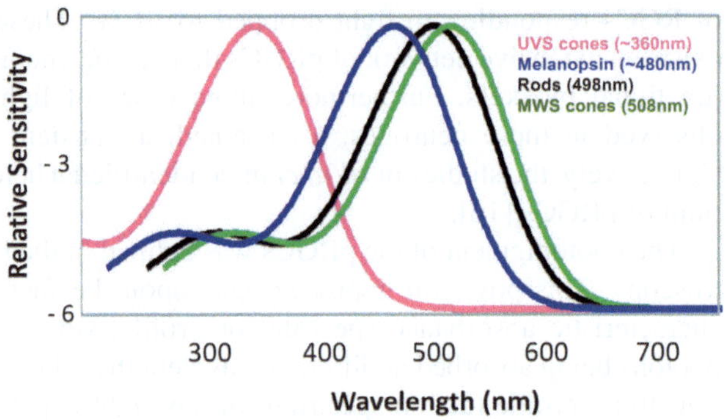

Fig. 3.1 Spectral sensitivity curves for the photopigments in the indicated photoreceptor cells in the mouse retina. The relative sensitivity of each photopigment is plotted as a function of wavelength of light

Even when input from the rods and cones is blocked, or rods and cones are ablated, pRGCs can still depolarise in response to light [15, 16]. However, the light-responsive properties of pRGCs differ markedly from rods and cones. For instance, the response of pRGCs to light is considerably slower than the rapid response of outer retinal rod and cone photoreceptors—following a light stimulus, pRGCs can remain tonically active for several minutes, whereas rods and cones deactivate almost immediately [16, 35]. Also, pRGCs are significantly less sensitive to light than rods and cones [15, 16, 20, 35], suggesting that melanopsin selectively detects high-intensity light. This reduced sensitivity may arise as a result of low photopigment density [33]. Consistent with this hypothesis, behavioural phase-shifting studies showed that melanopsin knockout mice show reduced phase shifts in response to high-intensity light relative to wild-type mice [40].

Mouse cones have two sensitivity peaks: short-wavelength cone photopigments in the UV range of the light spectrum (λ_{max} ~360 nm; UVS cones) and mid-wavelength cone photopigments with a λ_{max} at 508 nm (MWS cones) [41, 42]. The peak sensitivity of melanopsin is in the blue region of the spectrum (λ_{max} ~480 nm) [31, 43], and the peak sensitivity of rods is at 498 nm [44] (Fig. 3.1). Due to the lack of UV-absorbing pigments in the crystalline lens, the eyes of many rodents allow the transmission of UV radiation to the retina, including wavelengths as short as 300 nm [45].

The importance of pRGCs in the signalling of light to the SCN was shown by the finding that circadian photoentrainment is lost following their selective ablation [46–48]. However, since their original description, it has become clear that pRGCs are not a homogeneous population of cells, but instead consist of numerous functionally distinct cell types. To date, the classification of pRGC subtype has extended to include five classes M-1 to M-5, which are distinguished based upon the levels of melanopsin they express and the stratification of their dendrites within specific sublaminae of the inner plexiform layer (IPL) of the retina. Furthermore, the sensitivities, response latencies and central projections vary between the different subtypes, presumably relating to their differing regulatory roles in irradiance detection. For full details see [28].

3.3 Light Transduction to the SCN

Photic information can reach the SCN via three distinct afferent pathways. The most prominent pathway is via the RHT. This pathway originates in the retinal ganglion cells and projects monosynaptically to the SCN. The second pathway to the SCN is indirect via the geniculohypothalamic tract (GHT), which connects the retina to the SCN via the IGL and vLGN in the thalamus [49, 50]. The third pathway to the SCN is via serotonergic projections from the midbrain raphe cells [51].

The RHT appears to be the most important afferent pathway in photic entrainment. Axons originating from the retinal ganglion cells project monosynaptically via the optic nerve through the optic chiasm to the SCN. The RHT projects primarily to the ventral SCN [52]. Electrical stimulation of the RHT in vitro triggers the release of several neurotransmitters in the SCN, including glutamate and pituitary adenylate cyclase-activating peptide (PACAP) [53] (Fig. 3.2). Evidence from several groups all suggest that glutamate and PACAP mediate the effect of light on the SCN [54–57]. Moreover, glutamate is the primary neurotransmitter involved in light entrainment [58, 59]. The release of glutamate at the RHT increases the electrical activity of neurones within the SCN [60], and blocking the release of glutamate inhibits the SCN's response to optic nerve stimulation [57, 61, 62]. The in vitro night-time application of *N*-methyl-D-aspartate (NMDA) or glutamate can trigger a phase delay in the SCN, whereas the in vitro application of NMDA or glutamate during the day advances the SCN's rhythm [59, 63]. Finally, microinjecting glutamate or NMDA directly into the SCN in vivo causes a phase shift in behavioural activity, mimicking the effects of light [54, 64]. Collectively, these results support a primary role for glutamate in transducing light information to the SCN.

In the SCN, PACAP and glutamate are stored in the optic nerve terminals [65]. In vitro, PACAP application can phase shift the electrical rhythm of the SCN [66,

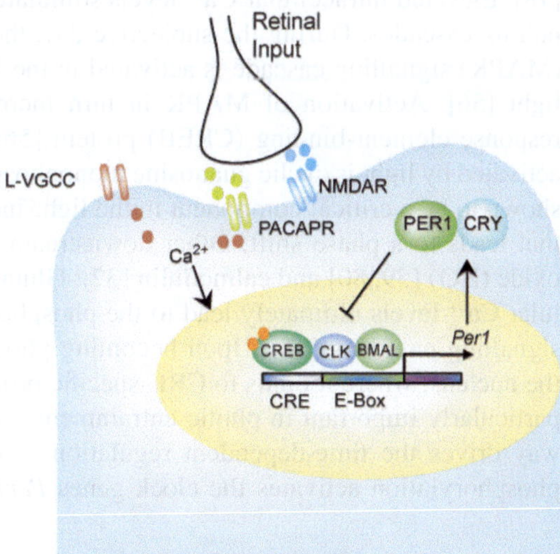

Fig. 3.2 Schematic overview of the light input pathway from the optic nerve to the SCN

67]. Moreover, microinjecting PACAP into the SCN in vivo can phase shift the endogenous rhythm of wheel-running activity [66, 68]. Interestingly, co-injecting PACAP blocks the late-night glutamate-induced phase advance, and PACAP antagonists enhance the glutamate-induced light-driven phase shift [69]. These findings provide compelling evidence that both PACAP and glutamate mediate the effect of light in the SCN. The modulatory effect of PACAP on this signalling is further supported by an in vitro study showing that PACAP enhances the glutamate-induced increase in intracellular calcium in SCN neurones [70]. Thus, the available evidence suggests that PACAP enhances the release of glutamate presynaptically and potentiates the effect of glutamate via PACAP receptors postsynaptically.

The effects of glutamate can be blocked both in vitro and in vivo by the application or injection, respectively, of NMDA receptor antagonists [59, 71, 72]. In addition, NMDA receptor antagonists block light-induced *c-fos* expression in the SCN [73]. These results suggest a critical role for NMDA receptors in the transmission of light signalling to the SCN. Interestingly, NMDA receptor activity is higher during the subjective night than during the subjective day [74], a finding that adds to our understanding of the time-dependent effects of light and glutamate on the SCN. In addition to NMDA receptors, the effects of glutamate may also be mediated via α-amino-3-hydroxy-5-methyl-4-isoxazolepropionic acid (AMPA) receptors [57, 75].

In the SCN, the light-induced release of glutamate and PACAP triggers the activation of postsynaptic NMDA and/or AMPA receptors and PACAP type 1 receptors, respectively. The activation of these receptors leads to an increase in the concentration of intracellular calcium [76]. Intracellular calcium (Ca^{2+}) levels are important for driving daily oscillations in clock gene expression, and blocking Ca^{2+} influx in SCN cultures abolishes the robust rhythms of *Per2* and *Bmal1* expression. NMDA-induced increases in Ca^{2+} levels are larger during the night than during the day [77], which corresponds with the intrinsic light-responsive rhythm of the SCN [78]. Elevated intracellular Ca^{2+} levels stimulate the activation of downstream signalling cascades. During the subjective day, the mitogen-activated protein kinase (MAPK) signalling cascade is activated in the SCN following a brief exposure to light [56]. Activation of MAPK in turn increases the activation of the cAMP response element-binding (CREB) protein [56]. Another signalling target that is activated by light is cyclic guanosine monophosphate (cGMP) [79], which has been shown to be a critical component in the light-induced expression of *Per1* and *Per2* that leads to a phase shift. Other downstream signalling pathways include nitric oxide (NO) [59, 80] and calmodulin [32]. Glutamate-induced increases in intracellular Ca^{2+} levels ultimately lead to the phosphorylation of CREB via intracellular signalling cascades [81]. Upon becoming phosphorylated, CREB translocates to the nucleus, where it binds to CRE-specific promoter regions [82]. Thus, CREB is particularly important in photic entrainment, as activation of this signalling pathway drives the time-dependent regulation of clock genes. For example, CREB phosphorylation activates the clock genes *Per1* and *Per2*, both of which contain a CRE-dependent promoter [82]. The timing of light-induced *Per1* and *Per2* expression corresponds with the timing in which light induces shifts in behavioural activity [30].

3.4 Light-Response Characteristics in the SCN

Both single-unit and multiunit electrophysiological recordings have shown that a subpopulation of SCN neurones responds to retinal illumination and stimulation of the RHT with changes in electrical activity [75, 78, 83–90] (Fig. 3.3a). In rodents such as rats, hamsters and mice, the majority of SCN neurones increase their electrical activity in response to light exposure or electrical stimulation of the RHT, whereas a small population of neurones responds with a suppression in SCN neuronal activity [75, 83, 84, 90, 91] or a transient response [91]. Interestingly, the population of light-suppressed SCN neurones is substantially larger in diurnal rodents such as the degu and the ground squirrel [92–94]. Approximately one-third of SCN neurones respond to light [75, 84, 90, 95], and these light-responsive cells are located predominantly in the terminal field of the RHT within the ventral SCN [84].

In response to a light pulse, a rapid light-induced transient overshoot in SCN electrical activity occurs when the light is first turned on, and this is followed by a sustained increase in activity for as long as the light stimulus remains [78, 87, 88, 90, 91, 96]. These response characteristics are similar to the "sustained" subclass of pRGCs [16]. After the light is turned off, the electrical activity of the SCN cells is transiently suppressed. The sustained electrical activity of SCN neurones is not observed in the visual areas of the brain, where transient dominate. However, the sustained responses are localised to other nonimage-forming brain areas such as the ventral lateral geniculate and intergeniculate leaflet [97], the raphe nucleus [98] and the pretectum [99], brain areas that are all strongly innervated by pRGCs.

The light-induced increases in SCN neuronal activity change during the course of the day. For example, long-term in vivo recordings showed that light induces a large increase in SCN neuronal activity during the subjective night, whereas light-induced neuronal activity increases to a much lesser extent during the day [78, 96]. In addition, the light-induced increase in neuronal activity is highly dependent on the intensity of the light [84, 85, 90, 91, 95] (Fig. 3.3b). Low-intensity light evokes relatively small changes in SCN electrical activity, whereas high-intensity light can saturate SCN electrical activity (i.e. a further increase in light intensity does not lead to a further increase in SCN activity). Light intensity–response curves in sustained SCN electrical activity are sigmoid shaped over a narrow working range [78, 88]. Moreover, the threshold light intensity for the intensity-response curve in SCN electrical activity is much higher than the light threshold for vision [84]. The sigmoid-shaped curves of the sustained responses in SCN electrical activity are quite similar to the phase-shift responses to various light intensities [85, 100].

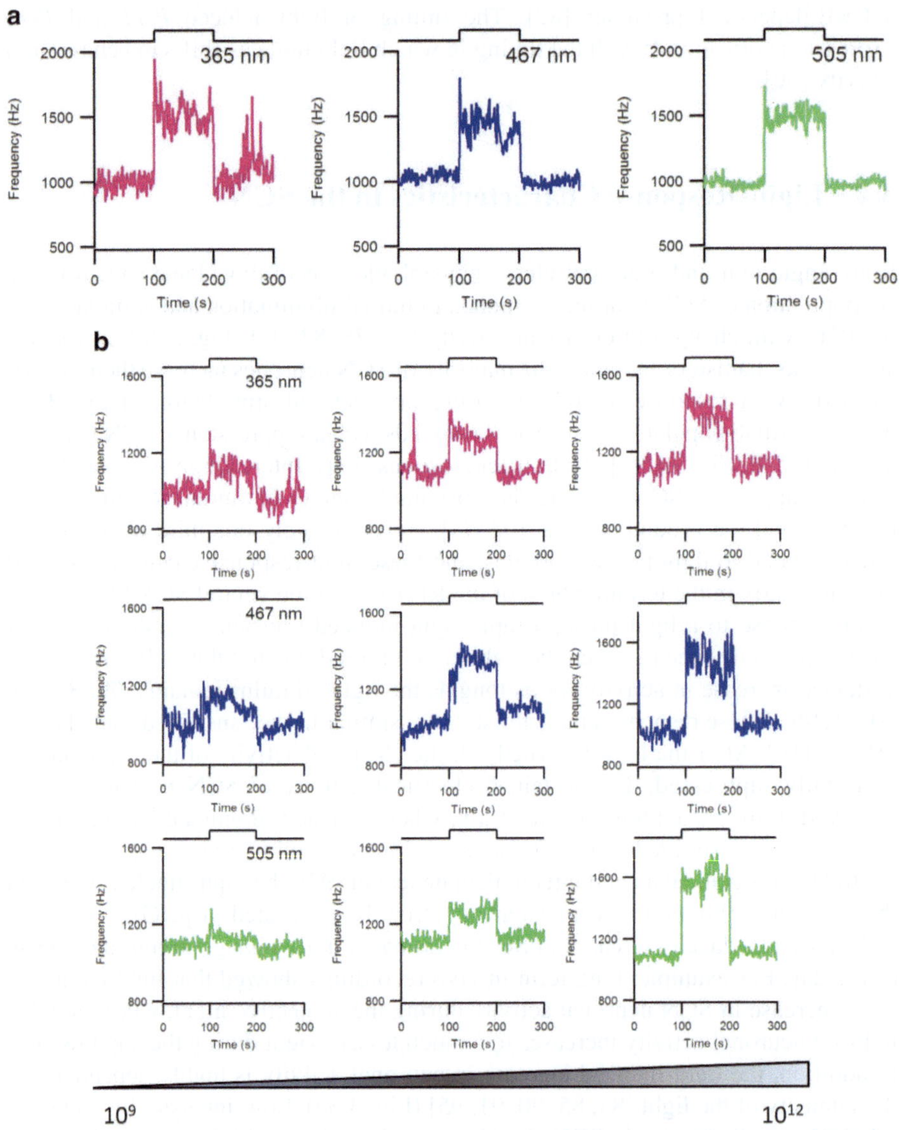

Fig. 3.3 Light-activated responses in the SCN. Activity was recorded using an implanted stationary electrode in an unanaesthetised mouse. (**a**) The examples show a typical fast-transient overshoot in electrical impulse frequency at the onset of the light pulse (1), sustained discharge plateaus during the light presentation (2) and a transient decrease at the end of the light pulse (3). After the transient decrease, the frequency returned to baseline. Similar responses were determined in response to UV (λ_{max} 365 nm), *blue* (λ_{max} 467 nm) and *green* (λ_{max} 505 nm) light. The duration of the pulse was 100 s. (**b**) Light-activated SCN electrical impulse frequency in response to various light intensities ranging from low-intensity (*top trace*) to high-intensity (*bottom trace*) white light. In response to all the various light intensities, a transient increase at the onset of the light pulse, a sustained discharge plateau during light exposure and a transient decrease at the end of the light pulse were recorded. Bin size is 1 s

3.5 Phase Response Curve

Depending on the phase of the circadian pacemaker in which light is perceived, the mammalian circadian clock can respond with a phase shift. These shifts can be evaluated best when animals are maintained in a constant environment, and short pulses of light are presented at specific phases in the circadian cycle (Figs. 3.4 and 3.5a). Such studies have revealed that both characteristically and essentially, the direction and magnitude of a phase shift are dependent upon the time of the pulse. Light exposure in the beginning of the active period (or the beginning of the "subjective night" in the case of nocturnal rodents) induces a delay in the behavioural activity rhythm, whereas light exposure at the end of the active period induces a phase advance. Light exposure during the resting phase of mammals does not induce a shift in activity; and as a result this phase is referred to as the "dead zone".

At each phase in the cycle, the magnitude of the shift is dependent upon the duration, intensity and wavelength of the light that is applied. In rodents, light pulses that last several minutes (*ca.* 5–10 min in duration) can induce a substantial phase shift in behavioural activity. If the duration of the light pulse is increased, the magnitude of the behavioural shift is increased correspondingly [101]. The relationship between the *duration* of light and the magnitude of the phase shift follows a logarithmic function [85, 102, 103], whereas the relationship between the *intensity* of a light pulse and the magnitude of the phase shift follows a sigmoidal function. The sensitivity of the shift to pulse duration was investigated by manipulating the duration of the light pulse whilst keeping the total number of photons constant. This experiment showed that the circadian system is more sensitive to 300-s stimuli than to either briefer (e.g. 30-s) or longer (e.g. 3,000 s) stimuli. Interesting, the intensity-response curve for phase shifting is similar to the intensity-response curve for acute SCN responses [85, 100, 102]. Moreover, an intensity threshold exists for low-intensity light; below this threshold, no phase shift occurs, and this threshold is

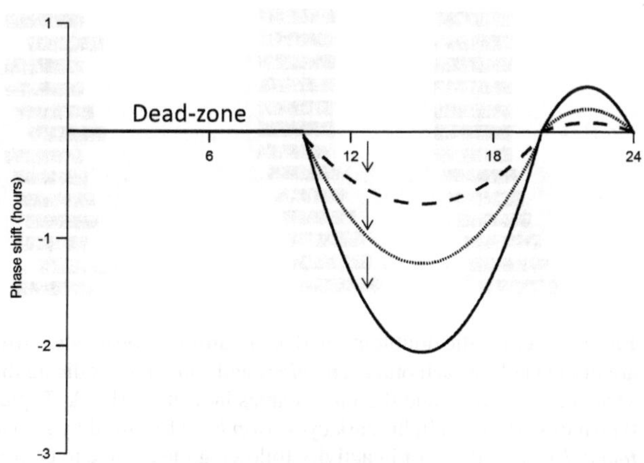

Fig. 3.4 Schematic phase response curve (PRC) of wheel-running activity of mice. Phase-shift magnitude and direction are plotted as a function of circadian time. Magnitude of the phase shift is affected over time by the intensity and duration of the light pulse as indicated by the *dotted lines*

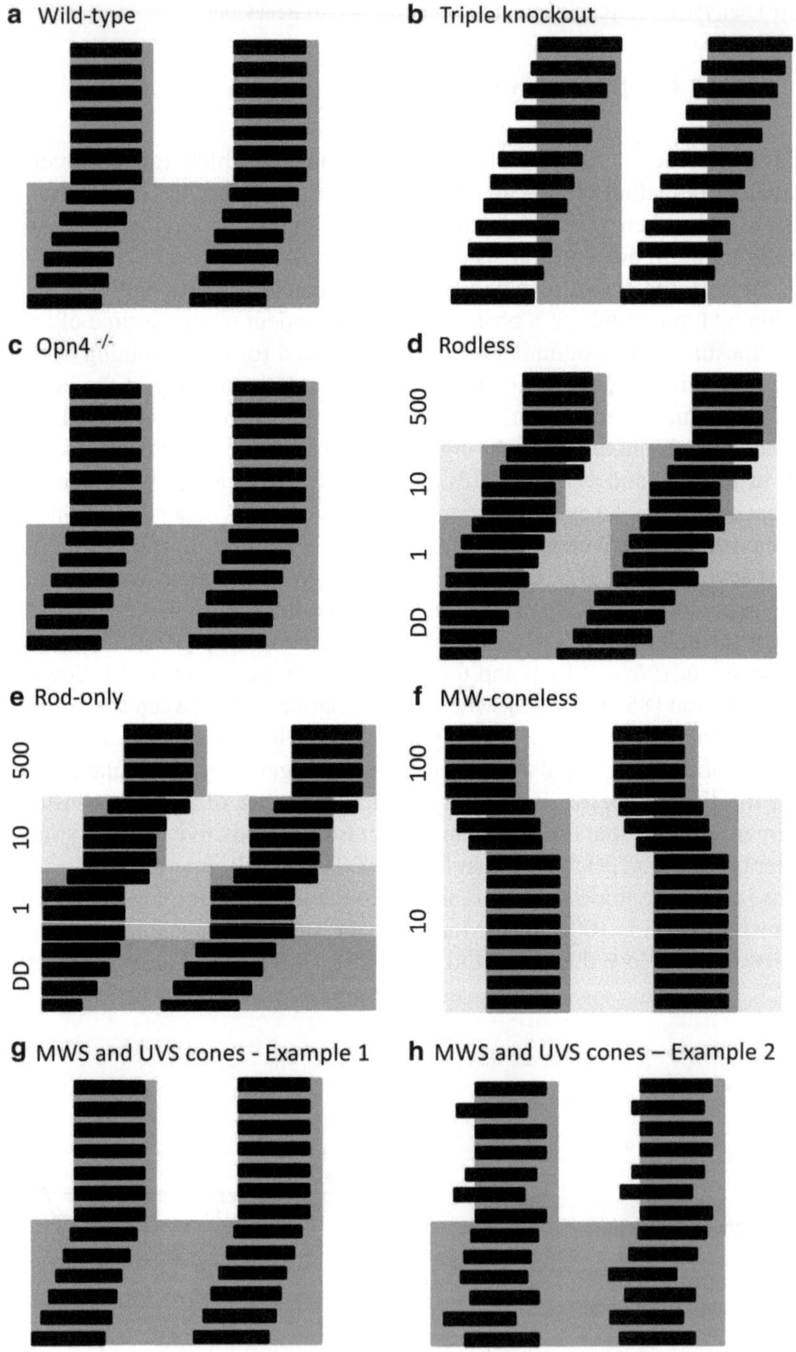

Fig. 3.5 Schematic double-plotted actograms of mouse wheel-running activity. Consecutive days are plotted below each other. The *black* and *white bars* indicate the light–dark cycle (the *unshaded areas* indicate light, and the *shades* areas indicate dark). **A**. Typical actogram recorded for a wild-type mouse during a light–dark cycle (*top half*) followed by a change to continuous darkness (*bottom half*). Note the shift in activity following the change to continuous darkness. (**b**) Actogram of a rod, cone and melanopsin triple knockout mouse during a light–dark cycle. Note that this mouse is unable to entrain to the light–dark cycle [39, 46]. C. Actogram of a melanopsin (Opn4) knockout mouse during a light–dark cycle. Opn4 knockout mice are able to entrain to a light–dark cycle, but

relatively high compared to image-forming vision. The effect of stimulus wavelength on the phase-shift magnitude is determined by the photopigments that contribute to entrainment, and this topic is considered in the final section.

3.6 Contribution of Photoreceptors to Photoentrainment

To determine the relative contributions of the inner and outer retinal photopigments to circadian photoentrainment, the phase-shifting capacity and entrainment properties of various retinal mouse models have been assessed (Table 3.1 and Fig. 3.5).

Table 3.1 Overview of the contribution of the photoreceptors to entrainment, phase-shifting capacity in behavioural activity, period lengthening in continuous light, re-entrainment after phase advance or delay, light-induced SCN clock gene expression and light-induced changes in SCN electrical activity

	Triple KO	Without ipRGCs	Without melanopsin	Only melanopsin	Cone only	MW-coneless	Rod only
Entrainment	None [39]	None [46]	Normal [40]	Normal [7]	Impaired [103]	Impaired [109]	Normal [107]
Phase shifting		None [46]	Reduced [40]	Normal [39]		Reduced [108, 109]	
Period change in constant light		None [46]	Reduced [40, 107]				Reduced [107]
Re-entrainment after phase delay or advance		None [46]			Varied [112]		Normal [107]
Light-induced SCN clock gene expression		Reduced [106]	Normal [111]			Reduced [108]	
SCN electrical activity			Normal [110] Normal [3, 96]	Reduced [3]			

Fig. 3.5 (continued) have a reduced phase-shifting capacity [40, 106]. (**d**) Actogram of a rodless mouse during various light–dark cycles with a range of light intensities. Light intensity varied from 0 lx (*darkest shading*) to 500 lx (*no shading*). Light intensities are indicated at the *y*-axis in lux. Rodless mice are unable to entrain at low light intensities, whilst they are still able to entrain and phase advance at higher light intensities [107]. (**e**) Actogram of a rod-only mouse with depolarised cones during various light–dark cycles with a range of light intensities. Light intensity varied from 0 (*darkest shading*) to 500 lx (*no shading*). Light intensities are indicated at the *y*-axis in lux. Rod-only mice are able to entrain at a wide range of light intensities [107]. (**f**) Actogram of a mid-wavelength-sensitive-coneless (MW-coneless) mouse during shifting light–dark cycles with light intensities from 10 to 100 lx. MW-coneless mice show a deficit in their phase-shifting capacity [109]. (**g, h**) Two examples of actograms of cone-only mice during a light–dark cycle (*top half*) followed by a change to continuous darkness. A minority of the cone-only mice are able to entrain (example 1), whilst others show deficits in normal circadian entrainment (example 2) [39, 40, 103, 107, 109]

Mice that lack all rods and cones (UVS and MWS) are still able to entrain to light and show normal phase shifts [9, 104, 105], demonstrating that melanopsin alone is sufficient for photoentrainment. However, mice that lack melanopsin but retain rods and cones can still entrain (Fig. 3.5c) and phase shift, although with reduced sensitivities. In addition, melanopsin-deficient mice show attenuated period lengthening when housed in continuous light (Table 3.1) [39, 40, 106]. The consensus from several independent lines of study all show that rods and cones can partially compensate for the loss of melanopsin.

These early findings provided compelling evidence that melanopsin is the dominant photopigment mediating photoentrainment, whilst classical photoreceptors can contribute at some level. More recent studies have begun to clarify the different contributions of the rods and cones. For example, rods have been shown to mediate entrainment at both low-intensity and high-intensity light [103, 107] (Fig. 3.5d, e). These studies suggested that two distinct rod pathways underlie these responses. At low light intensities, rods drive entrainment via the primary rod bipolar pathway. However, at high light intensities, cones are required, suggesting that rods signal to pRGCs via their electrical coupling to cones, forming a bright light rod–cone pathway [107]. This model is supported by additional studies showing that rods can mediate photoentrainment at very dim light intensities, but also at higher irradiances when cones would be expected to function [103]. These dynamic interactions within the retina presumably allow entrainment to different features of the dawn/dusk transition. For example, the rod bipolar pathway mediates entrainment to relatively long-duration, low-intensity light stimuli. By contrast the rod activation of the cone pathway, in conjunction with melanopsin activation, mediates the more familiar high-threshold phase-shifting responses. As cone photoreceptors adapt very quickly to sustained light stimuli, these photoreceptors were considered to play only a minor role in irradiance detection [103]. However, there has been a growing appreciation that both MWS and UVS cones can play an important role in photoentrainment [96, 108, 109] (Fig. 3.5f–h).

Light-induced phase shifts in behavioural activity are mirrored by light-induced changes in SCN neuronal activity, such that the magnitude of the phase shift in behaviour is directly correlated with light-induced changes in the impulse frequency of SCN neurones [85, 103]. In parallel with studies on circadian behaviour, the contribution of the various photoreceptor/photopigment classes to light responsiveness of the SCN has been studied at an electrophysiological level [83, 87, 88, 91, 96]. In an attempt to address the role of MWS cones in evoking changes in SCN impulse frequency, electrophysiological recordings were performed in mice in which the human "red cone" photopigment (λ_{max} ~560 nm) was genetically knocked in to the murine MWS cones. The resultant shift in the spectral sensitivity of the MWS cones from λ_{max} 508 nm to λ_{max} 556 nm, combined with different wavelength light stimuli, allowed the different contribution of MWS cones vs rods (λ_{max} 498 nm) to be elucidated. SCN recordings from anaesthetised mice suggested that the MWS cones are particularly important in transient on excitation, but not the sustained increase in firing [91]. Studies by other groups also suggested that cones are important for the fast-transient responses in SCN

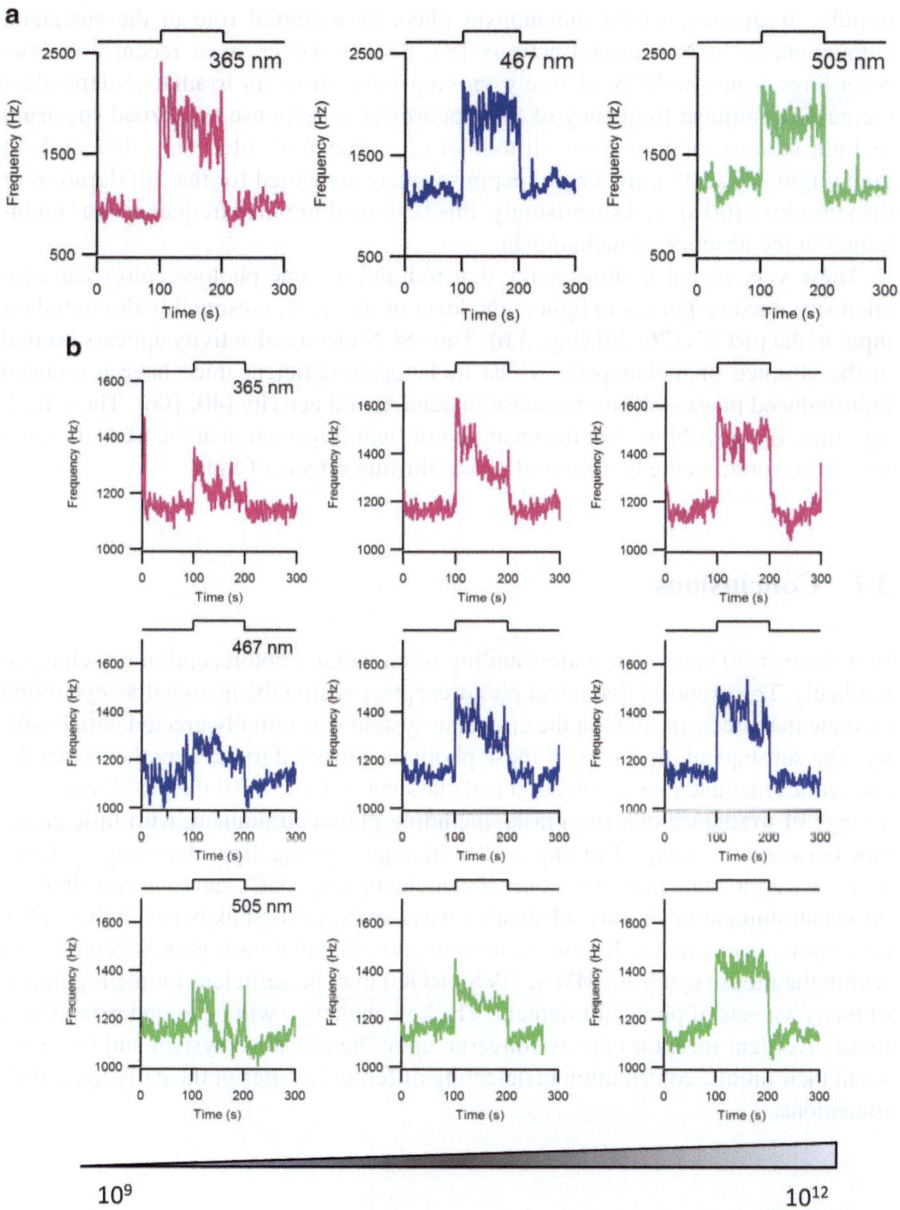

Fig. 3.6 (**a**) SCN electrical activity responses to UV (λ_{max} 365 nm), blue (λ_{max} 467 nm) and green (λ_{max} 505 nm) light in freely moving melanopsin deficient mice (*Opn4*$^{-/-}$). Graphs show representative SCN multiunit activity (MUA) traces in response to a 100 s light pulse. Light pulses are indicated above the graphs. Bin size is 1 s. SCN impulse frequency shows a sustained response during exposure to UV, blue and green light in *Opn4*$^{-/-}$ mice. (**b**) Light-induced increases in SCN impulse frequency in response to UV (λ_{max} 365 nm), blue (λ_{max} 467 nm) and green (λ_{max} 505 nm) light at three different light intensities in *Opn4*$^{-/-}$ mice. Mice were exposed to 100 s of light which is indicated in a step diagram above each graph. One representative trace of SCN electrical activity is shown per irradiance. Bin size is 1 s

impulse frequency, whilst melanopsin plays an essential role in the sustained component of SCN neuronal activity [87, 88]. However, most recently in vivo recordings from the SCN of freely moving mice show an irradiance-dependent increase in impulse frequency of SCN neurones in response to a broad spectrum of light stimuli ranging from ultraviolet (λ_{max} 365 nm), blue (λ_{max} 467 nm), to green light (λ_{max} 505 nm). These responses were sustained for the full duration of the stimulus (100 s), and surprisingly, this sustained impulse frequency was maintained in the absence of melanopsin.

These very recent findings show that rod and/or cone photoreceptors can also elicit sustained responses to light at the level of the SCN, presumably through their input to the pRGCs [26, 96] (Fig. 3.6). Thus SCN electrical activity appears normal in the absence of melanopsin, whilst melanopsin-deficient mice have a reduced light-induced phase-shifting response in behavioural activity [40, 106]. These findings raise the possibility that the encoding of light information at the SCN does not fully correspond to the behavioural phase-shifting effects of light.

3.7 Conclusions

In little over 20 years, our understanding of circadian photoreception has changed markedly. The proposal that novel photoreceptors within the mammalian eye might mediate the effects of light on the circadian system was initially greeted with hostility. The subsequent discovery of these photoreceptors led to the hypothesis that the rods/cones mediate image detection and classical vision, whilst the pRGCs regulate a range of irradiance detection tasks including photoentrainment, with little crosstalk between the image-forming and nonimage-forming light-detecting systems. Today we appreciate that rods, cones and melanopsin pRGCs can each contribute to photoentrainment in various substantive ways and the crosstalk between these photoreceptors is extensive. It now even seems possible that each photoreceptor class within the mouse eye (rod, MWS, UVS pRGC) may be sufficient for the regulation of many aspects of photoentrainment. The key challenge will be to understand how these divergent receptor classes converge upon the circadian system and to understand their unique contribution in detecting different features of the daily light–dark transitions.

References

1. Van den Pol AN. The hypothalamic suprachiasmatic nucleus of rat: intrinsic anatomy. J Comp Neurol. 1980;191(4):661–702. Epub 1980/06/15.
2. Abrahamson EE, Moore RY. Suprachiasmatic nucleus in the mouse: retinal innervation, intrinsic organization and efferent projections. Brain Res. 2001;916(1–2):172–91. Epub 2001/10/13.

3. van Diepen HC, Ramkisoesning A, et al. Irradiance encoding in the suprachiasmatic nuclei by rod and cone photoreceptors. FASEB J. 2013;27(10):4204–12.
4. Lowrey PL, Takahashi JS. Genetics of circadian rhythms in Mammalian model organisms. Adv Genet. 2011;74:175–230. Epub 2011/09/20.
5. Lamb TD, Pugh Jr EN. Phototransduction, dark adaptation, and rhodopsin regeneration the proctor lecture. Invest Ophthalmol Vis Sci. 2006;47(12):5137–52.
6. Soni BG, Philp AR, Knox BE, Foster RG. Novel retinal photoreceptors. Nature. 1998;394:27–8.
7. Foster RG, Provencio I, Hudson D, Fiske S, DeGrip W, Menaker M. Circadian photoreception in the retinally degenerate mouse (*rd/rd*). J Comp Physiol A. 1991;169(1):39–50.
8. Czeisler CA, Shanahan TL, Klerman EB, Martens H, Brotman DJ, Emens JS, et al. Suppression of melatonin secretion in some blind patients by exposure to bright light. N Engl J Med. 1995;332(1):6–11.
9. Freedman MS, Lucas RJ, Soni B, von Schantz M, Munoz M, David-Gray Z, et al. Regulation of mammalian circadian behavior by non-rod, non-cone, ocular photoreceptors. Science. 1999;284(5413):502–4. Epub 1999/04/16.
10. Provencio I, Foster RG. Circadian rhythms in mice can be regulated by photoreceptors with cone-like characteristics. Brain Res. 1995;694(1–2):183–90.
11. Provencio I, Wong S, Lederman AB, Argamaso SM, Foster RG. Visual and circadian responses to light in aged retinally degenerate mice. Vision Res. 1994;34(14):1799–806.
12. Foster RG. Shedding light on the biological clock. Neuron. 1998;20(5):829–32.
13. Foster RG. Keeping an eye on the time: the Cogan Lecture. Invest Ophthalmol Vis Sci. 2002;43(5):1286–98.
14. Foster RG, Hankins MW. Non-rod, non-cone photoreception in the vertebrates. Prog Retin Eye Res. 2002;21(6):507–27.
15. Berson DM, Dunn FA, Takao M. Phototransduction by retinal ganglion cells that set the circadian clock. Science. 2002;295:1070–3.
16. Sekaran S, Foster RG, Lucas RJ, Hankins MW. Calcium imaging reveals a network of intrinsically light-sensitive inner-retinal neurons. Curr Biol. 2003;13(15):1290–8. Epub 2003/08/09.
17. Peirson SN, Thompson S, Hankins MW, Foster RG. Mammalian photoentrainment: results, methods, and approaches. Methods Enzymol. 2005;393:697–726.
18. Lucas RJ, Douglas RH, Foster RG. Characterization of an ocular photopigment capable of driving pupillary constriction in mice. Nat Neurosci. 2001;4:621–6.
19. Hattar S, Lucas RJ, Mrosovsky N, Thompson S, Douglas RH, Hankins MW, et al. Melanopsin and rod-cone photoreceptive systems account for all major accessory visual functions in mice. Nature. 2003;424(6944):75–81.
20. Dacey DM, Liao HW, Peterson BB, Robinson FR, Smith VC, Pokorny J, et al. Melanopsin-expressing ganglion cells in primate retina signal colour and irradiance and project to the LGN. Nature. 2005;433(7027):749–54. Epub 2005/02/18.
21. Hankins MW, Lucas RJ. The primary visual pathway in humans is regulated according to long-term light exposure through the action of a non-classical photopigment. Curr Biol. 2002;12:191–8.
22. Zaidi FH, Hull JT, Peirson SN, Wulff K, Aeschbach D, Gooley JJ, et al. Short-wavelength light sensitivity of circadian, pupillary, and visual awareness in humans lacking an outer retina. Curr Biol. 2007;17(24):2122–8.
23. Provencio I, Jiang G, De Grip WJ, Hayes WP, Rollag MD. Melanopsin: An opsin in melanophores, brain, and eye. Proc Natl Acad Sci U S A. 1998;95(1):340–5.
24. Bellingham J, Whitmore D, Philp AR, Wells DJ, Foster RG. Zebrafish melanopsin: isolation, tissue localisation and phylogenetic position. Brain Res Mol Brain Res. 2002;107(2):128–36.
25. Provencio I, Rodriguez IR, Jiang G, Hayes WP, Moreira EF, Rollag MD. A novel human opsin in the inner retina. J Neurosci. 2000;20:600–5.

26. Semo M, Munoz Llamosas M, Foster RG, Jeffery G. Melanopsin (Opn4) positive cells in the cat retina are randomly distributed across the ganglion cell layer. Vis Neurosci. 2005; 22(1):111–6.
27. Pires SS, Shand J, Bellingham J, Arrese C, Turton M, Peirson S, et al. Isolation and characterization of melanopsin (Opn4) from the Australian marsupial Sminthopsis crassicaudata (fat-tailed dunnart). Proc Biol Sci. 2007;274(1627):2791–9.
28. Hughes S, Hankins MW, Foster RG, Peirson SN. Melanopsin phototransduction: slowly emerging from the dark. Prog Brain Res. 2012;199:19–40.
29. Newman LA, Walker MT, Brown RL, Cronin TW, Robinson PR. Melanopsin forms a functional short-wavelength photopigment. Biochemistry. 2003;42(44):12734–8.
30. Melyan Z, Tarttelin EE, Bellingham J, Lucas RJ, Hankins MW. Addition of human melanopsin renders mammalian cells photoresponsive. Nature. 2005;433(7027):741–5.
31. Panda S, Nayak SK, Campo B, Walker JR, Hogenesch JB, Jegla T. Illumination of the melanopsin signaling pathway. Science. 2005;307(5709):600–4. Epub 2005/02/01.
32. Isoldi MC, Rollag MD, Castrucci AM, Provencio I. Rhabdomeric phototransduction initiated by the vertebrate photopigment melanopsin. Proc Natl Acad Sci U S A. 2005;102(4): 1217–21.
33. Do MT, Kang SH, Xue T, Zhong H, Liao HW, Bergles DE, et al. Photon capture and signalling by melanopsin retinal ganglion cells. Nature. 2009;457(7227):281–7. Epub 2009/01/02.
34. Belenky MA, Smeraski CA, Provencio I, Sollars PJ, Pickard GE. Melanopsin retinal ganglion cells receive bipolar and amacrine cell synapses. J Comp Neurol. 2003;460(3):380–93. Epub 2003/04/15.
35. Wong KY, Dunn FA, Graham DM, Berson DM. Synaptic influences on rat ganglion-cell photoreceptors. J Physiol. 2007;582(Pt 1):279–96. Epub 2007/05/19.
36. Schmidt TM, Kofuji P. Differential cone pathway influence on intrinsically photosensitive retinal ganglion cell subtypes. J Neurosci. 2010;30(48):16262–71. Epub 2010/12/03.
37. Perez-Leon JA, Warren EJ, Allen CN, Robinson DW, Brown RL. Synaptic inputs to retinal ganglion cells that set the circadian clock. Eur J Neurosci. 2006;24(4):1117–23. Epub 2006/08/26.
38. Viney TJ, Balint K, Hillier D, Siegert S, Boldogkoi Z, Enquist LW, et al. Local retinal circuits of melanopsin-containing ganglion cells identified by transsynaptic viral tracing. Curr Biol. 2007;17(11):981–8. Epub 2007/05/26.
39. Hattar S, Lucas RJ, Mrosovsky N, Thompson S, Douglas RH, Hankins MW, et al. Melanopsin and rod-cone photoreceptive systems account for all major accessory visual functions in mice. Nature. 2003;424(6944):76–81. Epub 2003/06/17.
40. Panda S, Sato TK, Castrucci AM, Rollag MD, DeGrip WJ, Hogenesch JB, et al. Melanopsin (Opn4) requirement for normal light-induced circadian phase shifting. Science. 2002; 298(5601):2213–6. Epub 2002/12/14.
41. Jacobs GH, Neitz J, Deegan 2nd JF. Retinal receptors in rodents maximally sensitive to ultraviolet light. Nature. 1991;353(6345):655–6. Epub 1991/10/17.
42. Jacobs GH, Deegan 2nd JF. Sensitivity to ultraviolet light in the gerbil (Meriones unguiculatus): characteristics and mechanisms. Vision Res. 1994;34(11):1433–41. Epub 1994/06/01.
43. Qiu X, Kumbalasiri T, Carlson SM, Wong KY, Krishna V, Provencio I, et al. Induction of photosensitivity by heterologous expression of melanopsin. Nature. 2005;433(7027):745–9. Epub 2005/01/28.
44. Bridges CD. Visual pigments of some common laboratory mammals. Nature. 1959;184 Suppl 22:1727–8. Epub 1959/11/28.
45. Dillon J, Ortwerth BJ, Chignell CF, Reszka KJ. Electron paramagnetic resonance and spin trapping investigations of the photoreactivity of human lens proteins. Photochem Photobiol. 1999;69(2):259–64. Epub 1999/02/27.
46. Guler AD, Ecker JL, Lall GS, Haq S, Altimus CM, Liao HW, et al. Melanopsin cells are the principal conduits for rod-cone input to non-image-forming vision. Nature. 2008; 453(7191):102–5. Epub 2008/04/25.

47. Goz D, Studholme K, Lappi DA, Rollag MD, Provencio I, Morin LP. Targeted destruction of photosensitive retinal ganglion cells with a saporin conjugate alters the effects of light on mouse circadian rhythms. PLoS One. 2008;3(9):e3153. Epub 2008/09/06.
48. Hatori M, Le H, Vollmers C, Keding SR, Tanaka N, Buch T, et al. Inducible ablation of melanopsin-expressing retinal ganglion cells reveals their central role in non-image forming visual responses. PLoS One. 2008;3(6):e2451. Epub 2008/06/12.
49. Ibata Y, Okamura H, Tanaka M, Tamada Y, Hayashi S, Iijima N, et al. Functional morphology of the suprachiasmatic nucleus. Front Neuroendocrinol. 1999;20(3):241–68. Epub 1999/08/06.
50. Morin CL, Dolina S, Robertson RT, Ribak CE. An inbred epilepsy-prone substrain of BALB/c mice shows absence of the corpus callosum, an abnormal projection to the basal forebrain, and bilateral projections to the thalamus. Cereb Cortex. 1994;4(2):119–28. Epub 1994/03/01.
51. Moore RY, Halaris AE, Jones BE. Serotonin neurons of the midbrain raphe: ascending projections. J Comp Neurol. 1978;180(3):417–38. Epub 1978/08/01.
52. Moore RY. Entrainment pathways and the functional organization of the circadian system. Prog Brain Res. 1996;111:103–19. Epub 1996/01/01.
53. Hannibal J. Neurotransmitters of the retino-hypothalamic tract. Cell Tissue Res. 2002; 309(1):73–88. Epub 2002/07/12.
54. Colwell CS, Menaker M. NMDA as well as non-NMDA receptor antagonists can prevent the phase-shifting effects of light on the circadian system of the golden hamster. J Biol Rhythms. 1992;7(2):125–36. Epub 1992/01/01.
55. Ebling FJ. The role of glutamate in the photic regulation of the suprachiasmatic nucleus. Prog Neurobiol. 1996;50(2–3):109–32. Epub 1996/10/01.
56. Obrietan K, Impey S, Storm DR. Light and circadian rhythmicity regulate MAP kinase activation in the suprachiasmatic nuclei. Nat Neurosci. 1998;1(8):693–700. Epub 1999/04/10.
57. de Vries MJ, Treep JA, de Pauw ES, Meijer JH. The effects of electrical stimulation of the optic nerves and anterior optic chiasm on the circadian activity rhythm of the Syrian hamster: involvement of excitatory amino acids. Brain Res. 1994;642(1–2):206–12. Epub 1994/04/11.
58. Johnson RF, Moore RY, Morin LP. Loss of entrainment and anatomical plasticity after lesions of the hamster retinohypothalamic tract. Brain Res. 1988;460(2):297–313. Epub 1988/09/20.
59. Ding JM, Chen D, Weber ET, Faiman LE, Rea MA, Gillette MU. Resetting the biological clock: mediation of nocturnal circadian shifts by glutamate and NO. Science. 1994;266(5191): 1713–7. Epub 1994/12/09.
60. Shirakawa T, Moore RY. Responses of rat suprachiasmatic nucleus neurons to substance P and glutamate in vitro. Brain Res. 1994;642(1–2):213–20. Epub 1994/04/11.
61. Cahill GM, Menaker M. Kynurenic acid blocks suprachiasmatic nucleus responses to optic nerve stimulation. Brain Res. 1987;410(1):125–9. Epub 1987/04/28.
62. Shibata S, Liou SY, Ueki S. Influence of excitatory amino acid receptor antagonists and of baclofen on synaptic transmission in the optic nerve to the suprachiasmatic nucleus in slices of rat hypothalamus. Neuropharmacology. 1986;25(4):403–9. Epub 1986/04/01.
63. Shirakawa T, Moore RY. Glutamate shifts the phase of the circadian neuronal firing rhythm in the rat suprachiasmatic nucleus in vitro. Neurosci Lett. 1994;178(1):47–50. Epub 1994/08/29.
64. Mintz EM, Albers HE. Microinjection of NMDA into the SCN region mimics the phase shifting effect of light in hamsters. Brain Res. 1997;758(1–2):245–9. Epub 1997/05/30.
65. Hannibal J, Moller M, Ottersen OP, Fahrenkrug J. PACAP and glutamate are co-stored in the retinohypothalamic tract. J Comp Neurol. 2000;418(2):147–55. Epub 2000/03/04.
66. Harrington ME, Hoque S, Hall A, Golombek D, Biello S. Pituitary adenylate cyclase activating peptide phase shifts circadian rhythms in a manner similar to light. J Neurosci. 1999;19(15):6637–42. Epub 1999/07/22.
67. Hannibal J, Ding JM, Chen D, Fahrenkrug J, Larsen PJ, Gillette MU, et al. Pituitary adenylate cyclase-activating peptide (PACAP) in the retinohypothalamic tract: a potential daytime regulator of the biological clock. J Neurosci. 1997;17(7):2637–44. Epub 1997/04/01.
68. Piggins HD, Marchant EG, Goguen D, Rusak B. Phase-shifting effects of pituitary adenylate cyclase activating polypeptide on hamster wheel-running rhythms. Neurosci Lett. 2001; 305(1):25–8. Epub 2001/05/18.

69. Chen D, Buchanan GF, Ding JM, Hannibal J, Gillette MU. Pituitary adenylyl cyclase-activating peptide: a pivotal modulator of glutamatergic regulation of the suprachiasmatic circadian clock. Proc Natl Acad Sci U S A. 1999;96(23):13468–73. Epub 1999/11/11.
70. Kopp MD, Meissl H, Dehghani F, Korf HW. The pituitary adenylate cyclase-activating polypeptide modulates glutamatergic calcium signalling: investigations on rat suprachiasmatic nucleus neurons. J Neurochem. 2001;79(1):161–71. Epub 2001/10/12.
71. Mintz EM, Marvel CL, Gillespie CF, Price KM, Albers HE. Activation of NMDA receptors in the suprachiasmatic nucleus produces light-like phase shifts of the circadian clock in vivo. J Neurosci. 1999;19(12):5124–30. Epub 1999/06/15.
72. Colwell CS, Foster RG, Menaker M. NMDA receptor antagonists block the effects of light on circadian behavior in the mouse. Brain Res. 1991;554(1–2):105–10. Epub 1991/07/19.
73. Mikkelsen JD, Larsen PJ, Mick G, Vrang N, Ebling FJ, Maywood ES, et al. Gating of retinal inputs through the suprachiasmatic nucleus: role of excitatory neurotransmission. Neurochem Int. 1995;27(3):263–72. Epub 1995/09/01.
74. Pennartz CM, Hamstra R, Geurtsen AM. Enhanced NMDA receptor activity in retinal inputs to the rat suprachiasmatic nucleus during the subjective night. J Physiol. 2001;532(Pt 1): 181–94. Epub 2001/04/03.
75. Cui LN, Dyball RE. Synaptic input from the retina to the suprachiasmatic nucleus changes with the light–dark cycle in the Syrian hamster. J Physiol. 1996;497(Pt 2):483–93. Epub 1996/12/01.
76. Tominaga K, Geusz ME, Michel S, Inouye ST. Calcium imaging in organotypic cultures of the rat suprachiasmatic nucleus. Neuroreport. 1994;5(15):1901–5. Epub 1994/10/03.
77. Colwell CS. NMDA-evoked calcium transients and currents in the suprachiasmatic nucleus: gating by the circadian system. Eur J Neurosci. 2001;13(7):1420–8. Epub 2001/04/12.
78. Meijer JH, Watanabe K, Schaap J, Albus H, Detari L. Light responsiveness of the suprachiasmatic nucleus: long-term multiunit and single-unit recordings in freely moving rats. J Neurosci. 1998;18(21):9078–87. Epub 1998/10/24.
79. Prosser RA, McArthur AJ, Gillette MU. cGMP induces phase shifts of a mammalian circadian pacemaker at night, in antiphase to cAMP effects. Proc Natl Acad Sci U S A. 1989; 86(17):6812–5. Epub 1989/09/01.
80. Amir S. Blocking NMDA receptors or nitric oxide production disrupts light transmission to the suprachiasmatic nucleus. Brain Res. 1992;586(2):336–9. Epub 1992/07/24.
81. Ding JM, Faiman LE, Hurst WJ, Kuriashkina LR, Gillette MU. Resetting the biological clock: mediation of nocturnal CREB phosphorylation via light, glutamate, and nitric oxide. J Neurosci. 1997;17(2):667–75. Epub 1997/01/15.
82. Travnickova-Bendova Z, Cermakian N, Reppert SM, Sassone-Corsi P. Bimodal regulation of mPeriod promoters by CREB-dependent signaling and CLOCK/BMAL1 activity. Proc Natl Acad Sci U S A. 2002;99(11):7728–33. Epub 2002/05/29.
83. Aggelopoulos NC, Meissl H. Responses of neurones of the rat suprachiasmatic nucleus to retinal illumination under photopic and scotopic conditions. J Physiol. 2000;523(Pt 1): 211–22. Epub 2000/02/16.
84. Meijer JH, Groos GA, Rusak B. Luminance coding in a circadian pacemaker: the suprachiasmatic nucleus of the rat and the hamster. Brain Res. 1986;382(1):109–18. Epub 1986/09/10.
85. Meijer JH, Rusak B, Ganshirt G. The relation between light-induced discharge in the suprachiasmatic nucleus and phase shifts of hamster circadian rhythms. Brain Res. 1992;598 (1–2):257–63. Epub 1992/12/11.
86. Nakamura TJ, Fujimura K, Ebihara S, Shinohara K. Light response of the neuronal firing activity in the suprachiasmatic nucleus of mice. Neurosci Lett. 2004;371(2–3):244–8. Epub 2004/11/03.
87. Drouyer E, Rieux C, Hut RA, Cooper HM. Responses of suprachiasmatic nucleus neurons to light and dark adaptation: relative contributions of melanopsin and rod-cone inputs. J Neurosci. 2007;27(36):9623–31. Epub 2007/09/07.

88. Mure LS, Rieux C, Hattar S, Cooper HM. Melanopsin-dependent nonvisual responses: evidence for photopigment bistability in vivo. J Biol Rhythms. 2007;22(5):411–24. Epub 2007/09/19.
89. Groos GA, Meijer JH. Effects of illumination on suprachiasmatic nucleus electrical discharge. Ann N Y Acad Sci. 1985;453:134–46. Epub 1985/01/01.
90. Groos G, Mason R. Maintained discharge of rat suprachiasmatic neurons at different adaptation levels. Neurosci Lett. 1978;8(1):59–64. Epub 1978/04/01.
91. Brown TM, Wynne J, Piggins HD, Lucas RJ. Multiple hypothalamic cell populations encoding distinct visual information. J Physiol. 2011;589(Pt 5):1173–94. Epub 2011/01/13.
92. Meijer JH, Rusak B, Harrington ME. Photically responsive neurons in the hypothalamus of a diurnal ground squirrel. Brain Res. 1989;501(2):315–23. Epub 1989/11/06.
93. Jiao YY, Lee TM, Rusak B. Photic responses of suprachiasmatic area neurons in diurnal degus (Octodon degus) and nocturnal rats (Rattus norvegicus). Brain Res. 1999;817(1–2): 93–103. Epub 1999/01/16.
94. Jiao YY, Rusak B. Electrophysiology of optic nerve input to suprachiasmatic nucleus neurons in rats and degus. Brain Res. 2003;960(1–2):142–51. Epub 2002/12/31.
95. Nakamura T, Kawagoe Y, Matsuda T, Ueda Y, Koide H. Low-density lipoprotein apheresis in a patient with arteriosclerosis obliterans and light chain deposition disease. Clin Nephrol. 2004;61(6):429–33. Epub 2004/07/01.
96. van Oosterhout F, Fisher SP, van Diepen HC, Watson TS, Houben T, Vanderleest HT, et al. Ultraviolet light provides a major input to non-image-forming light detection in mice. Curr Biol. 2012;22(15):1397–402. Epub 2012/07/10.
97. Harrington ME. The ventral lateral geniculate nucleus and the intergeniculate leaflet: interrelated structures in the visual and circadian systems. Neurosci Biobehav Rev. 1997; 21(5):705–27. Epub 1997/11/14.
98. Mosko SS, Jacobs BL. Midbrain raphe neurons: spontaneous activity and response to light. Physiol Behav. 1974;13(4):589–93. Epub 1974/10/01.
99. Trejo LJ, Cicerone CM. Cells in the pretectal olivary nucleus are in the pathway for the direct light reflex of the pupil in the rat. Brain Res. 1984;300(1):49–62. Epub 1984/05/21.
100. Takahashi JS, DeCoursey PJ, Bauman L, Menaker M. Spectral sensitivity of a novel photoreceptive system mediating entrainment of mammalian circadian rhythms. Nature. 1984; 308(5955):186–8. Epub 1984/03/08.
101. Decoursey PJ. Phase control of activity in a rodent. Cold Spring Harb Symp Quant Biol. 1960;25:49–55. Epub 1960/01/01.
102. Nelson DE, Takahashi JS. Integration and saturation within the circadian photic entrainment pathway of hamsters. Am J Physiol. 1999;277(5 Pt 2):R1351–61. Epub 1999/11/24.
103. Lall GS, Revell VL, Momiji H, Al Enezi J, Altimus CM, Guler AD, et al. Distinct contributions of rod, cone, and melanopsin photoreceptors to encoding irradiance. Neuron. 2010; 66(3):417–28. Epub 2010/05/18.
104. Foster RG, Provencio I, Hudson D, Fiske S, De Grip W, Menaker M. Circadian photoreception in the retinally degenerate mouse (rd/rd). J Comp Physiol. 1991;169(1):39–50. Epub 1991/07/01.
105. Ebihara S, Tsuji K. Entrainment of the circadian activity rhythm to the light cycle: effective light intensity for a Zeitgeber in the retinal degenerate C3H mouse and the normal C57BL mouse. Physiol Behav. 1980;24(3):523–7. Epub 1980/03/01.
106. Ruby NF, Brennan TJ, Xie X, Cao V, Franken P, Heller HC, et al. Role of melanopsin in circadian responses to light. Science. 2002;298(5601):2211–3. Epub 2002/12/14.
107. Altimus CM, Guler AD, Alam NM, Arman AC, Prusky GT, Sampath AP, et al. Rod photoreceptors drive circadian photoentrainment across a wide range of light intensities. Nat Neurosci. 2010;13(9):1107–12. Epub 2010/08/17.
108. Dollet A, Albrecht U, Cooper HM, Dkhissi-Benyahya O. Cones are required for normal temporal responses to light of phase shifts and clock gene expression. Chronobiol Int. 2010;27(4):768–81. Epub 2010/06/22.

109. Dkhissi-Benyahya O, Gronfier C, De Vanssay W, Flamant F, Cooper HM. Modeling the role of mid-wavelength cones in circadian responses to light. Neuron. 2007;53(5):677–87. Epub 2007/03/03.
110. Tsai JW, Hannibal J, et al. Melanopsin as a sleep modulator: circadian gating of the direct effects of light on sleep and altered sleep homeostasis in Opn4(–/–) mice. PLoS Biol. 2009;7(6).
111. Lupi D, Semo M, et al. Impact of age and retinal degeneration on the light input to circadian brain structures. Neurobiol Aging. 2012;33(2):383–92.
112. Mrosovsky N, Hattar S. Impaired masking responses to light in melanopsin-knockout mice. Chronobiol Int. 2003;20(6):989–99.

Chapter 4
Role of Melatonin and Dopamine in the Regulation of Retinal Circadian Rhythms

Gianluca Tosini and P. Michael Iuvone

Abstract Melatonin and dopamine are two important players in the regulation of many physiological functions within the body. These neurohormones are also involved in the regulation of retinal physiology and its pathophysiology. Several studies have shown that melatonin synthesis in the retina primarily occurs during the night, whereas dopamine synthesis occurs during the day. We now know that melatonin and dopamine play opposing roles in the regulation of retinal physiology and visual processing in mammals; dopamine functions as a signal for light adaptation, while melatonin has dark-adaptive effects. This reciprocal feedback loop is believed to be a critical component of the regulation of circadian rhythms in the eye. Melatonin and dopamine exert their influence(s) by binding to G-protein-coupled receptors that are expressed by many different cells in the vertebrate retina. Although during the last 20 years we have acquired a substantial knowledge on the role of melatonin and dopamine in many different animal models, a clear role for the melatonin and dopamine systems in human retina is still missing. It is our hope that future genomic studies will be performed on human populations with ocular diseases to determine if polymorphisms in genes encoding enzymes—or receptors—involved in the melatonin and dopamine systems are associated with specific visual disorders.

Keywords Melatonin • Dopamine • Retina • Circadian • Photoreceptors • Melanopsin

G. Tosini, Ph.D. (✉)
Morehouse School of Medicine, Atlanta, GA 30310, USA
e-mail: gtosini@msm.edu

P.M. Iuvone, Ph.D.
Departments of Ophthalmology and Pharmacology, Emory University, Atlanta, GA, USA

G. Tosini et al. (eds.), *The Retina and Circadian Rhythms*, Springer Series in Vision Research 1, DOI 10.1007/978-1-4614-9613-7_4,

Melatonin and dopamine (DA) are two important players in the regulation of many physiological functions within the body. These neurohormones are involved in many aspects of retinal physiology and pathology [1, 2]. Our understanding of the roles that melatonin and dopamine play within the eye has significantly improved over the last 10 years. Much of the progress is due to the use of animal models in which the synthesis or the signaling of melatonin and dopamine has been affected by genetic manipulations. We now have solid experimental evidence that indicates that melatonin and dopamine play opposing roles in the regulation of retinal physiology and visual processing in mammals; dopamine functions as a humoral signal for light and facilitates photopic visual function, while melatonin has dark-adaptive effects. This reciprocal feedback loop, first identified in *Xenopus laevis* [3] but later in other nonmammalian vertebrates, is believed to be a critical component of the regulation of circadian rhythms in the eye. In this chapter, we will summarize the current knowledge about the role of melatonin and dopamine in the mammalian retina and more specifically on the contribution of these neurohormones to the regulation of retinal circadian rhythms.

4.1 Melatonin Synthesis and Metabolism

Melatonin is synthesized by the pineal gland and the retina of many vertebrate species via a well-defined biosynthetic pathway [4, 5]. Melatonin synthesis begins with the uptake of the amino acid tryptophan from the blood. Tryptophan is then sequentially converted to 5-hydroxytryptophan, serotonin, *N*-acetylserotonin, and melatonin by a series of enzymatic reactions (Fig. 4.1). In the retina, melatonin is almost exclusively produced by the photoreceptor cells [6, 7] although in some cases can be produced by other cell types situated within the inner nuclear layer [8] and by ganglion cells [9]. The amount of melatonin produced by the retina is much smaller than that produced by the pineal gland and retinal melatonin is thought to act as a local neuromodulator within the eye [10], although in some nonmammalian vertebrate species, retinal melatonin may also contribute to the levels of the hormone in the blood [11].

Several studies have shown that melatonin synthesis in the retina primarily occurs during the night and thus melatonin levels in the eye are high during the night and low during the day (reviewed in: [12]). In the majority of the species investigated thus far, melatonin synthesis in the retina is under the control of retinal circadian clocks since the retinae of fishes, amphibians, reptiles, birds, and mammals synthesize melatonin in rhythmic fashion when they are maintained in vitro under constant darkness (reviewed in: [5]). In addition, lesions of the suprachiasmatic nucleus (SCN) of the hypothalamus, i.e., the master circadian clock, do not abolish the circadian rhythm of *Aa-nat* transcription in the rat retina [13]. Although the above-mentioned experiment provided compelling evidence that the circadian rhythm in melatonin synthesis is generated within the mammalian retina, direct evidence could be only obtained by recording the circadian rhythm in melatonin

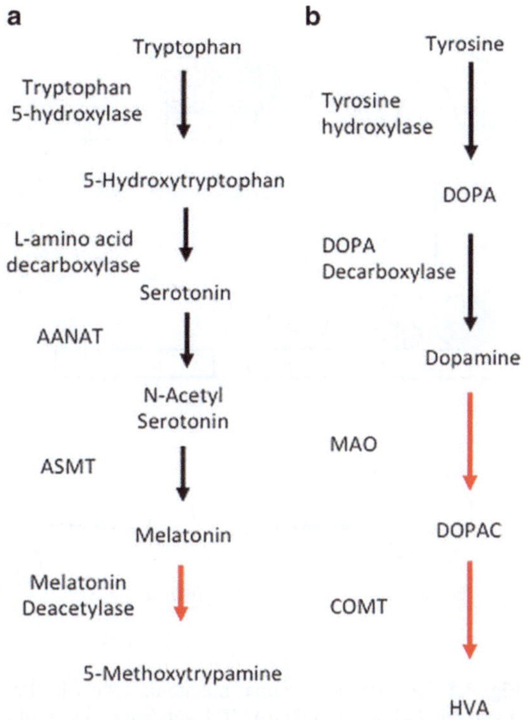

Fig. 4.1 Melatonin and dopamine biosynthesis. Melatonin is synthesized via four steps of a biochemical reaction that starts with the uptake of the circulating amino acid tryptophan. In the mammalian retina, the activity of melatonin deacetylase has not been demonstrated. Dopamine synthesis starts with the conversion of tyrosine to L-DOPA and then to dopamine. Retinal dopamine is then metabolized to DOPAC and HVA

release—or AANAT activity—in cultured retinae as reported for *Xenopus laevis* [14]. Such a direct demonstration was obtained by Tosini and Menaker [15], who reported that a circadian rhythm in melatonin release could be recorded for several days when the retinae were maintained in culture [15, 16]. Furthermore it was also reported that cultured retinae could be entrained to a new light:dark cycle in vitro [15, 16] and the circadian rhythm in melatonin release was temperature-compensated [17]. Interestingly, it has been also reported that, in the *tau* mutant hamster, the period of the circadian rhythm in melatonin synthesis of cultured retina is influenced by the *tau* gene in the same way that this gene influences the period of the circadian pacemaker located in the brain [15] suggesting that the molecular clockwork in the retina is identical to the molecular clockwork in the SCN.

Experimental evidence indicates that the circadian clock controlling melatonin synthesis is located within the photoreceptors. In *Xenopus*, chicken, and rat, rhythmic melatonin synthesis persists in retinae in which the inner retina has been destroyed by kainic acid treatment [6, 18–20] and in rodless (*rd*) mice, i.e., mice lacking rod photoreceptors, melatonin synthesis in cultured retina is no longer rhythmic [16]. Finally, it has been reported that, in the isolated photoreceptor layer, melatonin synthesis is still under the control of the circadian clock [6, 21]. However, it remains unclear if rhythmicity is produced by the rod photoreceptors or whether interactions among rods and cones are necessary to generate circadian rhythmicity in melatonin synthesis.

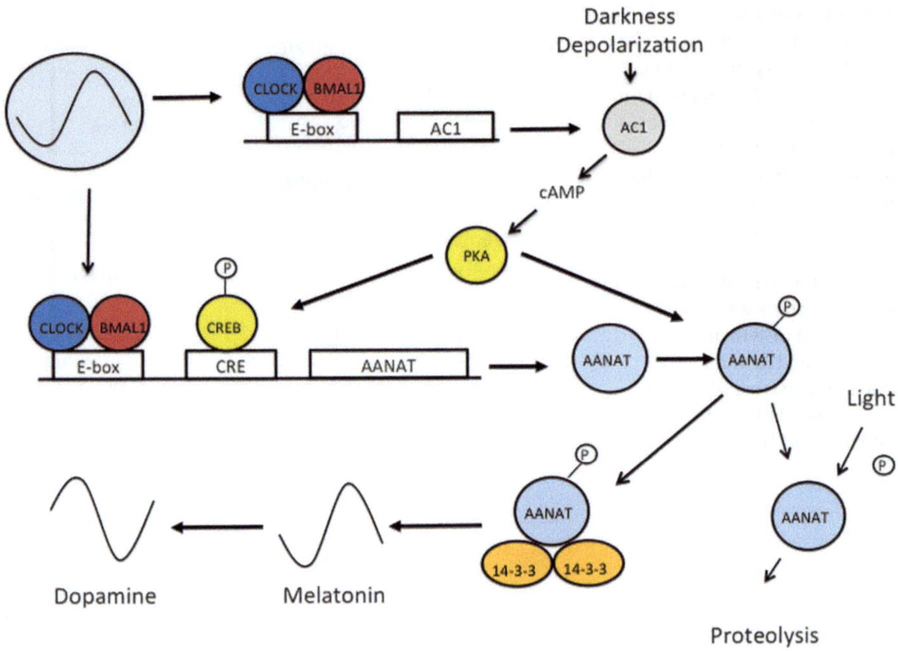

Fig. 4.2 Regulation of retinal melatonin levels in the eye by transcriptional and posttranscriptional regulation. At night cAMP levels are elevated, thus activating *Aanat* gene transcription and then AANAT phosphorylation. Phosphorylated AANAT (pAANAT) associates with 14-3-3 proteins, which activate and stabilize the enzyme resulting in increased conversion of serotonin to *N*-acetylserotonin, and then to melatonin. Light exposure at night decreases cAMP levels resulting in dephosphorylation of AANAT and its subsequent degradation by proteasomal degradation. The circadian clock controls melatonin levels by directly regulating *Aanat* transcription and by gating the cAMP signaling cascade. In constant darkness, the circadian rhythm in dopamine levels is driven by the circadian rhythm in melatonin levels (adapted from [1])

The key regulatory step in melatonin synthesis is catalyzed by arylalkylamine *N*-acetyltransferase (AANAT), which converts serotonin to *N*-acetylserotonin (Fig. 4.1). AANAT is subject to both transcriptional and posttranslational regulation [5] (Fig. 4.2). The control of the transcription of the *Aanat* gene in photoreceptors is under the direct control of light–dark cycles via the cAMP signaling cascade and controlled by the circadian clock [22, 23] (Fig. 4.2).

AANAT is also subject to posttranslational regulation (Fig. 4.2). Retinal AANAT is phosphorylated at night [24] thus promoting binding to 14:3:3 proteins, which stabilizes and activates the enzyme [24–26]. This process is also modulated by the retinal circadian clock by controlling the circadian expression of *Adcy1*, which encodes the type 1 Ca^{2+}/calmodulin-stimulated adenylyl cyclase (AC1) [23, 27]. The rhythm in AC1, in turn, generates circadian rhythms of cyclic AMP and PKA-dependent phosphorylation of AANAT [23, 27].

Posttranslational mechanisms ensure that melatonin levels are maintained at extremely low levels in the presence of light. For example, AANAT activity is abolished in animals maintained in constant light [28] and light exposure in the middle of the night induces a very rapid decrease in AANAT activity in the pineal gland and retina [4, 29, 30]. Light exposure rapidly decreases cAMP levels in photoreceptor cells [31–34] and promotes the dephosphorylation of AANAT, its dissociation from 14-3-3, and its degradation by proteasomal proteolysis [24, 25, 35–37]. This effect of light appears to be partially a direct effect on photoreceptor cells, combined with an effect of dopamine. Dopamine is released from amacrine and interplexiform cells in response to light and acts on dopamine D_4 receptors on the photoreceptor cells to further suppress cyclic AMP synthesis and Ca^{2+} levels and to inhibit melatonin biosynthesis [33, 38–41]. Such tight control of retinal melatonin levels suggests that high melatonin levels during the light phase may be deleterious for the photoreceptor cells [42, 43].

Another interesting aspect of retinal melatonin biology can be found in the regulation of its metabolism. In nonmammalian vertebrates, retinal melatonin is metabolized within the eye [44–46] via a well-defined pathway that involves melatonin deacetylation (see Fig. 4.1). This pathway seems to be absent in the mammalian retina [44, 47–49] and therefore it is not clear whether melatonin is metabolized in the retina of mammals.

While melatonin synthesis in the retina is well established in many mammalian and nonmammalian species, its synthesis in the retina of primates, including humans, has been questioned. Melatonin has been detected in human retina [47] and *Aanat* transcripts have been detected in human and macaque retinas [50, 51], but HIOMT (a.k.a. ASMT) transcripts and activity are barely detectable in human and macaque retinas [51, 52]. Nevertheless melatonin receptors are widely distributed in the human retina [53–56] and melatonin affects retinal functions (see below). Therefore, it is possible that in the primate retina melatonin of pineal origin modulates retinal functions.

4.2 Melatonin: Site of Action and Signaling

Melatonin exerts its influence by binding to G-protein-coupled receptors (GPCRs) named melatonin receptor type 1 (MT_1) and type 2 (MT_2). MT_1 and MT_2 receptors are both present in the vertebrate retina (reviewed in: [57], but see also Table 4.1). In rats, MT_1 receptors are found in the horizontal and amacrine cells within the inner plexiform layer, retinal ganglion cells (RGCs), and the retinal pigment epithelium (RPE) [58]. Dopaminergic neurons may also express MT_1 receptors [59], suggesting that melatonin can directly modulate the activity of these cells and is consistent with observations, in many species, that melatonin inhibits dopamine release [40, 60, 61]. In humans, melatonin receptors (MT_1 and MT_2) have been located on the rod photoreceptors and on RGCs [54–56, 62]. The colocalization of MT_1 and MT_2 receptors in photoreceptors suggests that these receptors may form heteromeric complexes in vivo, as demonstrated previously in vitro [63, 64].

Table 4.1 Distribution of melatonin and dopamine receptors in the vertebrate retina

	MT_1	MT_2	D_1-like	D_2-like
RPE	±	+	+	+
Photoreceptors	+	+	–	+
Outer plexiform layer	+	+	+	–
Bipolar cells	±	+	+	+
Horizontal cells	+	+	+	–
Amacrine cells	+	+	±	+
Inner plexiform layer	+	+	+	+
Ganglion cells	+	+	+	+
IpRGCs	+	–	–	+
Müller cells	+	+	+	–

+ Presence reported, ± presence uncertain, – no presence reported

In the mouse, MT_1 receptors have been localized to photoreceptors, inner retinal neurons, and RGCs [65, 66]. The fact that melatonin receptors are expressed on the same cells responsible for their synthesis raises the intriguing hypothesis that melatonin may feedback on the photoreceptors to regulate its own levels or to produce other autocrine effects. A few studies have also reported that melatonin receptors are present in the ciliary body [67–71]. The expression of melatonin receptors in the ciliary body processes has led to the hypothesis that melatonin may be involved in the regulation of intraocular pressure (IOP) and, indeed, several studies have shown that melatonin can modulate the IOP in various species [68, 70, 72, 73]. New experimental evidence in the mouse also indicates that melatonin modulates the daily rhythm in IOP [73] since removal of the MT_1 receptor significantly increased the mean IOP levels during the night.

4.3 Dopamine Synthesis and Regulation and in the Retina

Dopamine (DA) is the primary catecholamine in the retina, where this neuromodulator is synthesized and released from unique populations of cells in the inner nuclear layer that are either amacrine or interplexiform neurons [2]. The synthesis of DA occurs via the conversion of tyrosine to L-3,4-dihydroxyphenylalanine (L-DOPA) by tyrosine hydroxylase (TH), the rate limiting enzyme for DA synthesis, and the subsequent decarboxylation of L-DOPA by aromatic L-amino acid decarboxylase. Interestingly, it has been shown that dopaminergic neurons can synthesize dopamine throughout the entire cell (i.e., body and processes) [74]. Although the number of dopaminergic neurons in the retina is small (about 500 in the mouse retina [66]), the extensive length and ramification of the axons of these neurons can extend their action to almost the entire retina. Furthermore, DA may also have widespread effects within the retina via diffusion (volume conduction).

The daily changes in dopamine synthesis and release depend on the interactions between the photoreceptors and the dopaminergic neurons, where dopamine release is stimulated by light [75, 76]. Previous studies have also shown that dopamine release and dihydroxyphenylacetic acid (DOPAC) production are higher during the day than during the night [76] and, for this reason, it is believed that dopamine promotes light adaptation. However, it must be mentioned that dopamine release is also present in the darkness as demonstrated by several studies in which the animals were kept in darkness [66, 77]. Furthermore, it is worthwhile to mention that measuring total retinal dopamine content may not be the best method to assess the changes in the release of this neurotransmitter in the retina, since it includes the amount of dopamine present in the cell as well as extracellular DA. Therefore, a better way to assess the changes in dopamine level is to measure its release and reuptake or its metabolism (i.e., the level of DOPAC or homovanillic acid [HVA]). Methods that have been used to estimate retinal dopaminergic activity include measuring DA and/or DOPAC in the vitreous [78, 79], in vivo voltammetry [80], and in vivo microdialysis [81, 82]. Under most conditions, increases of dopamine release are accompanied by increases in DA synthesis, mediated by activation of TH [75, 76]. Thus, activation of TH, measured in vitro or in vivo, has been used as a measure of retinal dopaminergic activity. All of these methods consistently show that dopaminergic activity is elevated during the daytime or as a consequence of light exposure. However, not all retinal DA neurons appear to be light-sensitive. Electrophysiological recordings from mouse retinal DA neurons show heterogeneous responses, with some cells exhibiting light-independent activity [83].

4.4 Dopamine: Site of Action and Signaling

Dopamine receptors are GPCRs. There are five different subtypes (D_1–D_5) organized in two different families (D_1- and D_2-like). The D_1 family includes the D_1 and D_5 that are positively coupled with adenylyl cyclase, i.e., their activation increases the intracellular levels of cAMP. Although the D_1 and D_5 receptors have similar pharmacologies with respect to selective agonists and antagonists, the D_5 receptor has a much higher affinity for DA than does the D_1 receptor. The D_2 family includes the D_2, D_3, and D_4 subtypes and these receptors are negatively coupled with adenylyl cyclase, i.e., their activation decreases the intracellular levels of cAMP.

Most of the cells in the retina express dopamine receptors (Table 4.1), and therefore DA has profound effects in the regulation of many retinal functions. D_1 receptor immunoreactivity has been detected in the outer plexiform and inner plexiform layers and within the cell bodies of horizontal cells, some cone bipolar and amacrine cells, and perhaps in the ganglion cells [84]. D_5 receptor transcripts are expressed in cultured RPE [85], but the levels appear to be relatively low in the neural retina [86].

D_2 receptors (D_2R) are widely expressed in the retina and function as both postsynaptic receptors and autoreceptors that inhibit dopamine release. D_2 receptors are expressed by amacrine, bipolar, and ganglion cells [87, 88] and possibly by the intrinsically photosensitive RGCs (ipRGCs, [89]). D_2R immunoreactivity has been observed in Müller cells and on the basal surface of the bovine RPE, but not in other mammalian species [90]. More recently it has been reported that D_2R mRNA and protein are present in the human RPE [91] and the presence of D_2R transcripts has been reported for the human RPE-choroid. D_4 receptors are mostly expressed on the photoreceptors and transcripts for these receptors have been also detected in the interplexiform layer (IPL) and in ganglion cells [92]. D_3 receptors seem to be absent from the retina [86].

4.5 Role of Melatonin and Dopamine in the Regulation of Retinal Functions

Several studies have shown that melatonin and dopamine play opposing roles in the regulation of retinal adaptive physiology (reviewed in: [5, 12]). Dopamine functions as a humoral signal for light, producing light adaptive physiology, whereas melatonin produces dark-adaptive effects. Melatonin modulates the sensitivity of photoreceptors and second-order neurons at night when photopic input is low [93]. In some species, melatonin can affect glutamatergic transmission from cones to cone-driven bipolar cells [94] and may potentiate responses of ON bipolar cells to rod signals [95]. In *Xenopus laevis*, melatonin directly stimulates the responsiveness of rod photoreceptors [96]. Administration of exogenous melatonin in fishes and amphibians can increase the amplitude of the a- and b-wave of the scotopic electroretinogram (ERG [95, 96]). In chicken, administration of exogenous melatonin during the day reduces the amplitude of the b-wave [97], and chronic administration of melatonin abolishes rhythmicity of a-wave and b-wave implicit times and b-wave amplitude [98]. In humans, melatonin administration decreases the amplitude of the cone ERG [99] and the amplitude of the cone and mixed rod–cone response is negatively correlated with the concentration of endogenous melatonin [100].

In mice, administration of exogenous melatonin increases the amplitudes of a- and b-waves and lowers the scotopic threshold response (i.e., visual sensitivity) to levels observed at night under control conditions; removal of MT_1 receptors abolishes these effects [65]. In melatonin-proficient mice (C3H-$^{f+/+}$), removal of the MT_1 receptors abolished the daily rhythm in the amplitude of the dark- and light-adapted ERG [65]. Interestingly, the daily rhythm in the amplitude of the a- and b-wave of the ERGs also disappears during aging due to a reduced responsiveness to melatonin [101]. A decrease in the responsiveness to melatonin during aging has been observed also in other areas of the brain [102] and is probably due to a down-regulation of the melatonin receptors. In conclusion, it is evident that melatonin can affect visual functions, although its effects may be different in different species and/or cell types.

Finally, a few studies have suggested a role for melatonin in the regulation of disk shedding. Exogenous melatonin led to activation of disk shedding in *Xenopus* retina [103] and an increase in the frequency of large phagosomes in rat RPE cells [104]. However, it should be noted that circadian rhythms of disk shedding have been observed in mice that are genetically incapable of producing substantial quantities of melatonin [105].

The widespread distribution of dopamine receptors within the retina, together with the observation that dopamine can easily diffuse within the retina, indicates that this neuromodulator may modulate several retinal functions. Indeed, experimental evidence gathered over the last 20 years support this vision. Activation of D_2 or D_4 receptors present on the photoreceptor cells leads to a decrease of cAMP and inhibition of melatonin synthesis [40, 60], modulates the phosphorylation of phosducin [106], and probably affects the phosphorylation state of many proteins present in the photoreceptors, including rhodopsin [107]. Dopamine, via D_2-like receptors (presumably D_4R), can also affect rod–cone coupling via gap junctions [108, 109]. Activation of D_1-like receptors uncouples horizontal cells and, in dark-adapted retinas, depolarizes horizontal cells [2]. Activation of D_1 receptors also potentiates the effect of ionotropic glutamate receptor activation on horizontal cells [110]. Dopamine and D_1 receptor agonists induce the uncoupling of gap junctions in AII amacrine cells by cAMP-dependent phosphorylation of connexin proteins [111]. Activation of D_1-like receptors on starburst amacrine cells induces acetylcholine release [2]. Finally, dopamine modulates the function of RGCs including the newly discovered ipRGCs [89, 112, 113]. Emerging—and somewhat controversial—experimental evidence also suggests that these ipRGCs may be involved in the regulation of dopamine release and metabolism [114, 115]. Several lines of evidence also suggest that DA and its receptors are involved in the regulation of RPE functions. For example, inhibition of DA synthesis during the early part of the light phase induced a significant reduction of the morning peak of disk shedding and phagocytosis [116], and mice whose dopaminergic neurons have been destroyed by 1-methyl-4-phenyl-1,2,3,6-tetrahydropyridine (MPTP) accumulate a large number of residual bodies in the RPE [117]. However, it must be mentioned that the presence of dopamine receptors in the mammalian RPE is still controversial [88, 118].

4.6 Melatonin and Dopamine as a Key Regulator of Retinal Circadian Rhythms

In many species, the synthesis and release of both melatonin and dopamine are under circadian control, with melatonin released at night and dopamine during the day. Melatonin inhibits the release of dopamine through an action on melatonin receptors [60, 61, 119], and dopamine inhibits the synthesis and release of melatonin from photoreceptor cells by acting on D_2-like dopamine receptors [18, 40, 118]. Thus, the melatonin and dopamine secreting cells form a cellular feedback loop functioning to regulate circadian retinal physiology (Fig. 4.3). Although compelling molecular evidence suggests that dopaminergic neurons in the retina are likely to contain circadian clocks [120], the circadian rhythm of dopamine release and metabolism appears to be dependent on melatonin. Retinal dopamine content

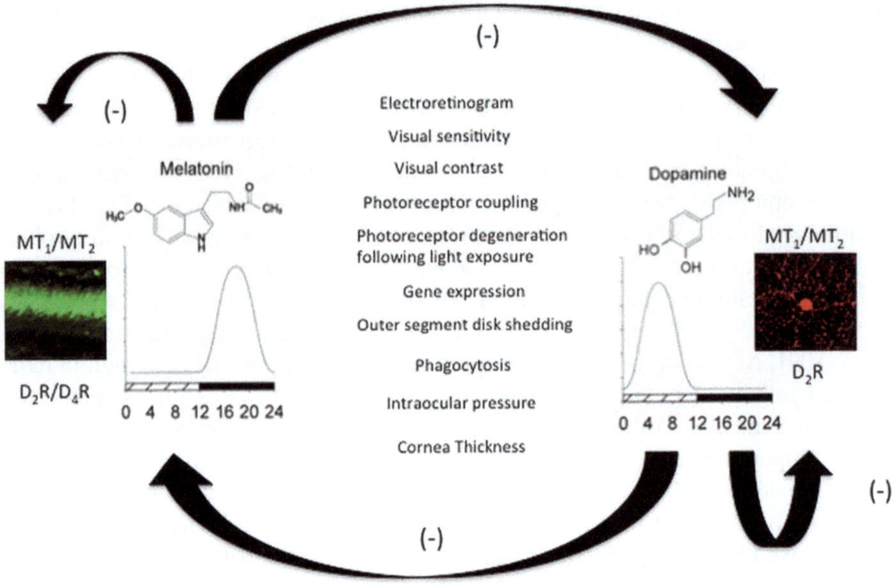

Fig. 4.3 Melatonin and dopamine modulate many retinal functions. Melatonin modulates dopaminergic neurons (paracrine role) in the retina and it can also modulate the functions of the photoreceptors where it is synthesized (autocrine role). Dopamine can also regulate melatonin synthesis by acting on D_2R and D_4R present on the melatonin synthesizing cells (photoreceptors) and feedback on dopaminergic cell via D_2R (autocrine role). Melatonin and dopamine receptors are expressed in many retinal cells (see Table 4.1) and therefore these neurohormones can modulate many different functions within the eye

and metabolism are circadian in mice that synthesize melatonin, but not in mice that are genetically incapable of synthesizing melatonin [76, 77, 106]; and daily injections of melatonin induce circadian rhythms of dopamine in retinas of mice that are unable to synthesize the neurohormone [77]. The role of melatonin in controlling DA rhythmicity is not unique to mice. For example, in pigeon and goldfish retinas, circadian clocks regulate rhythms of dopamine release via melatonin receptors [61, 82].

In *Xenopus laevis* DA, via D_2-like receptors, is involved in the entrainment of circadian rhythm of retinal melatonin synthesis [10, 121] via regulation of *Per2* mRNA expression [122, 123]. Recent experimental evidence also suggests that DA signaling plays an important role in the regulation of circadian rhythms in the mouse retina [106, 124–126] and these studies indicate that the modulation of circadian rhythms by DA within the retina involves multiple subtypes of DA receptors. Ruan et al. [124] demonstrated that the circadian rhythm in PER2:LUC bioluminescence observed within the inner retina is phase-shifted by activation of D_1R [124], whereas the entrainment of the circadian clock in the photoreceptors seems to be controlled by D_4R [106, 125]. Additional evidence for a role of DA signaling in circadian rhythm regulation in the mouse retina comes from a series of studies in which the

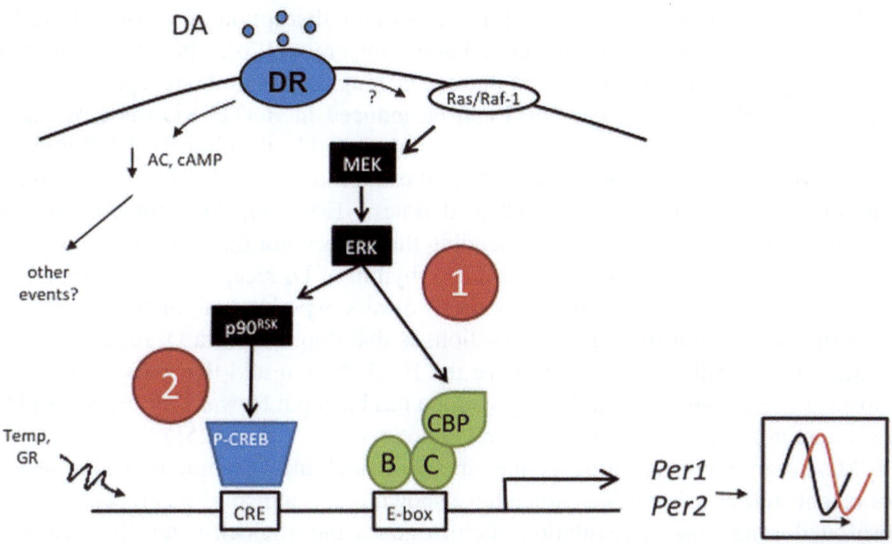

Fig. 4.4 Proposed model for the entrainment of the circadian clock by dopamine (DA) in the RPE. Dopamine receptor (DR) activation leads to activation of MEK, and then Erk1/2 (ERK) via a cAMP-independent pathway. Activation of *Period* (*Per*) *1* and *2* can occur by either (1) phosphorylation of CBP causing it to bind and activate BMAL1:CLOCK (B, C) or (2) phosphorylation of p90RSK, which in turn phosphorylates CREB at Ser133 thus activating *Per1/2* promoters. Adapted from [128] (Drawing by Dr. J.P. DeBruyne)

D_2R null mice were employed [127]. These D_2R null mice showed a significant reduction in the light masking response and light induction of *Per1* in the retina was dramatically reduced [127]. Moreover, CLOCK:BMAL1 transcriptional activity is augmented by activation of D_2R [128] and D_2R signaling also contributes to the stability of BMAL1 protein [129]. The precise molecular mechanisms by which dopamine drives the entrainment of the circadian clock are not fully understood; however, a number of studies support a model in which the rapid induction of the circadian clock genes *Period1* (*Per1*) and *Period2* (*Per2*) drives the resetting process [130, 131]. The induction of *Per1*mRNA is believed to be positively coupled with activation of cAMP response element-binding protein (CREB) located on the promoter region of the *Per1* [132, 133]. D_2-like receptors are negatively coupled with adenylyl cyclase (AC) and thereby lead to a decrease of cAMP levels. Thus, it is unlikely that D_2R activation induces *Per1* mRNA via the cAMP signaling pathway. *Per1* mRNA induction in the retina may be regulated by cAMP-independent signaling pathways since photic induction of *Per1* mRNA is dramatically decreased in $D_2R^{-/-}$ mice [126] and a functional CRE is not required for its induction [128] (see Fig. 4.4).

There is a circadian rhythm of b-wave amplitude of the ERG that is modulated by dopamine and dopamine D_4 receptors. In wild-type mice, the b-wave amplitude is higher in the daytime than at night and this difference persists in constant darkness [134, 135]. The ERG rhythm is disrupted in *Bmal1* and *Cry* mutant mice [134,

135], and in mice with a conditional retina-specific disruption of *Th* (RetTH-KO), that depletes dopamine from the retina but not the brain [126]. The effect of retinal dopamine depletion is phenocopied in mice lacking dopamine D_4 receptors and the circadian rhythm of photopic ERG can be induced in RetTH-KO mice by daily injection of a dopamine D_4 receptor agonist. Interestingly, the photopic ERG rhythm is observed in mice on a melatonin-deficient background, which do not show rhythmic dopamine metabolism in constant darkness. However, these mice express a circadian rhythm of the transcript encoding the D_4 receptor that persists in constant darkness [125]. We hypothesize that it is a rhythm of D_4 receptors that sustains the circadian ERG rhythm, but this hypothesis awaits experimental analysis. Another possible explanation for these observations is that dopamine via D_4 receptors may entrain the circadian clocks that drive the ERG rhythm and that without DA the clocks desynchronize. A similar explanation has been put forward for the role of D_4 receptors in rhythmic gene expression in photoreceptor cells [125].

Melatonin may also influence the circadian clock in the retina. Indeed, a series of recent studies using $MT_1^{-/-}$ mice have shown that removal of this receptor has a profound impact on the regulation of clock genes and of clock-controlled genes in many tissues. For example, Von Gall et al. [136] reported that rhythmic expression of the *Period1* (gene and protein) in the pituitary gland depends on melatonin via MT_1 signaling. Additional studies reported that the rhythmic expression of several clock genes (*Per1, Per 2, Bmal1, and Cry 1*) in the mouse *pars tuberalis* depends on MT_1 signaling [137]. A few studies have also suggested that melatonin influences the amplitude and the phase of *Per1, and Cry1 m*RNA in the mouse retina [138, 139].

As mentioned above, it appears that melatonin via MT_1 controls the daily rhythms of the scotopic and photopic ERGs in C3H mice [65], and other investigations performed with melatonin-deficient mice have shown that the daily rhythm in the amplitude of the scotopic ERGs does not persist in constant darkness, thus demonstrating a lack of circadian control [66, 134, 135]. Since melatonin is believed to be a key regulator of circadian functions within the retina, a recent study investigated whether there is circadian regulation of the photopic and scotopic ERGs in melatonin-proficient mice and found a lack of circadian control of the dark-adapted ERG [66]. On the other hand, a circadian rhythm in the photopic ERGs was observed in these mice, as in melatonin-deficient mice [134, 135]. Thus, melatonin may not be directly involved in the circadian regulation of the photopic ERGs in the mouse. However, the circadian regulation in the photopic ERG was abolished by targeted disruption of the MT_1 receptor, indicating that in melatonin-proficient mice MT_1 receptor signaling is required for the circadian regulation of the photic ERGs [66]. Interestingly, removal of the MT_1 receptors did not affect the circadian regulation of DA and DOPAC. Therefore, the loss of circadian regulation in the photopic ERGs observed in $MT_1^{-/-}$ mice is not a consequence of the loss in the circadian regulation of DA and DOPAC. A possible explanation for these results may be found in a previous study that demonstrated altered photopic ERG responses in mice lacking melanopsin (the photopigment present in the ipRGCs) [140]. Since MT_1 receptors are present on the ipRGCs [66], it is possible that the lack of MT_1 signaling may affect the circadian regulation of the photic ERG by acting on MT_1 receptors present on the ipRGCs.

Melatonin and dopamine have been implicated in the regulation of the disk shedding (see references above) although their role is not clear. For example, the circadian regulation of disk shedding does not appear to be different between melatonin-proficient mice (C3H/$f^{+/+}$) and melatonin-deficient mice (C57BL/6) since the disk shedding was rhythmic in both strains and was not affected by administration of exogenous melatonin [105]. In addition, since DA rhythms do not persist in C57BL/6 mice when exposed to constant darkness, it must be concluded that the circadian rhythm in dopamine metabolism is not involved in the regulation of the circadian rhythms in disk shedding [105]. However, a recent study has reported that in melatonin-proficient mice in which the melatonin receptors have been removed (MT_1 and MT_2 knock-down), the timing of the peak in the daily rhythm of phagocytosis occurs during the late night and not after the onset of light as occurs in wild-type mice thus suggesting that melatonin signaling may be involved in timing of the daily rhythm of shedding and phagocytosis [141]. Additional emerging evidence has shown that the RPE contains an autonomous circadian clock that controls the circadian rhythms in PER2:LUC bioluminescence rhythm [142]. Interestingly, this circadian rhythm in the RPE can be phase-shifted by exogenous dopamine but not by melatonin [143]. Further studies are required to fully understand the mechanisms by which melatonin and dopamine may regulate the circadian rhythms disk shedding and phagocytosis.

4.7 Conclusions

As we have described in the previous sections it is well established that melatonin and dopamine are important players in the regulation of many retinal functions and are the drivers of retinal circadian rhythms. Recent work using transgenic mice has greatly facilitated our understating of the mechanisms by which these two neurohormones regulate retinal physiology. Although extended knowledge now exists on the role of melatonin and dopamine in many different animal models, a clear role for the melatonin and dopamine systems in humans is still missing. It is our hope that future genomic studies will be performed on human populations with ocular diseases to determine if polymorphisms in genes encoding enzymes—or receptors—involved in the melatonin and dopamine systems are associated with specific visual disorders.

Acknowledgments Research in the authors' laboratories is supported by grants from the National Institutes of Health (R01 NS43459, R21 EY028821, R01 EY022216 (GT); R01 EY004864, P30 EY006360 (PMI)), and Research to Prevent Blindness, Inc. (RPB) (PMI). PMI is a recipient of Senior Scientific Investigator Award from RPB.

References

1. Tosini G, Baba K, Hwang CK, Iuvone PM. Melatonin: an underappreciated player in retinal physiology and pathophysiology. Exp Eye Res. 2012;103:82–9.
2. Witkovsky P. Dopamine and retinal function. Doc Ophthalmol. 2004;108:17–40.
3. Iuvone PM, Besharse JC. Dopamine receptor-mediated inhibition of serotonin N-acetyltransferase activity in retina. Brain Res. 1986;369:168–76.
4. Klein DC, Coon SL, Roseboom PH, Weller JL, Bernard M, et al. The melatonin rhythm-generating enzyme: molecular regulation of serotonin N-acetyltransferase in the pineal gland. Recent Prog Horm Res. 1997;52:307–57.
5. Iuvone PM, Tosini G, Haque R, Klein DC, Chaurasia SS. Circadian clocks, clock-controlled genes and melatonin biosynthesis in the retina. Prog Retin Eye Res. 2005;24:433–56.
6. Cahill GM, Besharse JC. Circadian clock functions localized in xenopus retinal photoreceptors. Neuron. 1993;10:573–7.
7. Liu C, Fukuhara C, Wessel JH, Iuvone PM, Tosini G. Localization of *Aa-nat* mRNA in the rat retina by fluorescence in situ hybridization and laser capture microdissection. Cell Tissue Res. 2004;315:197–1.
8. Sakamoto K, Liu C, Tosini G. Circadian rhythms in the retina of rats with photoreceptor degeneration. J Neurochem. 2004;90:1019–4.
9. Garbarino-Pico E, Carpentieri AR, Contin MA, Sarmiento MI, Brocco MA, et al. Retinal ganglion cells are autonomous circadian oscillators synthesizing N-acetylserotonin during the day. J Biol Chem. 2004;279:51172–81.
10. Cahill GM, Besharse JC. Resetting the circadian clock in cultured Xenopus eyecups: regulation of retinal melatonin rhythms by light and D2 dopamine receptors. J Neurosci. 1991; 11:2959–71.
11. Underwood H, Binkley S, Siopes T, Mosher K. Melatonin rhythms in the eyes, pineal bodies, and blood of Japanese quail (Coturnix coturnix japonica). Gen Comp Endocrinol. 1984;56:70–1.
12. Tosini G, Pozdeyev N, Sakamoto K, Iuvone PM. The circadian clock system in mammalian retina. Bioessays. 2008;30:624–3.
13. Sakamoto K, Oishi K, Shiraishi M, Hamano S, Otsuka H, et al. Two circadian oscillatory mechanisms in the mammalian retina. Neuroreport. 2000;11:3995–7.
14. Besharse JC, Iuvone PM. Circadian clock in Xenopus eye controlling retinal serotonin N-acetyltransferase. Nature. 1983;305:133–5.
15. Tosini G, Menaker M. Circadian rhythms in cultured mammalian retina. Science. 1996; 272:419–1.
16. Tosini G, Menaker M. The clock in the mouse retina: melatonin synthesis and photoreceptor degeneration. Brain Res. 1998;789:221–8.
17. Tosini G, Menaker M. Temperature compensation of retinal circadian oscillators in wild-type and *tau* mutant hamsters. Neuroreport. 1998;9:1001–5.
18. Zawilska JB, Iuvone PM. Melatonin synthesis in chicken retina: effect of kainic acid-induced lesions on the diurnal rhythm and D2-dopamine receptor-mediated regulation of serotonin N-acetyltransferase activity. Neurosci Lett. 1992;135:71–4.
19. Thomas KB, Tigges M, Iuvone PM. Melatonin synthesis and circadian tryptophan hydroxylase activity in chicken retina following destruction of serotonin immunoreactive amacrine and bipolar cells by kainic acid. Brain Res. 1993;601:303–7.
20. Sakamoto K, Liu C, Kasamatsu M, Iuvone PM, Tosini G. Intraocular injection of kainic acid does not abolish the circadian rhythm of arylalkylamine N-acetyltransferase mRNA in rat photoreceptors. Mol Vis. 2006;12:117–4.
21. Tosini G, Davidson AJ, Fukuhara C, Kasamatsu M, Castanon-Cervantes O. Localization of a circadian clock in mammalian photoreceptors. FASEB J. 2007;21:3866–71.
22. Chen W, Baler R. The rat arylalkylamine N-acetyltransferase E-box: differential use in a master vs. A slave oscillator. Brain Res Mol Brain Res. 2000;81:43–50.

23. Fukuhara C, Liu C, Ivanova TN, Chan GC-K, Storm DR, et al. Gating of the cAMP signaling cascade and melatonin synthesis by the circadian clock in mammalian retina. J Neurosci. 2004;24:1803–11.
24. Pozdeyev N, Taylor C, Haque R, Chaurasia SS, Visser A, et al. Photic regulation of arylalkylamine N-acetyltransferase binding to 14-3-3 proteins in retinal photoreceptor cells. J Neurosci. 2006;26:9153–1.
25. Ganguly S, Gastel JA, Weller JL, Schwartz C, Jaffe H, et al. Role of a pineal cAMP-operated arylalkylamine N-acetyltransferase/14-3-3-binding switch in melatonin synthesis. Proc Natl Acad Sci U S A. 2001;98:8083–8.
26. Obsil T, Ghirlando R, Klein DC, Ganguly S, Dyda F. Crystal structure of the 14-3-3zeta:serotonin N-acetyltransferase complex. A role for scaffolding in enzyme regulation. Cell. 2001;105:257–7.
27. Chaurasia SS, Haque R, Pozdeyev N, Jackson CR, Iuvone PM. Temporal coupling of cyclic AMP and Ca/calmodulin-stimulated adenylyl cyclase to the circadian clock in chick retinal photoreceptor cells. J Neurochem. 2006;99:1142–50.
28. Nowak JZ, Zurawska E, Zawilska J. Melatonin and its generating system in vertebrate retina: circadian rhythm, effect of environmental lighting and interaction with dopamine. Neurochem Int. 1989;14:397–6.
29. Klein DC, Weller JL. Rapid light-induced decrease in pineal serotonin N-acetyltransferase activity. Science. 1972;177:532–3.
30. Hamm HE, Takahashi JS, Menaker M. Light-induced decrease of serotonin N-acetyltransferase activity and melatonin in the chicken pineal gland and retina. Brain Res. 1983;266:287–3.
31. Orr HT, Lowry OH, Cohen AI, Ferrendelli JA. Distribution of 3′:5′-cyclic AMP and 3′:5′-cyclic GMP in rabbit retina in vivo: selective effects of dark and light adaptation and ischemia. Proc Natl Acad Sci U S A. 1976;73:4442–5.
32. DeVries GW, Cohen AI, Hall IA, Ferrendelli JA. Cyclic nucleotide levels in normal and biologically fractionated mouse retina: effects of light and dark adaptation. J Neurochem. 1978;31:1345–51.
33. Nir I, Harrison JM, Haque R, Low MJ, Grandy DK, et al. Dysfunctional light-evoked regulation of cAMP in photoreceptors and abnormal retinal adaptation in mice lacking dopamine D4 receptors. J Neurosci. 2002;22:2063–3.
34. Ivanova TN, Iuvone PM. Circadian rhythm and photic control of cAMP level in chick retinal cell cultures: a mechanism for coupling the circadian oscillator to the melatonin-synthesizing enzyme, arylalkylamine N-acetyltransferase, in photoreceptor cells. Brain Res. 2003; 991:96–03.
35. Gastel JA, Roseboom PH, Rinaldi PA, Weller JL, Klein DC. Melatonin production: proteasomal proteolysis in serotonin N-acetyltransferase regulation. Science. 1998;279:1358–60.
36. Fukuhara C, Dirden JC, Tosini G. Photic regulation of melatonin in rat retina and the role of proteasomal proteolysis. Neuroreport. 2001;12:3833–7.
37. Iuvone PM, Brown AD, Haque R, Weller J, Zawilska JB, et al. Retinal melatonin production: role of proteasomal proteolysis in circadian and photic control of arylalkylamine N-acetyltransferase. Invest Ophthalmol Vis Sci. 2002;43:564–72.
38. Cohen AI, Todd RD, Harmon S, O'Malley KL. Photoreceptors of mouse retinas possess D4 receptors coupled to adenylate cyclase. Proc Natl Acad Sci U S A. 1992;89:12093–7.
39. Zawilska JB, Derbiszewska T, Nowak JZ. Clozapine and other neuroleptic drugs antagonize the light-evoked suppression of melatonin biosynthesis in chick retina: involvement of the D4-like dopamine receptor. J Neural Transm Gen Sect. 1994;97:107–7.
40. Tosini G, Dirden JC. Dopamine Inhibits melatonin release in the mammalian retina: in vitro evidence. Neurosci Lett. 2000;286:119–2.
41. Ivanova TN, Alonso-Gomez AL, Iuvone PM. Dopamine D4 receptors regulate intracellular calcium concentration in cultured chicken cone photoreceptor cells: relationship to dopamine receptor-mediated inhibition of cAMP formation. Brain Res. 2008;1207:111–9.

42. Wiechmann AF, O'Steen WK. Melatonin increases photoreceptor susceptibility to light-induced damage. Invest Ophthalmol Vis Sci. 1992;33:1894–2.
43. Sugawara T, Sieving PA, Luvone PM, Bush RA. The melatonin antagonist luzindole protects retinal photoreceptors from light damage in the rat. Invest Ophthalmol Vis Sci. 1998; 39:2458–65.
44. Grace MS, Cahill GM, Besharse JC. Melatonin deacetylation: retinal vertebrate class distribution and Xenopus laevis tissue distribution. Brain Res. 1991;559:56–3.
45. Cahill GM, Besharse JC. Retinal melatonin is metabolized within the eye of Xenopus laevis. Proc Natl Acad Sci U S A. 1989;86:1098–02.
46. Li P, Pang SF, Tsang CW. Retinal 5-methoxytryptamine and 5-methoxyindole-3-acetic acid in the rat and quail: diurnal rhythms and interspecies differences. Biochem Biophys Res Commun. 1997;239:353–6.
47. Leino M. 6-Methoxy-tetrahydro-beta-carboline and melatonin in the human retina. Exp Eye Res. 1984;38:325–30.
48. Rogawski MA, Roth RH, Aghajanian GK. Melatonin: deacetylation to 5-methoxytryptamine by liver but not brain aryl acylamidase. J Neurochem. 1979;32:1219–6.
49. Hsu LL. Brain aryl acylamidase. Int J Biochem. 1982;14:1037–42.
50. Coon SL, Mazuruk K, Bernard M, Roseboom PH, Klein DC, Rodriguez IR. The human serotonin N-acetyltransferase (EC 2.3.1.87) gene (AANAT): structure, chromosomal localization, and tissue expression. Genomics. 1996;34:76–4.
51. Coon SL, Del Olmo E, Young 3rd WS, Klein DC. Melatonin synthesis enzymes in Macaca mulatta: focus on arylalkylamine N-acetyltransferase (EC 2.3.1.87). J Clin Endocrinol Metab. 2002;87:4699–6.
52. Rodriguez IR, Mazuruk K, Schoen TJ, Chader GJ. Structural analysis of the human hydroxyindole-O-methyltransferase gene. Presence of two distinct promoters. J Biol Chem. 1994;269:31969–77.
53. Reppert SM, Godson C, Mahle CD, Weaver DR, Slaugenhaupt SA, Gusella JF. Molecular characterization of a second melatonin receptor expressed in human retina and brain: the Mel1b melatonin receptor. Proc Natl Acad Sci U S A. 1995;92:8734–8.
54. Scher J, Wankiewicz E, Brown GM, Fujieda H. MT1 melatonin receptor in the human retina: expression and localization. Invest Ophthalmol Vis Sci. 2002;43:889–7.
55. Savaskan E, Wirz-Justice A, Olivieri G, Pache M, Kräuchi K, et al. Distribution of melatonin MT1 receptor immunoreactivity in human retina. J Histochem Cytochem. 2002;50:519–6.
56. Savaskan E, Jockers R, Ayoub M, Angeloni D, Fraschini F, et al. The MT2 melatonin receptor subtype is present in human retina and decreases in Alzheimer's disease. Curr Alzheimer Res. 2007;4:47–51.
57. Wiechmann AF, Summers JA. Circadian rhythms in the eye: the physiological significance of melatonin receptors in ocular tissues. Prog Retin Eye Res. 2008;27:137–60.
58. Fujieda H, Hamadanizadeh SA, Wankiewicz E, Pang SF, Brown GM. Expression of mt1 melatonin receptor in rat retina: evidence for multiple cell targets for melatonin. Neuroscience. 1999;93:793–9.
59. Fujieda H, Scher J, Hamadanizadeh SA, Wankiewicz E, Pang SF, Brown GM. Dopaminergic and GABAergic amacrine cells are direct targets of melatonin: immunocytochemical study of mt1 melatonin receptor in guinea pig retina. Vis Neurosci. 2000;17:63–70.
60. Dubocovich ML. Melatonin is a potent modulator of dopamine release in the retina. Nature. 1983;306:782–4.
61. Ribelayga C, Wang Y, Mangel SC. A circadian clock in the fish retina regulates dopamine release via activation of melatonin receptors. J Physiol. 2004;554:467–82.
62. Meyer P, Pache M, Loeffler KU, Brydon L, Jockers R, et al. Melatonin MT-1-receptor immunoreactivity in the human eye. Br J Ophthalmol. 2002;86:1053–7.
63. Ayoub MA, Couturier C, Lucas-Meunier E, Angers S, Fossier P, et al. Monitoring of ligand-independent dimerization and ligand-induced conformational changes of melatonin receptors in living cells by bioluminescence resonance energy transfer. J Biol Chem. 2002; 277:21522–8.

64. Ayoub MA, Levoye A, Delagrange P, Jockers R. Preferential formation of MT1/MT2 melatonin receptor heterodimers with distinct ligand interaction properties compared with MT2 homodimers. Mol Pharmacol. 2004;66:312–21.
65. Baba K, Pozdeyev N, Mazzoni F, Contreras-Alcantara S, Liu C, et al. Melatonin modulates visual function and cell viability in the mouse retina via the MT_1 melatonin receptor. Proc Natl Acad Sci U S A. 2009;106:15043–8.
66. Sengupta A, Baba K, Mazzoni F, Pozdeyev NV, Strettoi E, et al. Localization of melatonin receptor 1 in mouse retina and its role in the circadian regulation of the electroretinogram and dopamine levels. PLoS One. 2011;6:e24483.
67. Osborne NN, Chidlow G. The presence of functional melatonin receptors in the iris-ciliary processes of the rabbit eye. Exp Eye Res. 1994;59:3–9.
68. Pintor J, Martin L, Pelaez T, Hoyle CH, Peral A. Involvement of melatonin MT(3) receptors in the regulation of intraocular pressure in rabbits. Eur J Pharmacol. 2001;416:251–4.
69. Pintor J, Pelaez T, Hoyle CH, Peral A. Ocular hypotensive effects of melatonin receptor agonists in the rabbit: further evidence for an MT_3 receptor. Br J Pharmacol. 2003;138:831–6.
70. Wiechmann AF, Wirsig-Wiechmann CR. Melatonin receptor mRNA and protein expression in Xenopus laevis nonpigmented ciliary epithelial cells. Exp Eye Res. 2001;73:617–3.
71. Dortch-Carnes J, Tosini G. Melatonin receptor agonist-induced reduction of SNP-released nitric oxide and cGMP production in isolated human non-pigmented ciliary epithelial cells. Exp Eye Res. 2013;107:1–10.
72. Samples JR, Krause G, Lewy AJ. Effect of melatonin on intraocular pressure. Curr Eye Res. 1988;7:649–3.
73. Alcantara-Contreras S, Baba K, Tosini G. Removal of melatonin receptor type 1 increases intraocular pressure and retinal ganglion cells death in the mouse. Neurosci Lett. 2011;494:61–4.
74. Ehinger B, Florén I. Quantitation of the uptake of indoleamines and dopamine in the rabbit retina. Exp Eye Res. 1978;26:1–11.
75. Iuvone PM, Galli CL, Garrison-Gund CK, Neff NH. Light stimulates tyrosine hydroxylase activity and dopamine synthesis in retinal amacrine neurons. Science. 1978;202:901–2.
76. Nir I, Haque R, Iuvone PM. Diurnal metabolism of dopamine in dystrophic retinas of homozygous and heterozygous retinal degeneration slow (rds) mice. Brain Res. 2000;884:13–2.
77. Doyle SE, Grace MS, McIvor W, Menaker M. Circadian rhythms of dopamine in mouse retina: the role of melatonin. Vis Neurosci. 2002;19:593–601.
78. Stempels N, Tassignon MJ, Sarre S, Nguyen-Legros J. Microdialysis measurement of catecholamines in rabbit vitreous after retinal laser photocoagulation. Exp Eye Res. 1994; 59:433–9.
79. Megaw P, Morgan I, Boelen M. Vitreal dihydroxyphenylacetic acid (DOPAC) as an index of retinal dopamine release. J Neurochem. 2001;76:1636–4.
80. Witkovsky P, Nicholson C, Rice ME, Bohmaker K, Meller E. Extracellular dopamine concentration in the retina of the clawed frog, Xenopus laevis. Proc Natl Acad Sci U S A. 1993; 90:5667–1.
81. Adachi A, Nogi T, Ebihara S. Phase-relationship and mutual effects between circadian rhythms of ocular melatonin and dopamine in the pigeon. Brain Res. 1998;792:361–9.
82. Adachi A, Suzuki Y, Nogi T, Ebihara S. The relationship between ocular melatonin and dopamine rhythms in the pigeon: effects of melatonin inhibition on dopamine release. Brain Res. 1999;815:435.
83. Zhang DQ, Zhou TR, McMahon DG. Functional heterogeneity of retinal dopaminergic neurons underlying their multiple roles in vision. J Neurosci. 2007;27:69262–9.
84. Nguyen-Legros J, Simon A, Caillé I, Bloch B. Immunocytochemical localization of dopamine D1 receptors in the retina of mammals. Vis Neurosci. 1997;14:545–1.
85. Versaux-Botteri C, Giber JM, Nguyen-Legros J, Vernier P. Molecular identification of a dopamine D1b receptor in bovine retinal pigment epithelium. Neurosci Lett. 1995;237:9–2.
86. Jackson CR, Chaurasia SS, Zhou H, Haque R, Storm DR, Iuvone PM. Essential roles of dopamine D4 receptors and the type 1 adenylyl cyclase in photic control of cyclic AMP in photoreceptor cells. J Neurochem. 2009;109:148–57.

87. Derouiche A, Asan E. The dopamine D2 receptor subfamily in rat retina: ultrastructural immunogold and in situ hybridization studies. Eur J Neurosci. 1999;11:1391–02.
88. Nguyen-Legros J, Versaux-Botteri C, Vernier P. Dopamine receptor localization in the mammalian retina. Mol Neurobiol. 1999;19:181–04.
89. Sakamoto K, Liu C, Kasamatsu M, Pozdeyev NV, Iuvone PM, Tosini G. Dopamine regulates melanopsin mRNA expression in intrinsically photosensitive retinal ganglion cells. Eur J Neurosci. 2005;12:3129–6.
90. Wagner HJ, Luo BG, Ariano MA, Sibley DR, Stell BK. Localization of D2 dopamine receptors in vertebrates retinae with antipeptide antibodies. J Comp Neurol. 1993;331:469–81.
91. Dong F, An JH, Ren YP, Yan DS, Zhou XT, Lu F, Hu DN, Chen JF, Qu J. Expression of Dopamine receptor D2 and adenosine receptor A2A in human retinal pigment epithelium. Zhonghua Yan Ke Za Zhi. 2007;43:1110–3.
92. Kay JN, De la Huerta I, Kim IJ, Zhang Y, Yamagata M, et al. Retinal ganglion cells with distinct directional preferences differ in molecular identity, structure, and central projections. J Neurosci. 2011;31:7753–62.
93. Wiechmann AF, Yang XL, Wu SM, Hollyfield JG. Melatonin enhances horizontal cell sensitivity in salamander retina. Brain Res. 1988;453:377.
94. Huang H, Lee SC, Yang XL. Modulation by melatonin of glutamatergic synaptic transmission in the carp retina. J Physiol. 2005;569:857–71.
95. Ping Y, Huang H, Zhang XJ, Yang XL. Melatonin potentiates rod signals to ON type bipolar cells in fish retina. J Physiol. 2008;586:2683–4.
96. Wiechmann AF, Vrieze MJ, Dighe R, Hu Y. Direct modulation of rod photoreceptor responsiveness through a Mel(1c) melatonin receptor in transgenic Xenopus laevis retina. Invest Ophthalmol Vis Sci. 2003;44:4522–1.
97. Lu J, Zoran MJ, Cassone VM. Daily and circadian variation in the electroretinogram of the domestic fowl: effects of melatonin. J Comp Physiol A. 1995;177:299–6.
98. McGoogan JM, Cassone VM. Circadian regulation of chick electroretinogram: effects of pinealectomy and exogenous melatonin. Am J Physiol. 1999;277:R1418–27.
99. Gagne AM, Danilenko KV, Rosolen SG, Herbert M. Impact of oral melatonin on the electroretinogram. J Circadian Rhythm. 2009;7:14.
100. Rufiange M, Dumont M, Lachappelle P. Correlating retinal function with melatonin secretion in subject with an early or late circadian phase. Invest Ophthalmol Vis Sci. 2002;43:2491–9.
101. Baba K, Mazzoni F, Owino S, Contreras-Alcantara S, Strettoi E, Tosini G. Age-related changes in the daily rhythm of photoreceptor functioning and circuitry in a melatonin-proficient mouse strain. PLoS One. 2012;7:e37799.
102. Von Gall C, Weaver DR. Loss of responsiveness to melatonin in the aging mouse suprachiasmatic nucleus. Neurobiol Aging. 2008;29:464.
103. Besharse JC, Dunis DA. Methoxyindoles and photoreceptor metabolism: activation of rod shedding. Science. 1983;219:1341–3.
104. White MP, Fisher LJ. Effects of exogenous melatonin on circadian disc shedding in the albino rat retina. Vision Res. 1989;29:167–79.
105. Grace MS, Chiba A, Menaker M. Circadian control of photoreceptor outer segment membrane turnover in mice genetically incapable of melatonin synthesis. Vis Neurosci. 1999;16:909–18.
106. Pozdeyev NK, Tosini G, Ali F, Rozov S, Lee RH, Iuvone PM. Dopamine modulates diurnal and circadian rhythms of protein phosphorylation in photoreceptor cells of mouse retina. Eur J Neurosci. 2008;27:26691–700.
107. Udovichenko IP, Newton AC, Williams DS. Regulation of the phosphorylation state of rhodopsin by dopamine. J Biol Chem. 1998;273:7181–4.
108. Ribelayga C, Cao Y, Mangel SC. The circadian clock in the retina controls rod-cone coupling. Neuron. 2008;59:790–01.
109. Ribelayga C, Mangel SC. Identification of a circadian clock-controlled neural pathway in the rabbit retina. PLoS One. 2010;5:e11020.

110. Knapp AG, Dowling JE. Dopamine enhances excitatory amino acid-gated conductances in cultured retinal horizontal cells. Nature. 1987;325:437–9.
111. Hampson EC, Vaney DI, Weiler R. Dopaminergic modulation of gap junction permeability between amacrine cells in mammalian retina. J Neurosci. 1992;12:4911–22.
112. Vugler AA, Redgrave P, Semo M, Lawrence J, Greenwood J, Coffey PJ. Dopamine neurones form a discrete plexus with melanopsin cells in normal and degenerating retina. Exp Neurol. 2007;205:26–5.
113. Van Hook MJ, Wong KY, Berson DM. Dopaminergic modulation of ganglion-cell photoreceptors in rat. Eur J Neurosci. 2012;35:507–18.
114. Zhang DQ, Wong KY, Sollars PJ, Berson DM, Pickard GE, McMahon DG. Intraretinal signaling by ganglion cell photoreceptors to dopaminergic amacrine neurons. Proc Natl Acad Sci U S A. 2008;105:14181–6.
115. Cameron MA, Pozdeyev N, Vugler AA, Cooper H, Iuvone PM, Lucas RJ. Light regulation of retinal dopamine that is independent of melanopsin phototransduction. Eur J Neurosci. 2009;29:761–7.
116. Remé C, Wirz-Justice A, Rhyner A, Hofmann S. Circadian rhythm in the light response of rat retinal disk-shedding and autophagy. Brain Res. 1986;369:356.
117. Mariani AP, Neff NH, Hadjicostantinou M. 1-Methyl-4-phenyl-1,2,3,6-tetrahydropyridine (MPTP) treatment decreases dopamine and increases lipofuscin in mouse retina. Neurosci Lett. 1986;72:221–6.
118. Nguyen-Legros J, Chanut E, Versaux-Botteri C, Simon A, Trouvin JH. Dopamine inhibits melatonin synthesis in photoreceptor cells through a D2-like receptor subtype in the rat retina: biochemical and histochemical evidence. J Neurochem. 1996;67:2514–20.
119. Boatright JH, Rubim NM, Iuvone PM. Regulation of endogenous dopamine release in amphibian retina by melatonin: the role of GABA. Vis Neurosci. 1994;11:1013–8.
120. Ruan G-X, Zhang DQ, Zhou T, Yamazaki S, McMahon DG. The circadian organization of the mammalian retina. Proc Natl Acad Sci U S A. 2006;103:9703–8.
121. Hasegawa M, Cahill GM. A role for cyclic AMP in entrainment of the circadian oscillator in Xenopus retinal photoreceptors by dopamine but not by light. J Neurochem. 1999;72: 1812–20.
122. Steenhard BM, Besharse JC. Phase shifting the retinal circadian clock: xPer2 mRNA induction by light and dopamine. J Neurosci. 2000;20:8572–7.
123. Besharse JC, Zhuang M, Freeman K, Fogerty J. Regulation of photoreceptor Per1 and Per2 by light, dopamine and a circadian clock. Eur J Neurosci. 2004;20:167–4.
124. Ruan G-X, Allen GC, Yamazaki S, McMahon DC. An autonomous circadian clock in the inner mouse retina regulated by Dopamine and Gaba. PLoS Biol. 2008;6:e249.
125. Jackson CR, Chaurasia SS, Hwang CK, Iuvone PM. Dopamine D2 receptor activation controls circadian timing of the adenylyl cyclase 1/cyclic AMP signaling system in mouse retina. Eur J Neurosci. 2011;34:57–64.
126. Jackson CR, Ruan GX, Aseem F, Abey J, Gamble K, et al. Retinal dopamine mediates multiple dimensions of light-adapted vision. J Neurosci. 2012;32:9359–68.
127. Doi M, Yujnovsky I, Hirayama J, Malerba M, Tirotta E, Sassone-Corsi P, Borrelli E. Impaired light masking in dopamine D2 receptor-null mice. Nat Neurosci. 2006;9:732–4.
128. Yujnovsky I, Hyrayama J, Doi M, Borrelli E, Sassoni-Corsi P. Signaling mediated by dopamine D2 receptor potentiates circadian regulation by CLOCK:BMAL1. Proc Natl Acad Sci U S A. 2006;103:6386–1.
129. Sahar S, Zocchi L, Kinoshita C, Borrelli E, Sassone-Corsi P. Regulation of BMAL1 protein stability and circadian function by GSK3beta-mediated phosphorylation. PLoS One. 2010;5:e8561.
130. Akiyama M, Kouzu Y, Takahashi S, Wakamatsu H, Moriya T, et al. Inhibition of light- or glutamate-induced mPer1 expression represses the phase shifts into the mouse circadian locomotor and suprachiasmatic firing rhythms. J Neurosci. 1999;19:1115–21.

131. Albrecht U, Zheng B, Larkin D, Sun ZS, Lee CC. MPer1 and mper2 are essential for normal resetting of the circadian clock. J Biol Rhythms. 2001;16:100–4.
132. Obrietan K, Impey S, Storm DR. Light and circadian rhythmicity regulate MAP kinase activation in the suprachiasmatic nuclei. Nat Neurosci. 1998;1:693–00.
133. Obrietan K, Impey S, Smith D, Athos J, Storm DR. Circadian regulation of cAMP response element-mediated gene expression in the suprachiasmatic nuclei. J Biol Chem. 1999; 274:17748–6.
134. Storch KF, Paz C, Signorovitch J, Raviola E, Pawlyk B, Weitz CJ. Intrinsic circadian clock of the mammalian retina: importance for retinal processing of visual information. Cell. 2007; 130:730–41.
135. Cameron MA, Barnard AR, Hut RA, Bonnefont X, van der Horst GT, et al. Electroretinography of wild-type and Cry mutant mice reveals circadian tuning of photopic and mesopic retinal responses. J Biol Rhythms. 2008;23:489–01.
136. Von Gall C, Garabette ML, Kell CA, Frenzel S, Dehghani F, Schumm-Draeger PM, et al. Rhythmic gene expression in pituitary depends on heterologous sensitization by the neurohormone melatonin. Nat Neurosci. 2002;5:234–8.
137. Jilg A, Moek J, Weaver DR, Korf HW, Stehle JH, von Gall C. Rhythms in clock proteins in the mouse pars tuberalis depend on MT1 melatonin receptor signalling. Eur J Neurosci. 2005;22:2845–54.
138. Dinet V, Korf H-W. Impact of melatonin receptors on pCREB and clock-gene protein levels in the murine retina. Cell Tissue Res. 2007;330:29–4.
139. Dinet V, Ansari N, Torres-Farfan C, Korf H-W. Clock gene expression in the retina of melatonin-proficient (C3H) and melatonin-deficient (C57BL) mice. J Pineal Res. 2007; 42:83–1.
140. Hankins MW, Lucas RJ. The primary visual pathway in humans is regulated according to long-term light exposure through the action of a nonclassical photopigment. Curr Biol. 2002; 12:191–8.
141. Tosini G, Laurent V, Baba K, Hiragaki S, Contreras-Alcantara S, Hicks D. Modulation of the daily rhythms in disc shedding by melatonin in the mouse. ARVO abstract 2013; 306: 2648.
142. Baba K, Sengupta A, Tosini M, Contreras-Alcantara S, Tosini G. Circadian regulation of the PERIOD 2::LUCIFERASE bioluminescence rhythm in the mouse retinal pigment epithelium-choroid. Mol Vis. 2010;16:2605–11.
143. Baba K, Contreras-Alcantara S, Tosini G. Dopamine entrains the circadian rhythm of PER2::LUC bioluminescence in the mouse retinal pigment epithelium. ARVO abstract 2013; #2019.

Chapter 5
Circadian Organization of the Vertebrate Retina

Douglas G. McMahon

Abstract Our vision is different at different times of the day because our retina works differently at different times of day. These rhythms in the visual function are not just simple responses to the daily light–dark cycle but are in fact the overt expression of an endogenous, self-sustained circadian clock in the retina that drives many rhythms in retinal physiology and metabolism. Thus, the vertebrate retina is both a sensory organ and a 24-h biological clock. Here, we will explore the molecular, cellular, and neurochemical organization of the retinal circadian clock.

Keywords Circadian • Retina • Dopamine • Melatonin • Clock gene • Gene expression • Photoreceptor • Ganglion cell • Vision

The retina, in addition to elaborating the mechanisms for transduction and encoding of light into visual signals, contains a complete circadian clock system—clock genes that generate 24-h rhythms, an input pathway by which light synchronizes the cycling of the retinal clock to the environmental light/dark cycle, and neurochemical output pathways that transmit the clock's influence throughout the retina and into the rest of the brain. The retinal circadian clock drives intrinsic rhythms in retinal gene expression, synaptic communication, and metabolism that globally shape retinal function according to the time of the day. These endogenous 24-h rhythms bias retinal function in anticipation of the normal cycle of photopic and scotopic visual conditions that alternate with the predictable 24-h cycling of solar day and night. The retina is unique among vertebrate neural circadian clocks in containing

D.G. McMahon, Ph.D. (✉)
Department of Biological Sciences, Vanderbilt University, VU Station B,
Box 1634, Nashville, TN 37235, USA
e-mail: douglas.g.mcmahon@vanderbilt.edu

G. Tosini et al. (eds.), *The Retina and Circadian Rhythms*, Springer Series in Vision Research 1, DOI 10.1007/978-1-4614-9613-7_5,

all the functional elements of a clock system within a single neural tissue that is highly accessible, ordered and with well-defined circuitry.

While the vertebrate retina may be unique in its configuration as a circadian clock, its modulation of sensory function is emblematic of a broader biological phenomenon, the widespread influence of circadian clocks on sensory processes. Circadian gating of sensory function is a fundamental property of sensory organs and systems in a wide variety of species. Humans have demonstrated diurnal rhythms in overall visual sensitivity, with increased sensitivity to dim light stimuli at night [1], as well as enhanced temporal resolution of vision in the day [2]. These psychophysical measures are matched by corresponding rhythms in the amplitude and kinetics of the human electroretinogram (ERG), a summed measure of retinal electrical responses to light [3, 4], indicating that the retina itself likely drives or contributes to these visual rhythms. In addition, there are diurnal variations in odor responses [5] and in auditory temporal resolution [6] in humans as well. These sensory rhythms are present in other mammals (e.g., mouse ERG [7, 8], mouse olfactory response [9]). In invertebrates (Drosophila, Limulus, cockroaches), there are prominent circadian rhythms in the sensitivity of vision and olfaction, with changes in sensitivity of up to 6 log units from day to night [10–12]. These rhythms can be endogenously generated within the sensory tissue (e.g., Drosophila olfaction [13]) or via efferent control by central pacemakers (e.g., Limulus vision [11]).

That vertebrate retinas contain endogenous circadian clocks was first demonstrated by circadian melatonin secretion from isolated *Xenopus* frog retinas [14] and then extended to the mammalian retina with the demonstration by Tosini and colleagues of circadian rhythms in melatonin secretion by isolated hamster, mouse, and rat retinas [15–17]. In addition, circadian rhythms in melatonin synthesis, or gene expression, have been shown in isolated fish, reptile, and bird retinas [18–20], demonstrating that endogenous retinal clocks are widespread throughout the vertebrates.

Numerous aspects of retinal physiology and function are under the control of an intrinsic retinal circadian clock, including melatonin release [15, 16], dopamine synthesis [21, 22], gamma-aminobutyric acid (GABA) turnover rate and release [23], extracellular pH [24], ERG b-wave amplitude [7], rod disk shedding [25], and circadian clock gene expression [26–28]. In addition, the mammalian retinal clock and its outputs influence cell survival and growth processes in the eye including the susceptibility of photoreceptors to degeneration from light damage [29, 30], photoreceptor survival in animal models of retinal degeneration [31], photoreceptor and retinal ganglion cell survival in aging [32], and the degree of refractive errors in primate models of myopia [33]. In addition, circadian signals originating in the retina drive rhythms in the hypothalamic biological clock, even in the absence of light/dark cycles [34].

5.1 Molecular Organization of the Retinal Circadian Clock

5.1.1 Retinal Circadian Rhythms Are Generated by a Network of Clock Genes

Circadian rhythms in eukaryotic cells are generated by autoregulatory gene networks that spontaneously produce near-24-h cycles in gene and protein abundance through transcription–translation feedback loops [35]. In mammals, while more than a dozen genes contribute to this network, the core functional elements are six genes that directly participate in the negative feedback loop of the clock mechanism (Fig. 5.1). These genes are the *Period* genes *Per1* and *Per2*, the *Cryptochrome* genes *Cry1* and *Cry2*, and the transcription factors *Clock* and *Bmal1*. CLOCK and BMAL1 form a heterodimeric transcription complex that periodically drives the expression of two *Period* genes (*Per1*, *2*) and two *Cryptochrome* genes (*Cry1-2*). Once transcribed and translated, the resulting PER and CRY proteins also form heteromeric complexes that translocate back into the nucleus to suppress their own transcription by inhibiting CLOCK and BMAL1, forming the fundamental negative feedback loop at the core of the molecular circadian clock [35]. Gene knockout experiments, widely used to define the function roles of individual "clock genes," have indicated

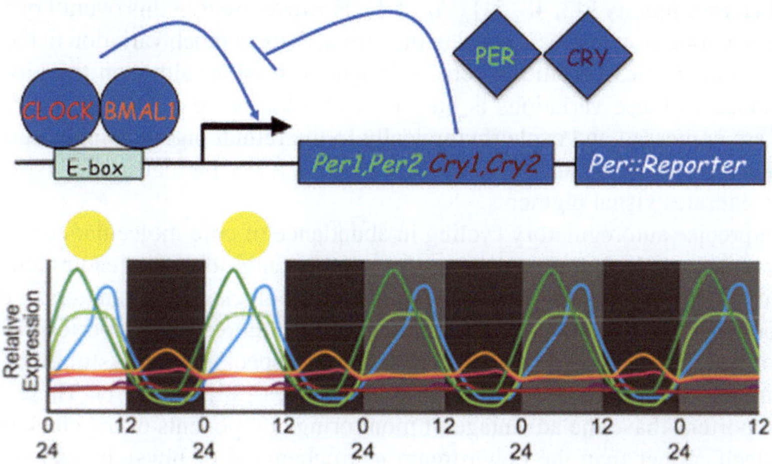

Fig. 5.1 Core circadian clock genes and circadian reporter. *Top*, simplified core circadian molecular feedback loop showing the six canonical core clock genes. CLOCK and BMAL1 form a complex that drives transcription of the *Period* (*Per*) and *Cryptochrome* (*Cry*) genes. PER and CRY proteins complex as a negative feedback element that inhibits CLOCK and BMAL1. *Per*::Reporter artificial genes that contain promoter elements from *Period* genes and make as their protein product a luminescent (e.g., firefly luciferase) or fluorescent (e.g., jellyfish green fluorescent protein) reporter molecule can be inserted by transgenesis and driven off of the clock molecular mechanism. *Bottom*, 24-h cycles in gene abundance typical of the clock genes in a light cycle or constant darkness—*Per* genes in *green*, *Cry* genes in *purple*, *Bmal1* in *orange*, *Clock* in *red*, *Per*::reporter in *blue*. Figure credit Christopher Ciarleglio

that there is partial redundancy among the *Per* genes and *Cry* genes and an absolute requirement for *Bmal1*, when assayed using disruption of behavioral activity/rest cycles as an endpoint [36–39]. Importantly, the clock gene network does not cycle in isolation, but in fact drives extensive, tissue-specific transcriptional networks of clock-controlled genes that are thought to be a fundamental basis rhythmic physiology and metabolism [35].

5.1.2 Clock Genes Are Widely Expressed in the Retina and RPE

The core circadian clock genes are widely expressed in the retina, with expression documented in all major retinal cell classes. Initial in situ hybridization studies that localized expression to particular retinal cell layers, but not to individual cell types, emphasized the high degree of expression of certain clock genes in the photoreceptor layer of nonmammalian vertebrates [40–45] and in the inner retinal layers of mammalian retinas [46–50]. Temporal profiles of clock gene expression showed that in retina, as in the clock nucleus of the brain, many of the clock genes were expressed with apparent 24-h rhythms in abundance, and many of these rhythmic variations persisted in constant darkness, fulfilling a requirement of endogenous circadian rhythmicity [43, 48, 51] (Fig. 5.1). However, within this overall picture of rhythmic expression of clock genes in the retina, there is much variation in the gene-specific and species-specific patterns of such expression, although the biological importance of these variations is not clear. In addition to the retina itself, clock genes are expressed and cycle rhythmically in the retinal pigment epithelium [52], which is in intimate contact with the retina and is a key partner in the visual cycle that regenerates visual pigments.

The precise autoregulatory cycling in abundance of core molecular components of the circadian clock makes them useful experimental tools for reading out clock function. Circadian gene/protein reporters in which clock gene promoters or chimeric clock proteins are linked with luminescence or fluorescence reporters that provide an optic signal of gene abundance have been especially successful as real-time readouts of circadian clock cycling in living tissues [53] (Fig. 5.1). These "clock gene reporters" have the advantage of monitoring components of the clock mechanism itself, rather than the downstream neurochemical or physiological rhythms. For real-time gene expression assays of the retina, transgene reporters that use bioluminescent firefly luciferase, and therefore do not require excitation by visible light, have proven particularly successful. Cultured retinas, isolated from transgenic mice harboring luciferase-based clock gene reporters (*Per1*::LUC; PER2::LUC), demonstrate robust, persistent circadian cycles in bioluminesce [27, 28, 54], both confirming the endogenous nature of the retinal circadian clock as indicated by assays of melatonin secretion and providing the means to examine the organization of the retinal circadian clock at the molecular level.

5.1.3 Genetic Disruption of the Clock Gene Network Disrupts Retinal Circadian Rhythms

The molecular organization of circadian clocks has been approached primarily by gene targeting (i.e., gene knockout) of clock genes followed by assays of behavioral or tissue rhythmicity. Individual clock genes were defined as essential for clock function if circadian rhythmicity was lost following knockout. Functional redundancy was explored by combining knockout mouse lines to produce double or even triple gene knockouts. In general, these types of experiments have demonstrated that the roles of individual clock genes in circadian rhythms generation vary across tissues, suggesting that there are tissue-specific variants of the core circadian clockworks [55, 56].

Which clock genes are essential for the expression of molecular circadian rhythms by the retina, and how does the molecular organization of the retinal circadian clock compare with other neural and nonneural clocks? Initial studies, using the ERG circadian rhythm as a functional assay, found that knockout of *Bmal1*, or of the two Cryptochrome genes, *Cry1* and *Cry2*, each ablated rhythmicity in the ERG [7, 8]. In addition, loss of *Bmal1* disrupted retinal circadian rhythms in gene expression, consistent with its role as a core clock gene [7]. Further studies, using the luciferase molecular rhythms assay from explanted PER2::LUC and *Per1*::LUC mouse retinas, showed that in addition to *Bmal1*, *Per1*, *Cry1*, and *Clock* are each required individually in order for the retina to express molecular circadian rhythms, whereas *Per2*, *Per3*, and *Cry2* were dispensible [54]. Interestingly, the suprachiasmatic nuclei (SCN), the hypothalamic "master" circadian clock, explanted from the same animals as the retinal explants, showed qualitatively similar but quantitatively different effects of the same gene knockout. Knockout of *Per1*, *Cry1*, or *Clock* resulted in decreased robustness of molecular SCN rhythms, but the disruption was less dramatic than the effect of those same gene knockouts on retinal molecular rhythms—only knockout of *Bmal1* alone can abrogate SCN molecular rhythms. Whereas the effects of gene knockout on the robustness of these two neural circadian clocks differed only in degree, the effects on the intrinsic period of molecular rhythms were divergent. Knockout of *Per1* or *Per3* shortened the circadian rhythm of retinas, while knockout of *Clock* lengthened retinal period. In contrast, *Per1* knockout SCN exhibited lengthened periods, while *Per3* and *Clock* knockout SCN were unaffected (Table 5.1). Thus, the retinal neural clock has a unique pattern of clock gene dependence, with the amplitude and robustness of molecular rhythms more vulnerable to gene knockout than the SCN clock, and with divergent regulation of rhythmic period by the *Period* genes.

One explanation for the increased penetrance of clock gene knockouts in the retina could be that neurochemical coupling of neurons is weaker in the retina than in the SCN, where it has been suggested that coupling buffers against genetic perturbation of the clock [55]. This suggests that the retinal clock may be more vulnerable to gene mutations and that there may be disturbances in retinal rhythms in genotypes in which SCN-driven rhythms appear normal. It should be noted however

Table 5.1 Divergent effects of clock gene knockout on the circadian period of retinal and SCN neural clocks

Gene/tissue	Retinal explant	SCN explant
Per1–/– Per1+/–	Arrhythmic/shorten	Lengthen
Per2–/–	No change	Shorten*
Per3–/–	Shorten	No change
Cry1–/–Cry2–/– or +/– Cry1–/–Cry2+/+	Arrhythmic/shorten	Shorten
Cry2–/–	Lengthen	Lengthen
Cry1+/–Cry2+/+, +–/	Shorten	No change
Cry1+/–Cry2–/–	Lengthen	No change
Clock–/– Clock+/–	Arrhythmic/lengthen	No change

Asterisk denotes results from [56]

that this tissue-level analysis leaves open the question of whether loss of rhythmicity in knockout retinas is due to cell-autonomous effects on cellular clocks or on the coupling of clock cells.

5.2 Cellular Organization of the Retinal Circadian Clock

5.2.1 Generation of Retinal Circadian Rhythms Is Distributed Among Many Cell Types

The forgoing experiments measured gene expression in whole tissue retinal explants—treating the retinal clock as a homogenous tissue. Of course, the retina is not at all homogeneous but consists of six classes of neurons—rods, cones, horizontal cells, bipolar cells, amacrine cells and ganglion cells—exquisitely ordered in three nuclear and two synaptic layers (Fig. 5.2). Which retinal cell types then actually generate circadian rhythms, i.e., which cells are the clock cells of the retina?

To address this question, two principal complementary approaches have been used: the cell-specific distribution of clock gene expression in the retina was mapped to determine which cell types expressed the genetic components of the circadian clock and the ability of cell types or retinal layers to generate circadian rhythms in isolation was examined. Circadian clock gene expression is concentrated in the outer nuclear layer of nonmammalian retinas, which contains the rods and cone photoreceptors. In addition, isolated *Xenopus* frog retinal photoreceptor layers continue to express circadian rhythms in melatonin synthesis [57], and isolated chicken photoreceptors express rhythms in opsin gene synthesis [58] and in ion channel activity [59]. Finally, photoreceptor-specific expression of a dominant negative form of the *Clock* gene, which breaks the circadian transcription/translation feedback loop, blocks the rhythms of melatonin synthesis in frog retinas [60], with most of this effect apparently due to abrogating clock cycling in rods [61]. Thus, in

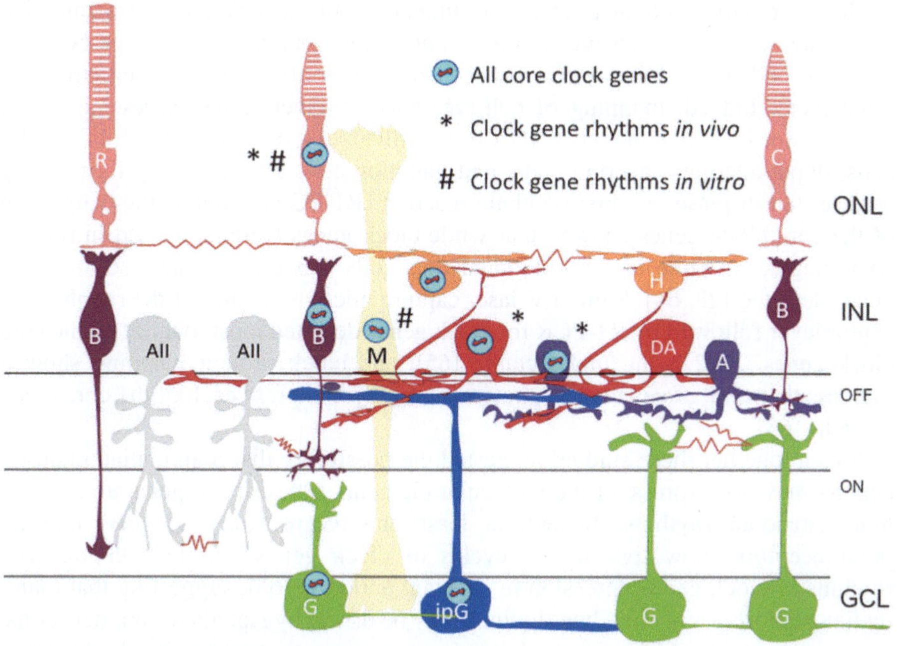

Fig. 5.2 Cell-specific expression of core clock genes in the retina. Diagram of mammalian retinal circuitry showing cell types in which there is coordinate expression of all six core clock genes. *ONL* outer nuclear layer, *INL* inner nuclear layer, *GCL* ganglion cell layer, *OFF* OFF sublamina of the inner plexiform layer, *ON* ON sublamina of the inner plexiform layer. Cell types, *R* rod photoreceptor, *C* cone photoreceptor, *H* horizontal cell, *B* bipolar cell, *DA* dopaminergic amacrine cell, *M* Müller cell, *AII* AII amacrine cell, *A* amacrine cell, *G* ganglion cell, *ipG* intrinsically photoreceptive ganglion cell. *Resistor* symbols indicate electrical synapses. *Oscillator* symbols identify cell populations in which cell-specific expression of all six core clock genes has been demonstrated. *Asterisks* denote cell populations in which clock gene rhythms have been observed from the in vivo retina. *Number* symbols denote cell populations in which clock gene rhythms have been observed in vitro in isolated cell populations. Note that for cones this was found in nonmammalian cones only

nonmammalian retinas, there is robust and convergent evidence that photoreceptors express the requisite clock components and express circadian rhythms in isolation from other cell types, both critical features of clock cells.

The organization of the mammalian retina is apparently more complex, with evidence for both photoreceptor and inner retinal clocks (Fig. 5.2). As with nonmammalian retinas, melatonin is rhythmically synthesized in retinal photoreceptors [62], and thus, rods and cones are putative clock cells in the mammal as well. However, at the level of in situ hybridization for clock gene transcripts and antibody detection of clock proteins, clock gene expression in the mammalian retina is concentrated in the inner nuclear layer, which contains the horizontal, bipolar, and amacrine cells and is less prominent in the photoreceptor and ganglion cell layers [28, 46, 47, 63]. These findings raise the question as to whether the cellular organization of the mammalian retinal circadian is fundamentally a different nonmammalian vertebrate, with circadian melatonin output of mammalian photoreceptors being driven by cells in the inner retina, rather than clock oscillations intrinsic to the photoreceptors. Indeed, mapping of cell-type-specific clock gene expression in the mouse retina by collecting the mRNA from individual rods, horizontal cells, bipolar cells, dopaminergic amacrine cells, and ganglion cells and then amplifying it by reverse transcriptase polymerase chain reaction (RT-PCR) to detect the expression of the core clock genes showed that while clock genes were expressed in rods to some degree, in none of the individual rods cells was the complete set of clock genes detected [28, 64]. Similarly, laser capture microdissection of the rat photoreceptor layer followed by RT-PCR for clock genes detected most, but not all the core clock genes, *Per2* being undetectable [65], and clock protein mapping showed that less than 5 % of cells in the photoreceptor layer express each of the core clock proteins [63].

Taken together these studies suggested the possibility that mammalian photoreceptors may not express all the molecular elements necessary to generate endogenous circadian rhythms. Indeed, at least one recent study suggests that rat photoreceptors show light-driven cycles of clock genes, but lack endogenous rhythms of clock gene expression in constant darkness [66], suggesting that mammalian photoreceptor circadian rhythms may be driven by signals from inner retinal clock neurons [28, 66], perhaps dopamine acting on photoreceptors rhythmically through D4-type dopamine receptors. However, there is also positive evidence indicating the ability of mammalian photoreceptors to generate intrinsic circadian rhythms. Rat photoreceptor layers, isolated from the inner retina by chemical lesioning, produce both *Per1*::Luc bioluminescence rhythms and melatonin rhythms [65]. One possibility is that the cone photoreceptors, which were not assayed extensively in the single cell gene expression studies, are the locus of rhythms generation in the photoreceptor layer of mammals (Fig. 5.2). A recent study on cell-type-specific clock protein expression indeed demonstrated prominent rhythms in cones, but not in rods [63].

In contrast to mammalian photoreceptors, in the inner retina, the cell types tested, horizontal cells, bipolar cells, dopaminergic amacrine cells, and ganglion cells all show some proportion of the individual cells of each of these cell types that express the full complement of core clock genes [28]. The cell type with the highest rate of clock gene expression is the dopaminergic amacrine cell [28, 64]. Since dopamine is secreted with a circadian rhythm, dopaminergic amacrine cells are candidate circadian clock neurons. They express key clock components at the protein level [49, 64, 67] and the circadian reporter construct, and the *Per1* and *Cry2* clock genes oscillate rhythmically in dopaminergic amacrine cells within the intact retina [49, 63] (Fig. 5.2), but the ability of these cells to generate rhythms intrinsically (i.e., in isolation from other retinal cell types) has not been tested.

Although dopaminergic amacrine neurons express all the requisite genetic components to be clock neurons, it is not yet clear if they indeed generate rhythms endogenously or require signaling from other retinal neurons. The retinal rhythm in dopamine secretion is blunted or lost in the absence of retinal melatonin synthesis due to genetic mutations [21], but not in the absence of melatonin signaling through MT1 melatonin receptors [68]. Whether melatonin signaling is required for circadian clock gene activity by dopamine neurons or just for dopamine secretion is not clear, as loss of dopamine output does not preclude persistent molecular circadian rhythms in the retina [69], and melatonin is not necessary for ongoing gene cycling in the retina [27].

Among other identified amacrine cell types, the nitric oxide synthase (NOS) expressing amacrine cells also express the *Per1*::GFP reporter rhythmically, as do the broad class of GABA neurotransmitter secreting amacrines, which include the dopamine neurons [70] (Fig. 5.2). In contrast, the glycinergic amacrine cell class does not express the *Per1*::GFP clock gene reporter rhythmically [70]. Thus, there is substantial evidence that the GABAergic class of retinal amacrine cells, which comprise the majority of amacrine cells in the mammalian retina including the dopamine neurons, express circadian rhythms in clock gene abundance in the intact retina.

The other inner nuclear layer neuronal cell types, horizontal cells, and bipolar cells also express the core circadian clock genes [28, 63]. Whether the gene transcripts cycle in these cells as well has not been tested. However, imaging of a bioluminescent reporter of PER2 has shown that the circadian cycling of gene abundance extends across the entire depth of the inner nuclear layer and is therefore likely to be produced by all three neuronal cell types of that layer—horizontal cells and bipolar cells as well as the amacrine cells in which there is direct evidence for gene cycling.

The core clock genes are also expressed in subsets of ganglion cells in the mammalian retina, including the intrinsically photoreceptive ganglion cells that serve as photoreceptors for the brain's biological clock and which provide intraretinal feedback to dopaminergic amacrine cells [49, 63]. These ganglion cells exhibit daily rhythms in the expression of the melanopsin pigment gene that are influenced by light and dopamine [71–73], suggesting that although they may be clock neurons generating endogenous circadian rhythms, expression of their photopigment responds to the external light cycle. Indeed, there are daily rhythms in the responsiveness of intrinsically photoreceptive ganglion cells to light [74] and in the amplitude of the pupillary light response driven by these cells [75], indicating functional circadian modulation of these ganglion cells. However, there have been no direct measurements of clock gene rhythms in these or other ganglion cell types to address the question endogenous vs. exogenous generation of these rhythms.

Finally, there is an additional cell type in the inner nuclear layer of retina that avidly expresses the full complement of core clock genes—the Müller glial cell. Müller glia in isolation from other cell types express clock gene rhythms and reporter gene rhythms, suggesting that they indeed possess endogenous circadian clocks that likely contribute to inner retinal gene rhythms (McMahon unpublished).

Is this inner retinal clock, contained within the inner nuclear layer, driven by photoreceptor clocks or is it an independent site of rhythms generation? Two lines of evidence suggest that it is an independent clock. First, retinas in which the photoreceptors have degenerated due to a genetic mutation, but in which the inner nuclear layer remains intact, continue to generate molecular circadian rhythms of the bioluminescent PER2 reporter gene and other clock genes as well [28]. In addition, retinas from a mouse strain in which melatonin synthesis is genetically disrupted also show robust circadian molecular rhythms similar to strains that can synthesize melatonin [27]. Thus, neither the physical presence of the photoreceptors nor their secreted circadian neurohormone melatonin is necessary for the expression of molecular rhythms by the inner retina.

To summarize then the findings on clock genes and cells in the retina—in non-mammalian vertebrate retinas—the photoreceptors contain endogenous gene clocks that synthesize melatonin rhythmically and control photoreceptor function through circadian expression of ion channels and visual pigments. Both rods and cones generate endogenous circadian rhythms, with the rods contributing the majority of melatonin synthesis. Clock genes are expressed at a lower level in the inner retina, and it is not clear if those cell types can generate endogenous rhythms.

In contrast, in the mammalian retina, there are likely multiple sites of molecular circadian rhythms generation within the retina (Fig. 5.2). There is a low level of clock gene expression in the photoreceptor layer, perhaps primarily limited to cones which make up a small percentage of the cells (e.g., ca. 2 % in the mouse retina). However, the cones themselves, and the photoreceptor layer in isolation, express molecular circadian rhythms. In addition, clock gene expression is concentrated within the inner nuclear layer, where the GABAergic class of amacrine cells, including in particular the dopaminergic amacrine cells, as well as the retinal Müller glia, can express circadian rhythms in clock gene expression. This inner retinal circadian clock generates molecular circadian rhythms independent of photoreceptors and melatonin and likely influences the photoreceptors and other retinal neurons through the rhythmic secretion of dopamine. There are also ganglion cells that express clock genes, including the ipRGCs that signal the central brain circadian clock and feedback to dopamine cells in the retina. Whether ganglion cells are endogenous clocks or driven by the photoreceptor or inner retinal clocks is not known.

5.3 Neurochemical Organization of the Retinal Circadian Clock

5.3.1 Intercellular Coupling May Not Be Necessary for Rhythms Generation

Studies of the molecular organization of the retinal circadian clock revealed that multiple cell types are apparently capable of generating and expressing circadian rhythms. How then are these multiple cell types organized into a tissue-level clock? Is neurochemical communication between cell types required for the retina to generate rhythms?

Of the major neurotransmitters and neuromodulators in the retina—glutamate, GABA, glycine, acetylcholine, dopamine, and melatonin—melatonin, dopamine, and GABA are secreted with a circadian rhythm [15, 21, 23, 76] and could serve as synchronizing signals throughout the retina. Cellular coupling through gap junctions is also prominent in retinal circuits and serves to synchronize clock cell populations in invertebrate retinal clocks [77]. In addition, in the SCN neural clock, spike-mediated neurotransmission and the neuropeptide vasoactive intestinal polypeptide (VIP) are required to maintain cellular synchronization and tissue-level gene rhythms [78–80]. Comprehensive testing of the necessity for retinal molecular rhythms generation of each of the major neurotransmitters, plus spike-mediated transmission and gap junctions with pharmacological agonists, antagonists, and genetic knockouts, revealed that none of these mechanisms of neurotransmission was individually necessary for tissue-level retinal molecular rhythms, as read out by the PER2 bioluminescence reporter gene [27]—that is, blockade or saturation of neurotransmission through glutamate, GABA, glycine, acetylcholine, dopamine, melatonin, gap junctions, or spikes did not block the production of molecular rhythms in explanted whole retinas. In addition, retinas explanted from mice in which VIP synthesis was knocked out also showed robust PER2 bioluminescence rhythms (Ruan and McMahon, unpublished).

Taken together, these results suggest that unlike the SCN neural clock in which cellular coupling is critical for maintenance of cellular and tissue-level rhythms, cellular coupling among clock cells in the retina may be minimal. The model for retinal rhythms generation most consonant with these data is that retinal clock cells are distributed throughout the retinal layers and generate molecular rhythms in a cell-autonomous manner. Consistent with this notion, the damping rate of the retinal clock in vitro is about 5× faster than that of the SCN clock [27], possibly due to individual cell clocks coming out of synchronization much more quickly. How does the retinal clock as a whole then maintain cell synchronization? One possibility is that the retina is normally exposed directly to the daily light–dark cycle, and this strong external synchronizing input is sufficient to maintain rhythmic organization in the absence of strong cellular coupling mechanisms.

5.3.2 *Entrainment of the Retinal Clock Involves Intraretinal Signaling by Dopamine*

While none of the major neurotransmitters of the retina were found to be necessary for rhythms generation, two in particular did have striking effects on other fundamental properties of the retinal circadian clock-altering dopamine levels resets the retinal circadian clock, while altering GABA levels change the amplitude of the clock gene oscillations. Dopamine is secreted in the retina with a circadian rhythm, with elevated levels in the daytime portion of the circadian cycle, and it is released in the retina in response to light in both mammalian and nonmammalian retinas. The sole source of dopamine in the retina is the dopaminergic amacrine cells of the

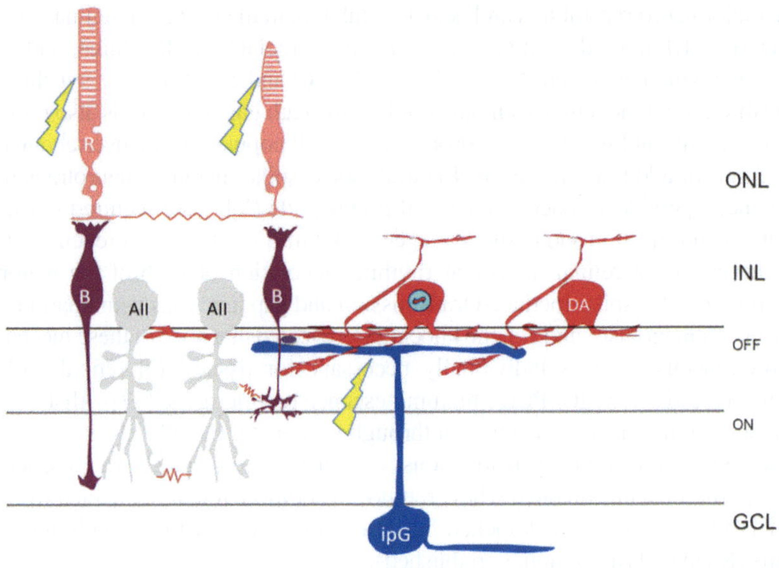

Fig. 5.3 Proposed circuitry for dopamine release mediating light entrainment. Cone and/or rod input is mediated through synaptic contacts of ON bipolar cells with dopaminergic amacrine cells, while the input of intrinsically photoreceptive ganglion cells to DA neurons takes place at appositions of these cells. *Lightning* symbols indicate cells transducing light signals. Abbreviations as in Fig. 5.2

inner nuclear layer, which receive light input from rods and cones through bipolar cells and from ipRGCs through retrograde intraretinal transmission [81, 82] (Fig. 5.3). Dopamine released from amacrine cells has widespread effects throughout the entire retina—with dopaminergic cells in many species elaborating interplexiform processes that innervate horizontal cells, bipolar cells, and photoreceptor terminals and evidence for volume transmission of dopamine by diffusion as well [83].

The receptors for dopamine are also widespread in retinal circuits—of particular relevance here is the expression of D2 family dopamine receptors (D2 or D4) on cone photoreceptors in a range of species [84, 85]. A role for retinal dopamine in light resetting of the retinal clock was first shown in *Xenopus*, in which blockade of dopaminergic transmission through D2-like receptors was shown to block light-induced phase shifts of melatonin secretion [85]. Since melatonin is primarily synthesized in the photoreceptors, this suggests a somewhat puzzling mechanism in which the clock within photoreceptors is not reset directly by the light signal transduced by the photoreceptors, but indirectly by feedback from dopaminergic amacrine cells. Indeed, light-induced dopamine release causes the induction of the *Period2* clock gene in Xenopus photoreceptors [41]. Similarly, in the mouse retina, blockade of dopaminergic signaling blunts light effects on circadian gene cycling, but in this species, D1 dopamine receptors play a more prominent role [27].

In addition, light-induced dopamine acutely stimulates the expression of *Period* clock genes in the retina, an effect that is blunted by knockout of D2 receptors [86].

Although it seems odd that rod and cone photoreceptors might, in a sense, ignore their own phototransduction for resetting of their clocks, one speculation is that the retrograde drive of the ipRGCs on dopamine cells [81, 82], which provides a much more sustained drive to dopamine neurons than rod/cone input, is a key factor in circadian entrainment of the retinal clock. This speculation is attractive as the ipRGCs would then serve entrainment of both the retinal and brain clocks, ensuring synchronization of these two neural oscillators [82]. Indeed, light-induced dopamine responses and dopamine rhythms have generally been found to be preserved in rod/cone degenerate retinas in which melanopsin ganglion cells could serve as the remaining photoreceptors [76, 87, 88], presumably due to melanopsin ipRGC drive to dopaminergic neurons (but see [89]). In any case, though the precise circuitry remains to be established, it is clear that dopamine acts as a critical input to the clock neurons in the retina, mediating the effects of light on clock phase.

5.3.3 GABA Signaling Influences the Amplitude of Retinal Circadian Rhythms

GABA is the principal fast inhibitory neurotransmitter in the retina, and the primary sources of GABA secretion are horizontal cells and amacrine cells. There are two major classes of GABA receptors: $GABA_A$ receptors that are ionotropic Cl-ion channels and $GABA_B$ receptors that are metabotropic g protein-coupled receptors. Isoforms of these two receptor classes are widely expressed in retinal circuits, and the retina is particularly rich in the *rho* isoform of $GABA_A$ receptor subunits, which is sometimes called the $GABA_C$ receptor [90]. GABA levels can fluctuate in the retina in a circadian rhythm, with elevated levels in the night portion of the circadian cycle [23].

Pharmacological blockade of $GABA_A$ and $GABA_C$, but not $GABA_B$ receptors, significantly increases the amplitude of PER2 bioluminescence rhythms by retinal explants, indicating that endogenous retinal $GABA_A$ neurotransmission, including $GABA_A$ *rho*-containing receptors, partially dampens the amplitude of gene cycling in the retinal circadian clock [27]. Indeed, increased stimulation of $GABA_A$ receptors with agonists greatly dampens the amplitude of the retinal circadian clock and at high doses actually stops the motion of the molecular clock. These effects can be partially reversed by counteracting the hyperpolarizing effects of GABA-induced Cl-flux on cell membrane potential and fully reversed when casein kinase epsilon (CKI_e), an enzyme that promotes the degradation of PERIOD proteins through regulating their phosphorylation, is also inhibited [27].

The function of the GABAergic effects on retinal clock gene cycling is less clear than that of dopamine, but one speculation is that whereas dopamine is a circadian signal for the day phase in the retina, GABA may be a circadian signal that reinforces the night phase. The nighttime peak in retinal GABA levels could be due to prolonged depolarization of GABA-secreting horizontal cells in the dark, resulting in prolonged secretion of GABA. Whereas, daytime circadian and light drive on dopamine secretion serve to reinforce the rise in the expression of *Period* clock genes during the day portion of the molecular circadian cycle, nighttime GABA-mediated stimulation of PERIOD protein degradation may reinforce the falling phase of the clock gene cycle that occurs at night, removing these negative feedback complexes and preparing retinal clock neurons for initiation of the next cycle of *Period* gene transcription at dawn. Interestingly, $GABA_A$ receptors have been found to physically associate with CKI_e in SCN clock neurons, although no function has been ascribed [91]. Thus, elevated $GABA_A$ signaling at night may reinforce the degradative portion of the *Period* clock gene cycle, in counter point to dopamine's reinforcement of the synthetic portion of the cycle during the day.

5.4 Outputs of the Retinal Circadian Clock

5.4.1 The Retinal Clock Makes Both Intraretinal and Extraretinal Outputs

Let us now consider the outputs of the retinal clock, the hands, rather than the gears, substances, and mechanisms by which the clock signals and influences its circadian rhythm throughout the retina and into the rest of the brain.

5.4.2 Melatonin

Primary among the output signals of the retinal clock is the neurohormone melatonin, the role of which is more fully elaborated elsewhere in this volume. Synthesized in photoreceptors, and secreted in the night phase of the retinal circadian cycle, melatonin is a key negative modulator of dopamine release [92]. It also contributes to regulation of retinal sensitivity as assayed by the ERG. Knockout of the MT1 dopamine receptor disrupts the circadian regulation of the light-adapted ERG, even though in this case retinal dopamine rhythms may be present [68]. In addition, it is a key trophic signal enhancing the survival of photoreceptors and of ganglion cells [32]. Finally, it is interesting to note that melatonin, though a powerful and pervasive output signal of the retinal circadian clock, does not appear to feedback on retinal rhythms generation. It is not necessary for molecular rhythms generation, nor does its application shift the phase of retinal rhythms [27]. It is a pure clock output signal.

5.4.3 Dopamine

In contrast, dopamine, which has clear input effects in that it can reset the retinal clock, also serves as an output, playing a complex dual role in retinal circadian organization. Retinal dopamine rhythms, which peak during the circadian day phase, have demonstrated effects on retinal cells and circuits, in particular through D2 family dopamine receptors. Mammalian cone photoreceptors express D4 dopamine receptors [84] with a circadian rhythm that peaks in the night phase [93]. At the level of the photoreceptors, dopamine acts to mediate light adaptation, decreasing the activation of adenyl cyclase type 1 and reducing levels of cAMP, as well as reducing the levels of the transduction modulating protein phosducin [94–97]. These effects are seen in response to light-adapted stimulation of dopamine release but are also driven by the intrinsic rhythms in dopamine signaling through D4-type dopamine receptors in the absence of light. Similarly, dopamine rhythms act on photoreceptor gap junctions through D2 family receptors to modulate rod–cone coupling in a circadian manner [98]. The relative sensitivity of dopamine receptor subtypes provides a dividing point between circadian and light-induced effects of dopamine. For example, whereas photoreceptor gap junction coupling is modulated through the more sensitive D2 family receptors in a circadian manner, horizontal cell gap junctions, which are modulated through less sensitive D1 type receptors, respond to the higher levels of dopamine achieved by light-induced dopamine release, but not to baseline circadian modulation [99, 100]. Thus, the D2 and D1 family receptors separate overall dopamine signaling in the retina into circadian and light adaptation streams as has been demonstrated recently in retinal dopamine knockout mice [69].

5.4.4 Gene Expression

One of the fundamental functions of the circadian clock gene network is to drive tissue-specific networks of clock controlled genes to underpin circadian physiology. Early work in the retina demonstrated light-induced as well as circadian clock control of opsin gene expression in the photoreceptors of nonmammalian vertebrates [58, 101], and recently this issue has been examined on a genomic scale to define the circadian transcriptome of the mammalian retina [7]. Using gene microarrays to characterize retinal gene expression across multiple time points in light–dark cycles (LD) and in constant darkness (DD), Storch et al. [7] found that ca. 3,000 genes cycled in a light-driven manner in LD and ca. 300 retinal genes continued to cycle under clock control in DD. The clock-controlled retinal genes came from classes that contributed to many basic neural and cellular functions such as synaptic transmission, photoreceptor signaling, intercellular communication, and regulation of the cytoskeleton and chromatin. Thus, the transcriptional regulation of retinal gene expression by the retinal circadian clock is a critical mechanism for its widespread control of retinal function and metabolism. Interestingly, the retinal clock mechanism also contributes to light regulation of gene expression, as knockout of the clock mechanism also disrupted rhythmic gene expression in LD cycles [7]. This widespread transcriptional control of retinal signaling and metabolism genes suggests that disruption of the retinal molecular clockworks is likely to contribute to retinal disease and pathology.

5.4.5 *Signaling by the Retinal Clock to the Brain*

The retina signals the daily light dark cycle to the hypothalamic SCN circadian clock through a specialized subset of retinal ganglion cells that express the photopigment melanopsin [102]. However, in addition to these light-driven daily signals, there is evidence for purely circadian signals from the retina to the brain that occur without light stimulation. In constant darkness, there are subregions of the SCN clock nuclei that express retina-driven circadian rhythms [34]. The retinorecipient zone of the SCN clock in hamsters shows circadian rhythms in the phosphorylation of the signaling kinase ERK, a molecule normally activated by retinal ganglion cell input to the SCN during light resetting of the clock. These rhythms in DD are ablated upon enucleation, suggesting that they originate in the retina, rather than in the SCN itself, and are transmitted to the SCN via the ipRGCs. In addition, retinal signals in DD also contribute to the maturation and functional organization of the brain's biological clock, as enucleation in DD during the first few weeks of life causes changes in the free-running period of locomotor rhythms in hamsters [103]. Interestingly, ipRGCs have been shown to express clock genes [63], and their light responses are modulated with a circadian rhythm and by dopamine [74, 104], suggesting the possibility that the retina-driven rhythms observed in the SCN originate in the ipRGCs themselves, or in retinal neurons with synaptic input to the melanopsin-expressing ipRGCs (e.g., dopamine amacrine cells).

5.5 The Retinal Clock's Role in Vision

Animals in natural terrestrial and aquatic environments are exposed to a predictable daily cycle of visual conditions—alternating between photopic conditions in the daylight and scotopic conditions at night, with mesopic transitions. Although there are mechanisms at many levels of the retina that adapt the photoreceptors and circuits in reaction to prevailing lighting conditions, the widespread evolution of retinal circadian clocks that match the solar day suggests that there is an advantage in tuning retinal function in anticipation of daily cycles in visual conditions. Indeed, for example, a circadian rhythm in tyrosine hydroxylase expression in retinal dopamine cells may ensure the synthetic capacity for an enhanced burst of dopamine synthesis at light onset [105].

Moving beyond the effects of the clock on individual cells and synapses, overall then, what is the global effect of the retinal circadian clock on visual function? The most consistent findings are that the retinal clock acts to bias the retina toward enhanced cone vision in the daytime and enhanced rod function in the nighttime. An extreme example of this effect that illustrates the more general point is the case of young zebrafish fry in which the cone synaptic ribbons are actually disassembled at night, preventing cone-mediated neurotransmission [106]. Since these newly hatched fish have not yet developed a significant number of rod photoreceptors, they are functionally blind at night during the first few days of free-swimming life. Synaptic transmission from their cones is turned off at night by circadian remodeling to the transmitter release apparatus, and they do not yet have sufficient functioning rods. Across a number of vertebrate species including fish, amphibians, birds, and mammals, the retinal circadian clock has been shown to drive shifts in the rod–cone balance of vision such that cone vision is favored in the day phase and rod vision is favored at night [98, 105, 107, 108].

Key to the circadian control of visual processing in the retina is circadian signaling by dopamine through D2 family receptors. The elevated circadian levels of dopamine in the day enhance light-adapted function of cones [94, 97] and uncouple cones from rods (the secondary rod pathway) and AII amacrine cells (the primary rod pathway) to isolate cone signals from the saturated rod output [98, 109]. Indeed, genetic depletion of retinal dopamine, by retina-specific knockout of the key dopamine synthesizing enzyme tyrosine hydroxylase (rTHKO), blunts the circadian rhythm in the light-adapted ERG, reducing the normal daytime elevation of this cone-driven response [69]. This effect is through D4 receptors, as it can be rescued by D4-specific agonists. Interestingly, a similar effect is observed on behaviorally measured contrast sensitivity in rTHKO mice, suggesting that this aspect of vision is likely under circadian control as well [96]. In these mice with greatly reduced retinal dopamine, the retina appears to be perpetually locked in the nighttime state (e.g., low amplitude photopic ERG and reduced contrast sensitivity). Human patients with parkinsonian degeneration of dopaminergic neurons exhibit similar visual deficits in ERG and contrast sensitivity, suggesting that depletion of retinal dopamine may play a role in the visual symptoms of this neurodegenerative disease [110–112]. In the rTHKO model, dopamine-specific changes in the functional adaptation of retinal ganglion cells are now being explored, and initial results reveal that the adaptational state of transient ON and OFF type ganglion cells in particular is subject to dopamine signaling (Risner and McMahon unpublished). Thus, the overall retinal sensitivity to photopic light, contrast sensitivity, and the function of specific classes of ganglion cells is under circadian control by dopamine signaling.

While the rTHKO experiments were performed in mice lacking the ability to secrete melatonin, and therefore demonstrating a melatonin-independent effect of dopamine in regulating photopic vision, experiments in melatonin-proficient mice demonstrated that signaling through the MT1 melatonin receptor is necessary for ERG rhythms, but not for dopamine rhythms [68]. In these mice, dopamine rhythms were not sufficient to drive ERG rhythms in the absence of melatonin signaling. In addition, melatonin-deficient mice lack circadian modulation on rod–cone coupling, likely due to disrupted dopamine circadian rhythms [98]. Thus, dopamine and melatonin have parallel roles in conveying the circadian clock output to visual function.

Finally, it is useful to distinguish the clock-driven changes in visual function from light-driven changes. In particular, light adaptation during the day evokes additional elevation of dopamine levels that stimulates D1 dopamine receptors, particularly on retinal horizontal cells [99, 100]. This leads to uncoupling of horizontal cell gap junctions and likely to shrinkage of inhibitory receptive field surrounds which may increase the spatial resolution of vision in bright light and high contrast. Indeed, in rTHKO mice, behaviorally measured visual acuity is decreased but can be rescued by D1-specific dopamine agonists. In addition, dopamine D1 stimulation of horizontal cells has been proposed to alter GABA release and enhance the function of rod bipolar cells, extending the range of rod function in response to background illumination [113]. Neither the scotopic ERG nor visual acuity has demonstrated circadian rhythms however, likely because these effects can only be invoked at D1 receptors by the high levels of dopamine that result from combined circadian drive and light-mediated release by photopic backgrounds presented during the day.

5.6 The Retinal Clock's Potential Role in Eye Disease

The widespread control of retinal signaling, metabolism, and gene expression exerted by the retinal circadian clock suggests that the retinal molecular clockworks, or its output signals, may contribute to retinal disease and pathology, as well as normal retinal function. This is an area just beginning to be explored at the molecular level, but previous studies have shown that there is a circadian modulation of photoreceptor vulnerability to light-induced degeneration [29], perhaps by maintenance of key K+ channel rhythms [114], and more recently, melatonin signaling through MT1 melatonin receptors has been shown to maintain photoreceptor and ganglion cell populations during aging [32]. Trophic signaling by the retinal clock and its outputs may also play a role in the regulation of eye growth and refractive errors. Disruption of retinal dopamine rhythms causes myopic changes in the chicken eye [115–119], and similar dopamine-dependent eye growth regulatory mechanisms likely operate in the mammalian retina [120]. Finally, circadian clock genes have been shown to regulate vascular endothelial growth factor (VEGF) signaling in tumorigenesis [121, 122], dopamine modulates VEGF receptor action on vascular endothelial cells [123, 124], and recent results that suggest that the *Period* genes may play a similar role in regulating vascularization signals in the retina in disease models of diabetic retinopathy [125]. Given that retinal clock gene expression is disrupted in proliferative neovascularizing diseases [126], this link may prove to be an important pathway for therapeutic intervention in these conditions.

5.7 Summary

The retina is unique among vertebrate neural circadian clocks in that it is both a primary sensory organ and a circadian pacemaker. It contains a complete circadian system—mechanisms for circadian phototransduction, rhythms generation, and rhythmic outputs to signal circadian time both within and downstream of the retina. Thus, the retina as an experimental system is highly suited to addressing fundamental questions of how neural circadian clocks are organized to generate and express circadian rhythms.

There are multiple sites within the retina of circadian rhythms generation (Fig. 5.3), with both the photoreceptor layers and inner retinal layers each shown capable of independently expressing circadian rhythms. The strength of circadian clock gene expression and the emphasis of rhythms expression are divergent across vertebrate retinas, with photoreceptors as the primary locus of rhythms generation in nonmammalian vertebrates, while in mammals, clock activity is most robust in the inner nuclear layer. Dopamine, secreted by amacrine cells in the inner nuclear layer, likely serves as an intraretinal signal from inner retinal clocks to the photoreceptors. On the cellular level, all the major retinal cell classes express circadian clock genes at some level, and the retinal rhythms are likely generated in a cell-autonomous fashion by cell populations throughout the retina. Cones and dopaminergic amacrine cells have robust and reproducible circadian rhythms in gene expression in the intact retina, while nonmammalian cones and mammalian Müller cells have been shown capable of circadian rhythmicity in isolation from other retinal cell types.

Cellular communication through the major neurotransmitter systems is not necessary for retinal rhythms generation; however, dopamine and GABA influence the phase and amplitude of retinal clock oscillations, respectively. Dopamine mediates in part the setting of the retinal clock by light, through positive induction of *Period* clock genes, whereas GABA, secreted in the dark phase, acts to decrease PER2 protein expression. These transmitter-driven influences on clock molecules apparently reinforce the autonomous transcription–translation cycling of clock genes that sees *Period* gene expression rising in the day and falling at night.

With respect to molecular organization of the retinal clock, it is comprised of the canonical clock genes, yet is much more sensitive to genetic disruption that of the central brain clock, the SCN (Fig. 5.2). The central brain circadian clock can more readily compensate for loss of individual clock genes compared to peripheral tissue circadian oscillators (e.g., liver or fibroblast), possibly because of strong inter-neural communication and the expression of *NPAS2*, a *Clock* paralog [55, 127]. In the SCN, the only single clock gene knockout able to ablate rhythmicity is *Bmal1*, whereas in peripheral tissue clocks, and in the retina *Bmal1*, *Per1*, *Cry1*, and *Clock* are all individually required for rhythms generation [7, 55, 128]. Thus, the retinal circadian clock is similar in its molecular constituents to other tissue circadian clocks but lacks the genetic and cellular buffering against perturbation that is characteristic of the SCN brain clock.

The retinal circadian clock is organized as apparently cell-autonomous clocks that are distributed among the retinal layers. It is likely that molecular rhythms generation occurs in multiple cell types, perhaps in subsets of each retinal cell type, and that retinal rhythms are primarily coordinated through common entrainment to the light cycle rather than through strong intercellular communication. The cell types that can serve as retinal clock cells remain to be defined, but retinal dopamine plays a unique role in that it is both an output of retinal circadian signals to retinal circuits and an intraretinal signal to reset the phase of the clock. Acting in large part through dopamine signaling, the retinal circadian clock reconfigures retinal circuits to enhance light-adapted cone-mediated visual during the day and dark-adapted rod-mediated visual signaling at night to preadapt vision to the temporal alternation of photopic and scotopic conditions in the natural environment. The extensive impact of circadian gene and metabolic regulation also impacts eye disease as well as normal visual function.

With multiple loci of rhythms generation and multiplex impacts on retinal function, the retinal circadian clock is a rich substrate for exploring the organization of neural circadian clocks. Key questions for the future include determining which of the several genetically competent cell populations in the retina are true clock cells, if and how their rhythms are coordinated, and the action of the circadian clock on specific retinal circuits and functional modules (e.g., ganglion cell subtypes), by which it modulates vision. Elucidation of these issues can shed light on the convergence of circadian and sensory function in the specialized senses of many organisms.

References

1. Bassi CJ, Powers MK. Daily fluctuations in the detectability of dim lights by humans. Physiol Behav. 1986;38(6):871–7.
2. Lotze M, Treutwein B, Roenneberg T. Daily rhythm of vigilance assessed by temporal resolution of the visual system. Vision Res. 2000;40(25):3467–73.
3. Nozaki S, Wakakura M, Ishikawa S. Circadian rhythm of human electroretinogram. Jpn J Ophthalmol. 1983;27(2):346–52.
4. Hankins MW, Jones RJ, Ruddock KH. Diurnal variation in the b-wave implicit time of the human electroretinogram. Vis Neurosci. 1998;15(1):55–67.
5. Nordin S, Lötsch J, Murphy C, Hummel T, Kobal G. Circadian rhythm and desensitization in chemosensory event-related potentials in response to odorous and painful stimuli. Psychophysiology. 2003;40(4):612–9.
6. Lotze M, Wittmann M, von Steinbüchel N, Pöppel E, Roenneberg T. Daily rhythm of temporal resolution in the auditory system. Cortex. 1999;35(1):89–100.
7. Storch K-F, Paz C, Signorovitch J, Raviola E, Pawlyk B, Li T, et al. Intrinsic circadian clock of the mammalian retina: importance for retinal processing of visual information. Cell. 2007; 130(4):730–41.
8. Cameron MA, Barnard AR, Hut RA, Bonnefont X, van der Horst GT, Hankins MW, et al. Electroretinography of wild-type and Cry mutant mice reveals circadian tuning of photopic and mesopic retinal responses. J Biol Rhythms. 2008;23(6):489–501.
9. Granados-Fuentes D, Tseng A, Herzog ED. A circadian clock in the olfactory bulb controls olfactory responsivity. J Neurosci. 2006;26(47):12219–25.

10. Krishnan B, Dryer SE, Hardin PE. Circadian rhythms in olfactory responses of Drosophila melanogaster. Nature. 1999;400(6742):375–8.
11. Barlow Jr RB. Circadian rhythms in the Limulus visual system. J Neurosci. 1983;3(4): 856–70.
12. Page TL, Koelling E. Circadian rhythm in olfactory response in the antennae controlled by the optic lobe in the cockroach. J Insect Physiol. 2003;49(7):697–707.
13. Tanoue S, Krishnan P, Krishnan B, Dryer SE, Hardin PE. Circadian clocks in antennal neurons are necessary and sufficient for olfaction rhythms in Drosophila. Curr Biol. 2004; 14(8):638–49.
14. Besharse JC, Iuvone PM. Circadian clock in Xenopus eye controlling retinal serotonin N-acetyltransferase. Nature. 1983;305(5930):133–5.
15. Tosini G, Menaker M. Circadian rhythms in cultured mammalian retina. Science. 1996; 272(5260):419–21.
16. Tosini G, Menaker M. The clock in the mouse retina: melatonin synthesis and photoreceptor degeneration. Brain Res. 1998;789(2):221–8.
17. Sakamoto K, Liu C, Tosini G. Circadian rhythms in the retina of rats with photoreceptor degeneration. J Neurochem. 2004;90(4):1019–24.
18. Kaneko M, Hernandez-Borsetti N, Cahill GM. Diversity of zebrafish peripheral oscillators revealed by luciferase reporting. Proc Natl Acad Sci U S A. 2006;103(39):14614–9.
19. Tosini G, Menaker M. Multioscillatory circadian organization in a vertebrate, Iguana iguana. J Neurosci. 1998;18(3):1105–14.
20. Steele CT, Tosini G, Siopes T, Underwood H. Time keeping by the quail's eye: circadian regulation of melatonin production. Gen Comp Endocrinol. 2006;145(3):232–6.
21. Doyle SE, Grace MS, McIvor W, Menaker M. Circadian rhythms of dopamine in mouse retina: the role of melatonin. Vis Neurosci. 2002;19(5):593–601.
22. Nir I, Haque R, Iuvone PM. Diurnal metabolism of dopamine in the mouse retina. Brain Res. 2000;870(1–2):118–25.
23. Jaliffa CO, Saenz D, Resnik E, Keller Sarmiento MI, Rosenstein RE. Circadian activity of the GABAergic system in the golden hamster retina. Brain Res. 2001;912(2):195–202.
24. Dmitriev AV, Mangel SC. Circadian clock regulation of pH in the rabbit retina. J Neurosci. 2001;21(8):2897–902.
25. Teirstein PS, Goldman AI, O'Brien PJ. Evidence for both local and central regulation of rat rod outer segment disc shedding. Invest Ophthalmol Vis Sci. 1980;19(11):1268–73.
26. Tosini G, Kasamatsu M, Sakamoto K. Clock gene expression in the rat retina: effects of lighting conditions and photoreceptor degeneration. Brain Res. 2007;1159:134–40.
27. Ruan G-X, Allen GC, Yamazaki S, McMahon DG. An autonomous circadian clock in the inner mouse retina regulated by dopamine and GABA. PLoS Biol. 2008;6(10):e249.
28. Ruan G-X, Zhang D-Q, Zhou T, Yamazaki S, McMahon DG. Circadian organization of the mammalian retina. Proc Natl Acad Sci U S A. 2006;103(25):9703–8.
29. Organisciak DT, Darrow RM, Barsalou L, Kutty RK, Wiggert B. Circadian-dependent retinal light damage in rats. Invest Ophthalmol Vis Sci. 2000;41(12):3694–701.
30. Grewal R, Organisciak D, Wong P. Factors underlying circadian dependent susceptibility to light induced retinal damage. Adv Exp Med Biol. 2006;572(3):411–6.
31. Ogilvie JM, Speck JD. Dopamine has a critical role in photoreceptor degeneration in the rd mouse. Neurobiol Dis. 2002;10(1):33–40.
32. Baba K, Pozdeyev N, Mazzoni F, Contreras-Alcantara S, Liu C, Kasamatsu M, et al. Melatonin modulates visual function and cell viability in the mouse retina via the MT1 melatonin receptor. Proc Natl Acad Sci U S A. 2009;106(35):15043–8.
33. Iuvone PM, Tigges M, Stone RA, Lambert S, Laties AM. Effects of apomorphine, a dopamine receptor agonist, on ocular refraction and axial elongation in a primate model of myopia. Invest Ophthalmol Vis Sci. 1991;32(5):1674–7.
34. Lee HS, Nelms JL, Nguyen M, Silver R, Lehman MN. The eye is necessary for a circadian rhythm in the suprachiasmatic nucleus. Nat Neurosci. 2003;6(2):111–2.

35. Takahashi JS, Hong HK, Ko CH, McDearmon EL. The genetics of mammalian circadian order and disorder: implications for physiology and disease. Nat Rev Genet. 2008; 9(10):764–75.
36. Bae K, Jin X, Maywood ES, Hastings MH, Reppert SM, Weaver DR. Differential functions of mPer1, mPer2, and mPer3 in the SCN circadian clock. Neuron. 2001;30(2):525–36.
37. Vitaterna MH, Selby CP, Todo T, Niwa H, Thompson C, Fruechte EM, et al. Differential regulation of mammalian period genes and circadian rhythmicity by cryptochromes 1 and 2. Proc Natl Acad Sci U S A. 1999;96(21):12114–9.
38. van der Horst GT, Muijtjens M, Kobayashi K, Takano R, Kanno S, Takao M, et al. Mammalian Cry1 and Cry2 are essential for maintenance of circadian rhythms. Nature. 1999; 398(6728):627–30.
39. Bunger MK, Wilsbacher LD, Moran SM, Clendenin C, Radcliffe LA, Hogenesch JB, et al. Mop3 is an essential component of the master circadian pacemaker in mammals. Cell. 2000; 103(7):1009–17.
40. Zhu H, LaRue S, Whiteley A, Steeves TD, Takahashi JS, Green CB. The Xenopus clock gene is constitutively expressed in retinal photoreceptors. Brain Res Mol Brain Res. 2000; 75(2):303–8.
41. Besharse JC, Zhuang M, Freeman K, Fogerty J. Regulation of photoreceptor Per1 and Per2 by light, dopamine and a circadian clock. Eur J Neurosci. 2004;20(1):167–74.
42. Zhuang M, Wang Y, Steenhard BM, Besharse JC. Differential regulation of two period genes in the Xenopus eye. Brain Res Mol Brain Res. 2000;82(1–2):52–64.
43. Zhu H, Green CB. Three cryptochromes are rhythmically expressed in Xenopus laevis retinal photoreceptors. Mol Vis. 2001;7:210–5.
44. Bailey MJ, Chong NW, Xiong J, Cassone VM. Chickens' Cry2: molecular analysis of an avian cryptochrome in retinal and pineal photoreceptors. FEBS Lett. 2002;513(2–3):169–74.
45. Haque R, Chaurasia SS, Wessel 3rd JH, Iuvone PM. Dual regulation of cryptochrome 1 mRNA expression in chicken retina by light and circadian oscillators. Neuroreport. 2002; 13(17):2247–51.
46. Gekakis N, Staknis D, Nguyen HB, Davis FC, Wilsbacher LD, King DP, et al. Role of the CLOCK protein in the mammalian circadian mechanism. Science. 1998;280(5369):1564–9.
47. Miyamoto Y, Sancar A. Vitamin B2-based blue-light photoreceptors in the retinohypothalamic tract as the photoactive pigments for setting the circadian clock in mammals. Proc Natl Acad Sci U S A. 1998;95(11):6097–102.
48. Namihira M, Honma S, Abe H, Masubuchi S, Ikeda M, Honmaca K. Circadian pattern, light responsiveness and localization of rPer1 and rPer2 gene expression in the rat retina. Neuroreport. 2001;12(3):471–5.
49. Witkovsky P, Veisenberger E, LeSauter J, Yan L, Johnson M, Zhang D-Q, et al. Cellular location and circadian rhythm of expression of the biological clock gene Period 1 in the mouse retina. J Neurosci. 2003;23(20):7670–6.
50. Thompson CL, Rickman CB, Shaw SJ, Ebright JN, Kelly U, Sancar A, et al. Expression of the blue-light receptor cryptochrome in the human retina. Invest Ophthalmol Vis Sci. 2003;44(10):4515–21.
51. Namihira M, Honma S, Abe H, Tanahashi Y, Ikeda M, Honma K. Circadian rhythms and light responsiveness of mammalian clock gene, Clock and BMAL1, transcripts in the rat retina. Neurosci Lett. 1999;271(1):1–4.
52. Baba K, Sengupta A, Tosini M, Contreras-Alcantara S, Tosini G. Circadian regulation of the PERIOD 2::LUCIFERASE bioluminescence rhythm in the mouse retinal pigment epithelium-choroid. Mol Vis. 2010;16:2605–11.
53. Hastings MH, Reddy AB, McMahon DG, Maywood ES. Analysis of circadian mechanisms in the suprachiasmatic nucleus by transgenesis and biolistic transfection. Methods Enzymol. 2005;393:579–92.
54. Ruan G-X, Gamble KL, Risner ML, Young LA, McMahon DG. Divergent roles of clock genes in retinal and suprachiasmatic nucleus circadian oscillators. PLoS One. 2012; 7(6):e38985.

55. Liu AC, Welsh DK, Ko CH, Tran HG, Zhang EE, Priest AA, et al. Intercellular coupling confers robustness against mutations in the SCN circadian clock network. Cell. 2007;129(3): 605–16.
56. Pendergast JS, Friday RC, Yamazaki S. Distinct functions of Period2 and Period3 in the mouse circadian system revealed by in vitro analysis. PLoS One. 2010;5(1):e8552.
57. Cahill GM, Besharse JC. Circadian clock functions localized in Xenopus retinal photoreceptors. Neuron. 1993;10(4):573–7.
58. Pierce ME, Sheshberadaran H, Zhang Z, Fox LE, Applebury ML, Takahashi JS. Circadian regulation of iodopsin gene expression in embryonic photoreceptors in retinal cell culture. Neuron. 1993;10(4):579–84.
59. Ko GY, Ko ML, Dryer SE. Circadian regulation of cGMP-gated channels of vertebrate cone photoreceptors: role of cAMP and Ras. J Neurosci. 2004;24(6):1296–304.
60. Hayasaka N, LaRue SI, Green CB. In vivo disruption of Xenopus CLOCK in the retinal photoreceptor cells abolishes circadian melatonin rhythmicity without affecting its production levels. J Neurosci. 2002;22(5):1600–7.
61. Hayasaka N, LaRue SI, Green CB. Differential contribution of rod and cone circadian clocks in driving retinal melatonin rhythms in Xenopus. PLoS One. 2010;5(12):e15599.
62. Tosini G. Melatonin circadian rhythm in the retina of mammals. Chronobiol Int. 2000; 17(5):599–612.
63. Liu X, Zhang Z, Ribelayga CP. Heterogeneous expression of the core circadian clock proteins among neuronal cell types in mouse retina. PLoS One. 2012;7(11):e50602.
64. Dorenbos R, Contini M, Hirasawa H, Gustincich S, Raviola E. Expression of circadian clock genes in retinal dopaminergic cells. Vis Neurosci. 2007;24(4):573–80.
65. Tosini G, Davidson AJ, Fukuhara C, Kasamatsu M, Castanon-Cervantes O. Localization of a circadian clock in mammalian photoreceptors. FASEB J. 2007;21(14):3866–71.
66. Sandu C, Hicks D, Felder-Schmittbuhl M-P. Rat photoreceptor circadian oscillator strongly relies on lighting conditions. Eur J Neurosci. 2011;34(3):507–16.
67. Gustincich S, Contini M, Gariboldi M, Puopolo M, Kadota K, Bono H, et al. Gene discovery in genetically labeled single dopaminergic neurons of the retina. Proc Natl Acad Sci U S A. 2004;101(14):5069–74.
68. Sengupta A, Baba K, Mazzoni F, Pozdeyev NV, Strettoi E, Iuvone PM, et al. Localization of melatonin receptor 1 in mouse retina and its role in the circadian regulation of the electroretinogram and dopamine levels. PLoS One. 2011;6(9):e24483.
69. Jackson CR, Ruan G-X, Aseem F, Abey J, Gamble K, Stanwood G, et al. Retinal dopamine mediates multiple dimensions of light-adapted vision. J Neurosci. 2012;32(27):9359–68.
70. Zhang D-Q, Zhou T, Ruan G-X, McMahon DG. Circadian rhythm of Period1 clock gene expression in NOS amacrine cells of the mouse retina. Brain Res. 2005;1050(1–2):101–9.
71. Sakamoto K, Liu C, Kasamatsu M, Pozdeyev NV, Iuvone PM, Tosini G. Dopamine regulates melanopsin mRNA expression in intrinsically photosensitive retinal ganglion cells. Eur J Neurosci. 2005;22(12):3129–36.
72. Sakamoto K, Liu C, Tosini G. Classical photoreceptors regulate melanopsin mRNA levels in the rat retina. J Neurosci. 2004;24(43):9693–7.
73. Mathes A, Engel L, Holthues H, Wolloscheck T, Spessert R. Daily profile in melanopsin transcripts depends on seasonal lighting conditions in the rat retina. J Neuroendocrinol. 2007;19(12):952–7.
74. Weng S, Wong KY, Berson DM. Circadian modulation of melanopsin-driven light response in rat ganglion-cell photoreceptors. J Biol Rhythms. 2009;24(5):391–402.
75. Zele AJ, Feigl B, Smith SS, Markwell EL. The circadian response of intrinsically photosensitive retinal ganglion cells. PLoS One. 2011;6(3):e17860.
76. Doyle SE, McIvor WE, Menaker M. Circadian rhythmicity in dopamine content of mammalian retina: role of the photoreceptors. J Neurochem. 2002;83(1):211–9.
77. Block GD, McMahon DG. Cellular analysis of the Bulla ocular circadian pacemaker system III. Localization of the circadian pacemaker. J Comp Physiol A. 1984;155:387–95.

78. Yamaguchi S, Isejima H, Matsuo T, Okura R, Yagita K, Kobayashi M, et al. Synchronization of cellular clocks in the suprachiasmatic nucleus. Science. 2003;302(5649):1408–12.
79. Aton SJ, Colwell CS, Harmar AJ, Waschek J, Herzog ED. Vasoactive intestinal polypeptide mediates circadian rhythmicity and synchrony in mammalian clock neurons. Nat Neurosci. 2005;8(4):476–83.
80. Maywood ES, Reddy AB, Wong GK, O'Neill JS, O'Brien JA, McMahon DG, et al. Synchronization and maintenance of timekeeping in suprachiasmatic circadian clock cells by neuropeptidergic signaling. Curr Biol. 2006;16(6):599–605.
81. Zhang D-Q, Belenky MA, Sollars PJ, Pickard GE, McMahon DG. Melanopsin mediates retrograde visual signaling in the retina. PLoS One. 2012;7(8):e42647.
82. Zhang D-Q, Wong KY, Sollars PJ, Berson DM, Pickard GE, McMahon DG. Intraretinal signaling by ganglion cell photoreceptors to dopaminergic amacrine neurons. Proc Natl Acad Sci U S A. 2008;105(37):14181–6.
83. Witkovsky P. Dopamine and retinal function. Doc Ophthalmol. 2004;108(1):17–40.
84. Cohen AI, Todd RD, Harmon S, O'Malley KL. Photoreceptors of mouse retinas possess D4 receptors coupled to adenylate cyclase. Proc Natl Acad Sci U S A. 1992;89(24):12093–7.
85. Cahill GM, Besharse JC. Resetting the circadian clock in cultured Xenopus eyecups: regulation of retinal melatonin rhythms by light and D2 dopamine receptors. J Neurosci. 1991; 11(10):2959–71.
86. Yujnovsky I, Hirayama J, Doi M, Borrelli E, Sassone-Corsi P. Signaling mediated by the dopamine D2 receptor potentiates circadian regulation by CLOCK:BMAL1. Proc Natl Acad Sci U S A. 2006;103(16):6386–91.
87. Morgan WW, Kamp CW. Dopaminergic amacrine neurons of rat retinas with photoreceptor degeneration continue to respond to light. Life Sci. 1980;26(19):1619–26.
88. Vugler AA, Redgrave P, Hewson-Stoate NJ, Greenwood J, Coffey PJ. Constant illumination causes spatially discrete dopamine depletion in the normal and degenerate retina. J Chem Neuroanat. 2007;33(1):9–22.
89. Cameron MA, Pozdeyev N, Vugler AA, Cooper H, Iuvone PM, Lucas RJ. Light regulation of retinal dopamine that is independent of melanopsin phototransduction. Eur J Neurosci. 2009;29(4):761–7.
90. Qian H, Dowling JE. Novel GABA responses from rod-driven retinal horizontal cells. Nature. 1993;361(6408):162–4.
91. Ning K, Li L, Liao M, Liu B, Mielke JG, Chen Y, et al. Circadian regulation of GABAA receptor function by CKI epsilon-CKI delta in the rat suprachiasmatic nuclei. Nat Neurosci. 2004;7(5):489–90.
92. Dubocovich ML. Melatonin is a potent modulator of dopamine release in the retina. Nature. 1983;306(5945):782–4.
93. Klitten LL, Rath MF, Coon SL, Kim J-S, Klein DC, Møller M. Localization and regulation of dopamine receptor D4 expression in the adult and developing rat retina. Exp Eye Res. 2008;87(5):471–7.
94. Nir I, Harrison JM, Haque R, Low MJ, Grandy DK, Rubinstein M, et al. Dysfunctional light-evoked regulation of cAMP in photoreceptors and abnormal retinal adaptation in mice lacking dopamine D4 receptors. J Neurosci. 2002;22(6):2063–73.
95. Pozdeyev N, Tosini G, Li L, Ali F, Rozov S, Lee RH, et al. Dopamine modulates diurnal and circadian rhythms of protein phosphorylation in photoreceptor cells of mouse retina. Eur J Neurosci. 2008;27(10):2691–700.
96. Jackson CR, Chaurasia SS, Hwang CK, Iuvone PM. Dopamine D4 receptor activation controls circadian timing of the adenylyl cyclase 1/cyclic AMP signaling system in mouse retina. Eur J Neurosci. 2011;34(1):57–64.
97. Jackson CR, Chaurasia SS, Zhou H, Haque R, Storm DR, Iuvone PM. Essential roles of dopamine D4 receptors and the type 1 adenylyl cyclase in photic control of cyclic AMP in photoreceptor cells. J Neurochem. 2009;109(1):148–57.
98. Ribelayga C, Cao Y, Mangel SC. The circadian clock in the retina controls rod-cone coupling. Neuron. 2008;59(5):790–801.

99. Ribelayga C, Mangel SC. Absence of circadian clock regulation of horizontal cell gap junctional coupling reveals two dopamine systems in the goldfish retina. J Comp Neurol. 2003; 467(2):243–53.
100. Ribelayga C, Mangel SC. Tracer coupling between fish rod horizontal cells: modulation by light and dopamine but not the retinal circadian clock. Vis Neurosci. 2007;24(3):333–44.
101. Korenbrot JI, Fernald RD. Circadian rhythm and light regulate opsin mRNA in rod photoreceptors. Nature. 1989;337(6206):454–7.
102. Berson DM. Strange vision: ganglion cells as circadian photoreceptors. Trends Neurosci. 2003;26(6):314–20.
103. Yamazaki S, Alones V, Menaker M. Interaction of the retina with suprachiasmatic pacemakers in the control of circadian behavior. J Biol Rhythms. 2002;17(4):315–29.
104. Van Hook MJ, Wong KY, Berson DM. Dopaminergic modulation of ganglion-cell photoreceptors in rat. Eur J Neurosci. 2012;35(4):507–18.
105. Manglapus MK, Iuvone PM, Underwood H, Pierce ME, Barlow RB. Dopamine mediates circadian rhythms of rod-cone dominance in the Japanese quail retina. J Neurosci. 1999; 19(10):4132–41.
106. Emran F, Rihel J, Adolph AR, Dowling JE. Zebrafish larvae lose vision at night. Proc Natl Acad Sci U S A. 2010;107(13):6034–9.
107. Wang Y, Mangel SC. A circadian clock regulates rod and cone input to fish retinal cone horizontal cells. Proc Natl Acad Sci U S A. 1996;93(10):4655–60.
108. Krizaj D, Gabriel R, Owen WG, Witkovsky P. Dopamine D2 receptor-mediated modulation of rod-cone coupling in the Xenopus retina. J Comp Neurol. 1998;398(4):529–38.
109. Hampson EC, Vaney DI, Weiler R. Dopaminergic modulation of gap junction permeability between amacrine cells in mammalian retina. J Neurosci. 1992;12(12):4911–22.
110. Ingster-Moati I, Le Coz P, Albuisson E, Fromont G, Pierron C, Grall Y, et al. [Static contrast sensitivity in idiopathic Parkinson disease]. Rev Neurol (Paris). 1996;152(12):738–43.
111. Haug BA, Trenkwalder C, Arden GB, Oertel WH, Paulus W. Visual thresholds to low-contrast pattern displacement, color contrast, and luminance contrast stimuli in Parkinson's disease. Mov Disord. 1994;9(5):563–70.
112. Ikeda H, Head GM, Ellis CJ. Electrophysiological signs of retinal dopamine deficiency in recently diagnosed Parkinson's disease and a follow up study. Vision Res. 1994;34(19): 2629–38.
113. Herrmann R, Heflin SJ, Hammond T, Lee B, Wang J, Gainetdinov RR, et al. Rod vision is controlled by dopamine-dependent sensitization of rod bipolar cells by GABA. Neuron. 2011;72(1):101–10.
114. Hölter P, Kunst S, Wolloscheck T, Kelleher DK, Sticht C, Wolfrum U, et al. The retinal clock drives the expression of Kcnv2, a channel essential for visual function and cone survival. Invest Ophthalmol Vis Sci. 2012;53(11):6947–54.
115. Bartmann M, Schaeffel F, Hagel G, Zrenner E. Constant light affects retinal dopamine levels and blocks deprivation myopia but not lens-induced refractive errors in chickens. Vis Neurosci. 1994;11(2):199–208.
116. Feldkaemper M, Diether S, Kleine G, Schaeffel F. Interactions of spatial and luminance information in the retina of chickens during myopia development. Exp Eye Res. 1999; 68(1):105–15.
117. Guo SS, Sivak JG, Callender MG, Diehl-Jones B. Retinal dopamine and lens-induced refractive errors in chicks. Curr Eye Res. 1995;14(5):385–9.
118. Rohrer B, Spira AW, Stell WK. Apomorphine blocks form-deprivation myopia in chickens by a dopamine D2-receptor mechanism acting in retina or pigmented epithelium. Vis Neurosci. 1993;10(3):447–53.
119. Weiss S, Schaeffel F. Diurnal growth rhythms in the chicken eye: relation to myopia development and retinal dopamine levels. J Comp Physiol A. 1993;172(3):263–70.
120. Stone RA, Pardue MT, Iuvone PM, Khurana TS. Pharmacology of myopia and potential role for intrinsic retinal circadian rhythms. Exp Eye Res. 2013;114:35–47.

121. Koyanagi S, Kuramoto Y, Nakagawa H, Aramaki H, Ohdo S, Soeda S, et al. A molecular mechanism regulating circadian expression of vascular endothelial growth factor in tumor cells. Cancer Res. 2003;63(21):7277–83.
122. Chilov D, Hofer T, Bauer C, Wenger RH, Gassmann M. Hypoxia affects expression of circadian genes PER1 and CLOCK in mouse brain. FASEB J. 2001;15(14):2613–22.
123. Sarkar C, Chakroborty D, Mitra RB, Banerjee S, Dasgupta PS, Basu S. Dopamine in vivo inhibits VEGF-induced phosphorylation of VEGFR-2, MAPK, and focal adhesion kinase in endothelial cells. Am J Physiol Heart Circ Physiol. 2004;287(4):H1554–60.
124. Sinha S, Vohra PK, Bhattacharya R, Dutta S, Sinha S, Mukhopadhyay D. Dopamine regulates phosphorylation of VEGF receptor 2 by engaging Src-homology-2-domain-containing protein tyrosine phosphatase 2. J Cell Sci. 2009;122(Pt 18):3385–92.
125. Bhatwadekar AD, Yan Y, Qi X, Thinschmidt JS, Neu MB, Li Calzi S, et al. Per2 mutation recapitulates the vascular phenotype of diabetes in the retina and bone marrow. Diabetes. 2013;62(1):273–82.
126. Busik JV, Tikhonenko M, Bhatwadekar A, Opreanu M, Yakubova N, Caballero S, et al. Diabetic retinopathy is associated with bone marrow neuropathy and a depressed peripheral clock. J Exp Med. 2009;206(13):2897–906.
127. Debruyne JP, Noton E, Lambert CM, Maywood ES, Weaver DR, Reppert SM. A clock shock: mouse CLOCK is not required for circadian oscillator function. Neuron. 2006;50(3): 465–77.
128. DeBruyne JP, Weaver DR, Reppert SM. Peripheral circadian oscillators require CLOCK. Curr Biol. 2007;17(14):R538–9.

Chapter 6
Rhythmicity of the Retinal Pigment Epithelium

Linda Ruggiero and Silvia C. Finnemann

Abstract The retinal pigment epithelium (RPE) forms the outermost layer of the retina. Its interactions with outer segment portions of rod and cone photoreceptors are essential for vision. Among the many essential support functions of the RPE for photoreceptors, the contribution of the RPE in photoreceptor outer segment renewal is strictly rhythmic occurring once a day for both rods and cones. RPE cells contribute to shedding of distal outer segment tips of photoreceptors and are solely responsible for the concomitant clearance of shed debris by phagocytosis. Here, we discuss current knowledge of how the diurnal rhythm is established and how the rhythm translates into synchronized activation of the phagocytic machinery of the RPE.

Keywords Retinal pigment epithelium • Outer segment renewal • Shedding • Phagocytosis • Rhythmic clearance • Phosphatidylserine exposure

The intimate reciprocal relationship that exists between the neural retina and the underlying retinal pigment epithelium (RPE) in the vertebrate retina is crucial for vision. Specifically, photoreceptor neurons and RPE cells are so dependent upon each other that they can be considered as a single functional unit (Fig. 6.1). Their joint functions and interactions are strongly influenced by diurnal light–dark cycles and the circadian rhythm. Recent research suggests that diurnal rhythmicity of the RPE itself is critical for photoreceptor support and retinal function. Our knowledge to date of the molecular mechanisms governing RPE rhythmicity is the subject of this chapter.

L. Ruggiero, Ph.D. • S.C. Finnemann, Ph.D. (✉)
Department of Biological Sciences, Center for Cancer, Genetic Diseases and Gene Regulation, Fordham University, Larkin Hall,
441 East Fordham Road, Bronx, NY 10458, USA
e-mail: pixietail@gmail.com; finnemann@fordham.edu

G. Tosini et al. (eds.), *The Retina and Circadian Rhythms*, Springer Series in Vision Research 1, DOI 10.1007/978-1-4614-9613-7_6,

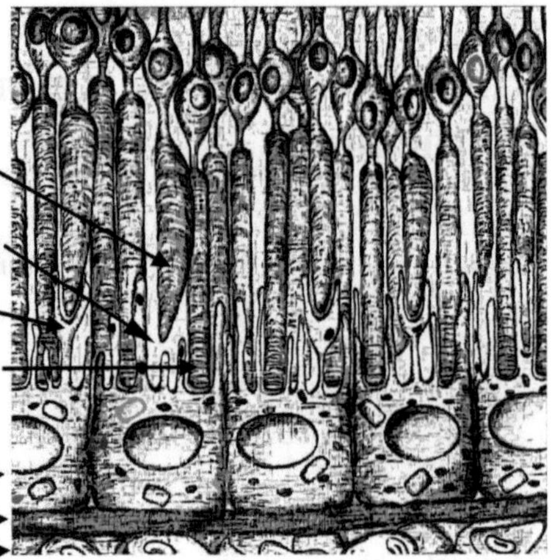

Fig. 6.1 The RPE forms the outermost layer of the retina. Each RPE cell extends apical microvillar processes (MV) into the interphotoreceptor matrix (IPM). These microvilli ensheath both cone and rod photoreceptor outer segments (C-OS, R-OS). At their basolateral surface, RPE cells interact with Bruch's membrane (BM), which separates the RPE from the vascularized choroid (CH)

6.1 Maintenance of Retinal Health and Function by Activities of the RPE

Like the retinal neurons they serve, RPE cells are post-mitotic in the mammalian eye. They must therefore continue to fulfill each one of their support tasks for the retina efficiently for sustained function. RPE cells form a polarized monolayer that lies adjacent to the rod and cone photoreceptors of the outer retina. At the apical surface, RPE cells extend highly elongated microvilli that reach into the subretinal spaces and ensheath the outer segment portions of rods and cones. Specific cell surface receptor proteins promote recognition, adhesion, and signaling of the RPE upon ligation by outer segment surface proteins or by extracellular glycoproteins residing in the interphotoreceptor matrix of the subretinal space. Tight junctions between RPE cells generate the outer blood–retinal barrier. Para- and transcellular transport across the RPE and trans-RPE cell migration are tightly regulated to maintain the subretinal space as an immune-privileged, avascular site. The basolateral plasma membrane of RPE cells faces Bruch's membrane, which consists of five layers of extracellular matrix with defined composition and elasticity and which separates the RPE from the choroidal vasculature. In addition to controlling access to the retina by way of barrier formation, RPE cells produce and release numerous trophic factors in a vectorial manner that are required for the survival and maintenance of photoreceptors both during development and throughout the lifetime of an individual (reviewed by Strauss [1]). There are no published studies to date that address the question whether or not the barrier and trophic functions of the RPE vary in a circadian rhythm.

In contrast, the subcellular localization of cytoplasmic melanosome pigment granules varies with time of day in wild-type mice entrained to a 12 h light and 12 h dark (12:12 LD) cycle [2]. Melanin pigment functions to absorb light and dissipates radical oxygen species thereby preventing photooxidative damage in the RPE [3]. Diurnal melanosome redistribution is modest in distance when compared to what is known from nonmammalian species such as teleosts but it is significant and accomplished in an active process that depends on Rab27a GTPase regulation of myosinVa motor mobility along the F-actin cytoskeleton in apical microvilli. Whether or not diurnal organelle transport in the RPE is subject to circadian regulation has not yet been investigated.

Two major tasks of the RPE are essential to maintain photoreceptor phototransduction activity. The first function is the participation of RPE cells in the visual cycle of chromophore bleaching and regeneration as first outlined in 1935 [4]. Rod and cone photoreceptor outer segments comprise stacks of membranous disks containing high concentrations of the photopigments rhodopsin or cone opsins, respectively. Light isomerizes the retinal chromophore of opsins from the 11-cis to the all-trans configuration stimulating phototransduction. Bleached all-trans retinal must be re-isomerized in an enzymatic process for reuse. Bleached rhodopsin releases its chromophore for delivery by specialized binding proteins to the adjacent RPE, which possesses the RPE65 isomerase that regenerates 11-cis retinal for transport back to rods and reassociation with opsin [5, 6]. Cone chromophore re-isomerization is less well understood but may utilize RPE65 in some species but non-RPE, non-RPE65 pathways have also been proposed (recently reviewed in [7]). Besides regenerating bleached retinal for recycling, RPE cells also take up vitamin A from the circulation and thus continuously replenish supplies for photoreceptors. RPE cells may store large amounts of retinylesters in specialized storage organelles [8]. The activity of the RPE in the visual cycle is not constant but necessarily varies with photoreceptor activity and therefore with light conditions. Circadian variation in any of the cycle's components or activities or in vitamin A uptake and processing remains to be studied.

The second function through which RPE cells maintain photoreceptor outer segment functionality is their role in the process of outer segment renewal that was first described in 1967–1968 [9, 10]. Photoreceptors constantly grow their outer segment portions by assembling new membrane disks at the proximal end of the outer segment. At the distal tip, they shed packets of disks yielding outer segment fragments that are about 1 μm in diameter in mouse retina. These photoreceptor outer segment fragments, referred to as "POS" from here on, are promptly removed from the retina by the RPE by phagocytosis, a three-phase mechanism that includes recognition, engulfment, and digestion. Shedding of POS is not entirely intrinsic to photoreceptors but a process to which RPE cells contribute: shed POS have never been observed unengulfed in between outer segments and RPE apical surface in the subretinal space [11]. Instead, disk packets have been found closely surrounded by apical RPE extensions, and interactions by RPE microvilli may aid in detaching POS. Furthermore, shedding is abnormal in rodent animal models with phagocytic defects of the RPE [12–14]. In the mammalian retina, the processes of POS shedding and phagocytosis are thus co-regulated and must be examined together. They occur in a strict diurnal rhythm that is influenced by the light–dark cycle and by circadian elements as follows.

6.2 A Conserved Role of the RPE in Outer Segment Renewal of Both Rods and Cones

While the precise contributions of rhythmic outer segment turnover to photoreceptor function in vision remain speculative, it is generally assumed that outer segment components become damaged and dysfunctional during normal levels of light exposure and that routine turnover of the entire outer segment structure is an efficient way of ensuring lifelong visual function [15]. This view is also supported by the fact that diurnal POS shedding and POS phagocytosis by the RPE are conserved across vertebrate species. Because of the prevalence of rod photoreceptors in common experimental animal models (97.2 % in mouse retina [16]) the majority of studies have examined rods and rod outer segment–RPE interactions. Pulse-chase experiments first demonstrated that radioactive amino acids were incorporated into the proximal end of the rod outer segment in the rat retina, then migrated to the photoreceptor's apical tip, and were eventually found in phagosomal and later phagolysosomal inclusion bodies in the RPE [17], thus elucidating the critical role of RPE cells in the renewal of rod outer segments. Later studies reported the same turnover process in mouse, cat, and frog retina.

When Young conducted his original studies on photoreceptor renewal, he did not observe the same synchronized turnover of radioactivity in cones that he observed in rods, leading him to conclude that cones do not regenerate their outer segments [18]. However, when studies on human fovea, a cone-rich region of the retina devoid of rod photoreceptors, showed the presence of phagosomes within the RPE, his theory was questioned [19]. It was also found that in cat and human retina, extensions of the RPE's apical membrane ensheathed cone outer segments [20]. This close proximity could allow for cone–RPE interactions, similar to those observed in rods. Further examination soon demonstrated the presence of phagosomes within these membranous extensions of the RPE, as well as within the RPE cell body, demonstrating that cones shed POS for subsequent phagocytosis by the RPE [21]. Additional studies in squirrels and rhesus monkeys confirmed that mammalian cones, like rods, yield shed POS that are phagocytosed by the RPE [22–24]. Today, it is generally accepted that highly similar processes of rod and cone photoreceptor outer segment renewal, that involve phagocytosis by the RPE, take place in all vertebrate species. However, rod and cone POS turnovers differ markedly in their diurnal rhythms.

6.3 Distinct Diurnal Rhythms of RPE Phagocytosis of Rod and Cone POS

Rod POS shedding and RPE phagocytosis display a strict diurnal rhythm. In rats maintained on a light cycle of 12 h of light followed by 12 h of darkness (12:12 LD), phagosome numbers in the RPE are higher shortly after light onset than at all other times of day that were tested [25]. Follow-up studies to this seminal observation determined this rhythm to be circadian. When rats are maintained in constant darkness (DD), the rhythm in shedding/phagocytosis is sustained. The peak in phagosome number displayed a period close to 24 h, suggesting that this rhythm is endogenous. Furthermore, entrainment of POS shedding and RPE phagocytosis occurs in rats shifted from one cyclic lighting regime to another. In these animals, the rhythm of POS phagocytosis shifts to coincide with the new lighting conditions, and the peak in RPE phagosome number occurs shortly after the new time of light onset. The circadian regulation of POS renewal has been further characterized by observing disk shedding/phagocytosis in animals maintained under altered lighting regimes. The initial studies demonstrating the persistence of POS renewal in DD were conducted on rats maintained in constant conditions for up to 3 days. Subsequent studies examined animals kept in DD for longer periods and showed that while the rhythm of shedding/phagocytosis is sustained after 12 days in DD, its period becomes slightly longer, and the timing of peak phagosome number within the RPE was broadened [26]. In contrast, in rats kept in constant light (LL), the rhythmic peak of shedding/phagocytosis is lost, suggesting that LL disturbs the regulation of the POS renewal rhythm. Transfer of animals from LL to LD can restore this rhythm [27]. Taken together, these studies conclusively demonstrate that the diurnal rhythm of rod POS shedding and RPE phagocytosis in rat retina is under circadian regulation.

In the cat, the rhythm of rod POS shedding and RPE phagocytosis is similar to that of rats and mice, peaking approximately 2 h after light onset [28]. In the frog *Rana pipiens*, rod POS renewal peaks also occur 1–2 h after light onset but (unlike in *Xenopus laevis*) they do not persist in DD [29]. In chickens reared on a 12:12 LD cycle, POS are also shed and phagocytosed by the RPE during the early light phase [30]. In addition, diurnal rhythms have been observed in goldfish, with rod POS shedding/phagocytosis peaking 2–4 h into the light phase [31]. However, the rhythm of POS shedding and RPE phagocytosis in goldfish does not appear to be circadian as it is readily lost in DD. Taken together, in all species examined rod POS shedding and RPE phagocytosis occur in a diurnal burst shortly after light onset. Light and the dark/light transition as well as circadian mechanisms regulate this rhythm. The relative importance of light changes and circadian effects varies considerably by species.

While shedding of cones has been studied much less extensively than shedding of rods, cone POS shedding and RPE phagocytosis have proven diurnal in all instances studied. The generation and characterization of the neural retina leucine zipper gene (Nrl) knockout mouse has shown that this mouse provides a murine model useful in the study of cones [32]. Loss of the transcription factor Nrl in these mice precludes rod photoreceptor differentiation yielding an all-cone photoreceptor retina [33].

When Nrl knockout mice are maintained on 12:12 LD, cone POS phagosome content in the RPE exhibits a peak 1 h after light onset. The rhythm was sustained when the mice were shifted to DD, suggesting circadian regulation, although the amplitude of the peak in phagosome number was reduced [34]. It is important to note, however, that Nrl knockout retina shows obvious signs of pathology such as retinal rosettes at early ages. This raises concern that there may be abnormalities in retinal and photoreceptor physiology and that therefore experimental insight obtained from studying Nrl knockout cones may not be directly applicable to turnover of cones in general. The Nile rat, *Arvicanthis ansorgei*, is a particularly exciting experimental model organism for the study of cone POS renewal. This rodent is active during the day, like humans but unlike mice and rats, and its retina by nature contains tenfold more cone photoreceptors than mice [35, 36]. When these rats are maintained in cyclic light, both rod and cone POS phagosome content of the RPE peaks approximately 1 h after light onset. In DD, these rhythms persist, suggesting that cone POS shedding and RPE phagocytosis are under circadian control [37]. Like the Nile rat, some other species shed and renew their cone POS with a rhythm similar to that of rods. In the cone-rich tree shrew, both cone POS shedding and RPE phagocytosis peak shortly after light onset [38]. In the cat, RPE cells possess peak numbers of cone phagosomes early during the light phase as well. Still other species, however, display different profiles of cone POS renewal. In lizards and chickens, RPE cells harbor the greatest number of cone POS phagosomes approximately 2 h after onset of darkness. In goldfish, RPE cells show peak cone POS phagosome load at approximately 4 h after onset of darkness [39]. In ground squirrels, cone POS shedding occurs even later during the dark phase, with maximum RPE phagosome content observed 7 h after dark onset [31, 40, 41]. Taken together, the timing of cone POS shedding and RPE phagocytosis varies widely among different species.

6.4 Local and Central Mechanisms Controlling the Phagocytosis Rhythm of the RPE

Much evidence suggests both local regulation in the eye and central control in the brain of rhythmic POS renewal. In mammals, the brain clock is located in the suprachiasmatic nucleus (SCN) of the hypothalamus. Circadian rhythms are generated as a result of intrinsic physiological properties of SCN neurons. When isolated in culture or slice preparations, these cells display oscillations in firing rates that persist for several days ex vivo [42, 43]. The endogenous rhythmic activity of the SCN is generated by a feedback loop of clock gene expression in individual cells. Cells with clock gene activity are present in many if not most tissues and organs outside of the SCN as well, and these "peripheral clocks" display sustained oscillations when isolated from the rest of the organism. Because rhythms in peripheral tissues may be out of phase with rhythms of the SCN, the SCN is believed to act as the "master clock" by using information from the environment to synchronize peripheral clocks within the organism (reviewed in [44]). Light input from the retina is the most potent entraining stimulus for the circadian clock in mammals. Retinal projections to the brain via the optic nerve allow for synchronization of the neuronal activity of SCN cells [45]. Outputs of the SCN, in turn, project to other parts of the hypothalamus, where they induce rhythms by releasing signaling molecules and by producing rhythmic firing patterns [46–48]. Once the cells of the SCN are synchronized with environmental light changes, photoentrainment of physiological rhythms is established.

The mammalian retina displays persistent rhythmic activity even when isolated from the brain. POS shedding and RPE phagocytosis in rats with transected nerves continue diurnally despite loss of synaptic connections between the eye and the brain, suggesting that this rhythm is generated and controlled locally in the eye [49]. However, this rhythm cannot be reset by light unless the optic nerve remains intact [50]. Thus, the circadian renewal of POS involves both local control within the retina and central regulation by the brain.

The circadian mechanisms operating in the retina and RPE are complex. Both retinal neurons and RPE cells display clock activity when isolated from the brain. It remains unclear as to whether rhythms regulating POS shedding and RPE phagocytosis are generated within retinal neurons, RPE cells, or both. Clock activity within retinal photoreceptors is responsible for controlling retinal melatonin synthesis in mammals [51–53]. Clock genes, such as period-2, have proven to be useful markers for tracking rhythmic cell activity. In cultured mouse retina lacking photoreceptors but expressing a period-2-luciferase (PER2-LUC) fusion protein, a circadian rhythm in PER2-LUC activity is present, lending support to clock activity of inner retinal cells [54, 55]. The RPE has also been reported to express clock genes and exhibits circadian rhythms in adenylyl cyclase activity and in Bmal1-luciferase expression [56, 57]. Moreover, Period-2 mRNA is cyclic in mouse RPE in vivo and a circadian rhythm in PER2-LUC bioluminescence can be recorded from cultured RPE cells. Further in vitro studies demonstrate that the PER2-LUC rhythm bioluminescence in the retina can be phase-shifted by light [58]. However, light does not phase-shift the rhythm in the RPE, suggesting that the RPE cannot respond to photic input directly but rather depends on input from the retina [59].

Researchers have postulated that melatonin and/or dopamine may be involved in the control of circadian POS shedding and RPE phagocytosis. Production and release of melatonin and dopamine, two regulatory signals of great importance in retinal physiology, are under circadian control in the retina [60–64]. The activity and expression profiles of dopamine and melatonin suggest that these molecules serve as chemical analogs of day and night, respectively. Retinal dopamine levels are highest during the light phase in contrast to melatonin whose levels peak during the dark period. Dopamine secretion from amacrine cells of the inner retina inhibits the production and release of retinal melatonin from photoreceptors. Melatonin, conversely, inhibits dopamine release [65–67]. In vitro studies on *Xenopus* eyecups have shown that application of exogenous melatonin stimulates a burst of POS shedding [68]. However, rhythmic disk shedding and RPE phagocytosis was observed in mice incapable of melatonin production, suggesting that this rhythm does not strictly require melatonin action in vivo [69].

Besides melatonin and dopamine ligands for G-protein-coupled receptors, a number of other molecules and signaling pathways have been implicated in the control of the rhythm of POS renewal. Experiments studying *Xenopus* eyecups have shown that application of the excitatory amino acids L-aspartate, L-glutamate, and kainic acid, potassium chloride, or ouabain can induce rod disk shedding [70]. While these compounds were only tested on isolated frog retina, they provide valuable hints into the potential mechanisms underlying the circadian regulation of POS renewal. Taken together, the mechanisms underlying the circadian and non-circadian regulation of the diurnal rhythm of POS shedding and RPE phagocytosis are complex. Future research is needed to fully understand them and to tease apart central and local, retinal and RPE contributions.

6.5 The Molecular Mechanisms Governing Rhythmic RPE Phagocytosis

The cell and molecular mechanisms underlying circadian regulation of shedding/phagocytosis are complex. However, cell culture models of polarized RPE that retains phagocytic activity and the availability of genetically modified animals have enabled the identification of key molecules and signaling pathways. RPE cells utilize a molecular mechanism for POS phagocytosis that is similar to the mechanism for uptake of apoptotic cells used by other professional phagocytic cell types like macrophages and dendritic cells. Indeed, POS and apoptotic cells compete for recognition by both RPE and macrophages in culture [71].

Phagocytosis of spent POS by RPE cells is a relatively slow process that occurs in distinct phases: recognition of and binding to POS by RPE cells, engulfment of shed POS, and lysosomal digestion of spent POS. Binding to POS occurs via a saturable, receptor-mediated recognition and attachment event [72] and is governed by αvβ5 integrin, the only integrin receptor expressed at the apical surface of RPE cells [73]. Complex formation at the cell surface with the tetraspanin CD81 enhances integrin binding activity toward POS that are opsonized with integrin ligand [74]. The secreted glycoprotein Milk Fat Globule-E8 (MFG-E8) is the physiological ligand for αvβ5 integrin during retinal phagocytosis [75]. MFG-E8 is expressed in the retina and RPE of mouse and rat, and it binds αvβ5 integrin via an RGD motif [75–77]. Studies of knockout mice lacking either αvβ5 integrin receptors ($\beta5^{-/-}$ mice) or MFG-E8 (MFG-E8$^{-/-}$ mice) have revealed that this receptor–ligand pair is essential for the diurnal rhythm of POS uptake. If maintained in 12:12 LD, both $\beta5^{-/-}$ and MFG-E8$^{-/-}$ mice lack the rhythmic peak in shedding and RPE phagocytosis after light onset that is characteristic to wild-type animals [14, 75]. Rather, these animals showed a constant load of POS phagosomes in the RPE at all times tested throughout the 24 h time period. These data imply that αvβ5 integrin ligation synchronizes RPE phagocytosis with environmental lighting conditions. Studies on the trigger of integrin activation after light onset remain largely obscure. We recently found that at the whole retina level, MFG-E8 protein levels are significantly higher around light onset as compared to several hours later when phagocytosis has stopped (Fig. 6.2). We speculate that the light cycle and/or circadian rhythm triggered molecular pathways that govern MFG-E8 synthesis and that this plays a role in rhythmic integrin activation after light onset. As αvβ5 integrin is expressed exclusively on the

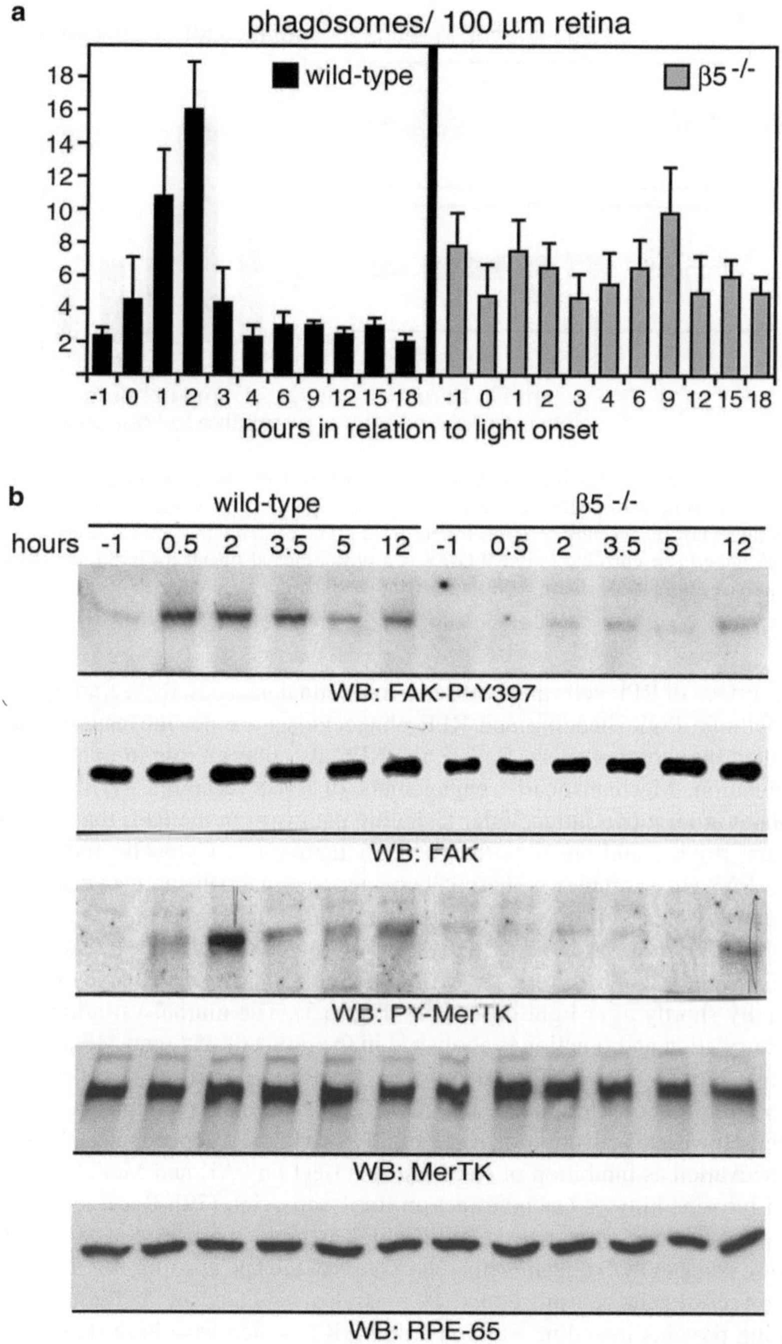

Fig. 6.2 The diurnal rhythm of RPE phagocytosis involves a daily burst of FAK and MerTK signaling after light onset and depends on αvβ5 integrin. (**a**) $\beta5^{-/-}$ mouse retina lacks the characteristic rhythm of RPE phagocytosis that follows light onset in wild-type mouse retina. *Bars* represent mean numbers of phagosomes ±SD ($n=3$). (**b**) Increased tyrosine phosphorylation of FAK and MerTK coincides with the rhythm of RPE phagocytosis while protein levels remain constant. © Nandrot et al., 2004. Originally published in J Exp Med. 200:1539–45 [14]

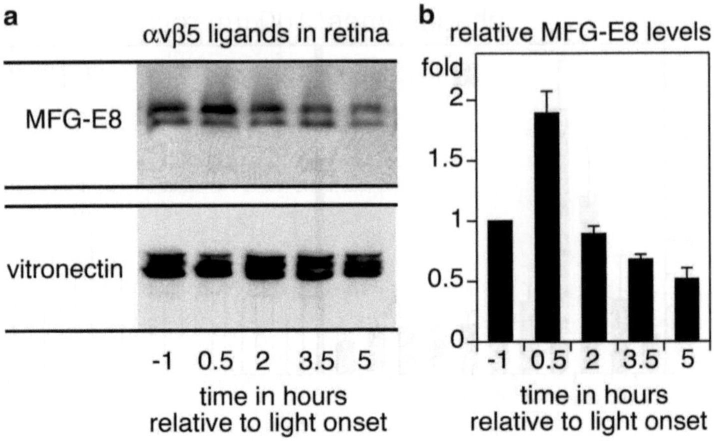

Fig. 6.3 Increased MFG-E8 levels in the retina coincide with the rhythmic phagocytic function of RPE cells. (**a, b**) Levels of MFG-E8 peak shortly after light onset in mouse retina while levels of vitronectin do not change at time points tested. Both MFG-E8 and vitronectin are capable of binding αvβ5 integrin in vitro but only MFG-E8 is a physiological ligand for αvβ5 receptors at the apical surface of the RPE, where it initiates rhythmic signaling and POS clearance

apical surface of RPE cells and is absent from retinal neurons, these findings suggest that rhythmic POS shedding and RPE phagocytosis are not intrinsic mechanisms controlled by photoreceptors. Rather, the RPE also plays a role in this process and its regulation. Mechanistically, engagement of αvβ5 receptors by MFG-E8-POS stimulates at least two intracellular signaling pathways in the RPE that both follow a diurnal rhythm and are required for POS uptake [1]. Cytosolic focal adhesion kinase (FAK) in a complex with αvβ5 becomes enzymatically active after ligation of αvβ5 and dissociates from the receptor complex ultimately causing a peak in tyrosine phosphorylation and likely activation of Mer tyrosine kinase (MerTK) [78]. Both FAK and MerTK in retina of wild-type mice in 12:12 LD display brief bursts of activity shortly after light onset [14] (Fig. 6.3). The diurnal variation of kinase phosphorylation and function is abolished in the retina of $\beta5^{-/-}$ and MFG-E8$^{-/-}$ mice [2]. Like MerTK and FAK, activation of the RhoA family GTPase Rac1 peaks shortly after light onset in mouse and rat RPE in vivo in an αvβ5 integrin-dependent manner. However, activation of Rac1 by POS-αvβ5 is independent of MerTK and FAK activation as inhibition of Rac1 has no effect on FAK and MerTK and inhibition of tyrosine kinases has no effect on Rac1 activation [79]. Rac1 activation ultimately targets the F-actin cytoskeleton, which assembles in a special phagocytic cup structure beneath POS that are tethered to αvβ5 at the apical surface of the RPE [80]. F-actin recruitment beneath bound POS is also regulated by additional cytosolic signaling proteins including PI3kinase and AKT, which have both shared and distinct functions in phagocytic cup assembly [81]. AKT affects not only F-actin but also recruitment of the F-actin-binding protein and molecular motor myosin-II to the phagocytic cup. Myosin-II motor activity is required for POS engulfment [82].

The rhythmicity of PI3kinase and AKT signaling is currently under investigation. Another actin-binding protein, annexin-2, shows an increase in phosphorylation shortly after light onset, suggesting a diurnal rhythm in its activity that coincides with peak phagocytic activity by the RPE [83]. Notably, AKT increases levels of F-actin and myosin-2 but not of annexin-2 in phagocytic cups. Lack of annexin-2 in mice does not eliminate the rhythm of POS shedding and RPE phagocytosis. Rather, POS processing is delayed possibly related to a shift in peak phosphorylation of FAK. One can thus speculate that annexin-2 activities may affect the rhythm generated by the MFG-E8-αvβ5 signaling pathways.

Engulfment of POS is followed by lysosomal degradation of spent photoreceptor disks within the RPE. This process relies on the activity of lysosomal digestive enzymes, such as the protease cathepsin D. Cathepsin D is specifically required for the cleavage of the rod photopigment rhodopsin by far the most abundant protein component of rod POS [84, 85]. Interestingly, in rabbits maintained in LD, a diurnal rhythm in cathepsin D activity peaks 1.5 h after light onset, at the same time as peak numbers of undigested POS phagosomes reside in the RPE [86]. Electron microscopy studies showed diurnal changes in cathepsin D also in rat retina [87]. In the RPE of entrained rats, opsin may be immunolabeled in early phagosomes that are devoid of cathepsin D. However, shortly after cathepsin D is detected in maturing phagosomes, opsin levels sharply decline suggesting that cathepsin D delivery to POS phagosomes is synchronized to cause a burst of opsin degradation in RPE phagolysosomes.

Taken together, numerous proteins that play a role in RPE phagocytosis of shed POS change activity or in their level of expression to coincide with the uptake rhythm. The ligation of the integrin receptor αvβ5 by its ligand MFG-E8 is required for the rhythm itself and responsible for inducing rhythmic tyrosine kinase signaling. Elevated MFG-E8 levels in the retina at the time of phagocytosis may contribute to the synchronized ligation and signaling of αvβ5. Other studies on the regulatory elements upstream of integrin ligation and the photoreceptor contribution to synchronized shedding and POS clearance will be facilitated by the availability of new methodology that can image shedding rod tips in live mouse retina within minutes of dissection as discussed next.

6.6 Dependence of the Rhythm of POS Shedding on the Rhythm of RPE Phagocytosis

MFG-E8 is a globular protein with several conserved functional domains. Besides an RGD motif for integrin ligation, MFG-E8 also possesses a binding site for anionic phospholipids. The best characterized "eat-me" signal of apoptotic cells that triggers opsonization and clearance phagocytosis via integrins among other pathways is the exposure of phosphatidylserine (PS). In cells in the early stages of apoptosis, PS, which resides in the inner leaflet of the plasma membrane in healthy cells, is actively externalized to the outer leaflet. Apoptotic cells and POS compete for binding to RPE cells in culture, which use αvβ5 for particle binding. It had thus

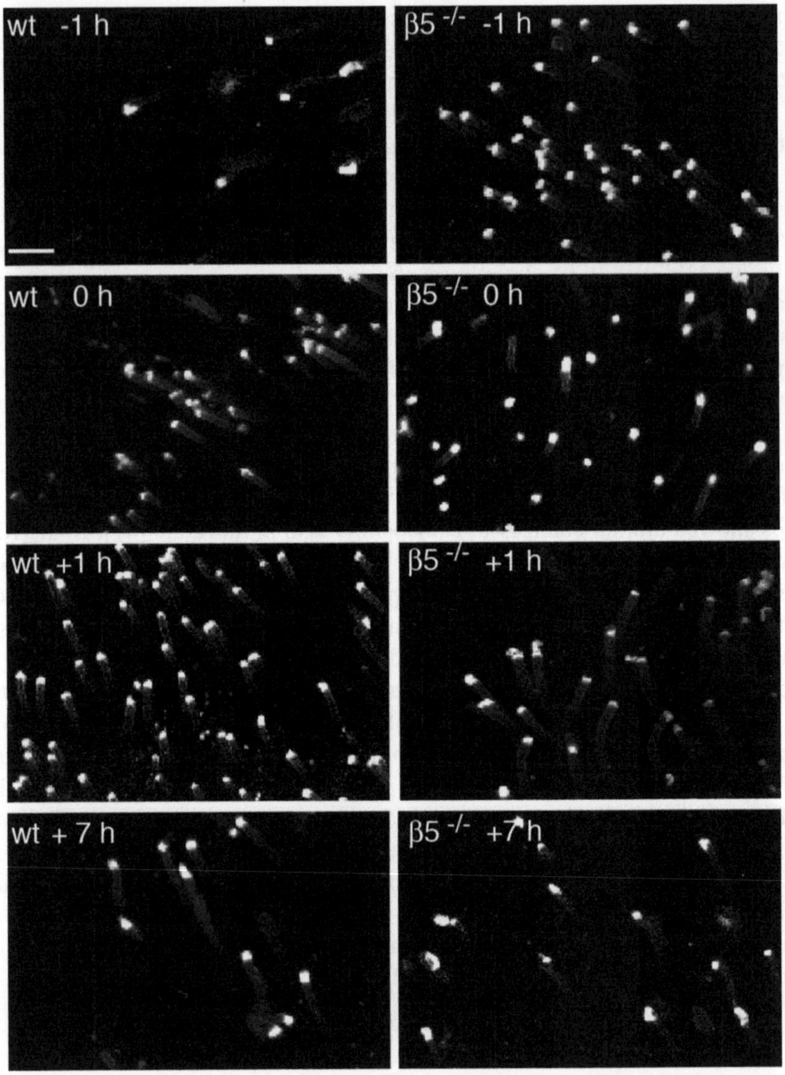

Fig. 6.4 The diurnal rhythm in frequency of PS-marked rod outer segment tips requires αvβ5 integrin. *Scale bar*, 10 μm. © Ruggiero et al., 2012. Originally published in Proc Natl Acad Sci U S A. 109(21):8145–8 [88]

long been proposed that the PS serves as an "eat-me" signal for POS as well. However, as shedding cannot be mimicked ex vivo, it remained unclear if outer segments expose PS in the intact retina to initiate diurnal RPE phagocytosis.

Using a novel fluorescent biosensor for PS, we recently demonstrated that rod outer segment tips externalize PS exclusively at their distal tips with discrete boundaries [88]. Frequency of these PS-marked tips is strictly rhythmic in mice in 12:12 LD with highest numbers immediately after light onset (Fig. 6.4).

Moreover, PS-marked tips elongate significantly at light onset. Interestingly, $\beta 5^{-/-}$ and MFG-E8$^{-/-}$ mice do not show this daily rhythm of tip formation and elongation in accordance with their loss of phagocytic rhythm. It is our interpretation that PS externalization serves as an "eat-me" signal on rod tips at light onset and triggers POS shedding and RPE phagocytosis. In mice lacking the integrin-dependent phagocytosis rhythm of the RPE, the rhythm of PS exposure is lost indicating that the diurnal rhythm of shedding and RPE phagocytosis is not intrinsic to rod photoreceptors but requires activities of the RPE as well.

6.7 Importance of the Diurnal RPE Phagocytic Rhythm for Retinal Health

Lack of POS phagocytosis or digestion leads to photoreceptor dystrophy and blindness. The lack of engulfment activity of the RPE in rats deficient in MerTK causes dramatic and early onset retinal degeneration [25]. Mutations in the MERTK gene have been identified as the cause of retinitis pigmentosa in human patients [89–91]. Mice lacking the lysosomal protease cathepsin D develop postnatal retinal atrophy [92], and transgenic mice expressing a mutant variant of cathepsin D exhibit gradual debris accumulation, age-related RPE atrophy, and photoreceptor degeneration [85, 93]. Unlike for loss-of-phagocytosis models, it is not immediately intuitive that lack of a daily rhythm of POS shedding and RPE phagocytosis must impair retinal or RPE functions. Indeed, $\beta 5^{-/-}$ RPE and retina are morphologically normal and electroretinogram recordings show no abnormalities until animals are ~4 months of age. Thus, the lack of diurnal rhythm of POS turnover does not impair retinal function in the short term. However, mice lose cone and rod photoreceptor function starting in middle age, at ~6 months of age. By 12 months of age, rod and cone functions are severely diminished. In aged animals, we also observe accumulation of lipofuscin-like autofluorescent material in RPE lysosomes [14]. As lipofuscin in the human eye has been shown to contain oxidized lipids and to act as a photooxidizer, we examined the oxidative burden of $\beta 5^{-/-}$ retina and RPE with age [94]. Strikingly, $\beta 5^{-/-}$ RPE cells, but not retinal neurons, increasingly accumulate oxidized proteins and lipids, which destabilizes the F-actin cytoskeleton. Thus, loss of rhythmicity of RPE phagocytosis, the primary defect in $\beta 5^{-/-}$ mice, leads to elevated oxidative stress and damage with age. Dietary enrichment with natural antioxidants is sufficient to prevent oxidative damage and lipofuscin accumulation, and maintains photoreceptor function, but has no effect on the phagocytosis rhythm of the RPE. We, therefore, conclude that the rhythm of RPE phagocytosis that depends on $\alpha v\beta 5$ integrin is important for retinal health. More studies are needed to understand its molecular regulation upstream of MFG-E8 and how rhythmicity prevents oxidative damage.

6.8 Summary

The strict diurnal rhythm of RPE phagocytosis is highly conserved among vertebrates. Its regulation varies depending on species. In mammals, the phagocytosis rhythm is responsive to light changes and controlled by clock mechanisms both locally in the retina and centrally in the brain. Synchronization of RPE phagocytosis in the retina fails to take place if the RPE lacks the integrin receptor αvβ5 or its ligand MFG-E8, which opsonizes shedding outer segment tips exposing PS. How clock mechanisms target the integrin receptor pathway remains to be studied.

To date, our understanding of the role of rhythmicity in RPE activities is largely restricted to the RPE's function in photoreceptor outer segment renewal. In the future, greater availability of experimental models and more sensitive methodology will increasingly allow examining RPE functionality in situ. We anticipate that such studies will uncover roles for synchronization and rhythmicity in additional RPE activities other than phagocytosis.

Acknowledgment This work was supported by NIH grant EY013295.

References

1. Strauss O. The retinal pigment epithelium in visual function. Physiol Rev. 2005;85(3): 845–81.
2. Futter CE, Ramalho JS, Jaissle GB, Seeliger MW, Seabra MC. The role of Rab27a in the regulation of melanosome distribution within retinal pigment epithelial cells. Mol Biol Cell. 2004;15(5):2264–75.
3. Burke JM, Kaczara P, Skumatz CM, Zareba M, Raciti MW, Sarna T. Dynamic analyses reveal cytoprotection by RPE melanosomes against non-photic stress. Mol Vis. 2011;17:2864–77.
4. Wald G. Carotenoids and the visual cycle. J Gen Physiol. 1935;19(2):351–71.
5. Moiseyev G, Chen Y, Takahashi Y, Wu BX, Ma JX. RPE65 is the isomerohydrolase in the retinoid visual cycle. Proc Natl Acad Sci U S A. 2005;102(35):12413–8.
6. Redmond TM, Poliakov E, Yu S, Tsai JY, Lu Z, Gentleman S. Mutation of key residues of RPE65 abolishes its enzymatic role as isomerohydrolase in the visual cycle. Proc Natl Acad Sci U S A. 2005;102(38):13658–63.
7. Tang PH, Kono M, Koutalos Y, Ablonczy Z, Crouch RK. New insights into retinoid metabolism and cycling within the retina. Prog Retin Eye Res. 2013;32:48–63.
8. Imanishi Y, Batten ML, Piston DW, Baehr W, Palczewski K. Noninvasive two-photon imaging reveals retinyl ester storage structures in the eye. J Cell Biol. 2004;164(3):373–83.
9. Young RW, Bok D. Participation of the retinal pigment epithelium in the rod outer segment renewal process. J Cell Biol. 1969;42(2):392–403.
10. Young RW. The renewal of photoreceptor cell outer segments. J Cell Biol. 1967;33(1):61–72.
11. Besharse J, Defoe D. The role of the retinal pigment epithelium in photoreceptor membrane turnover. In: Marmor MF, Wolfensberger TJ, editors. The retinal pigment epithelium. New York: Oxford University Press; 1998. p. 152–72.
12. Bok D, Hall MO. The role of the pigment epithelium in the etiology of inherited retinal dystrophy in the rat. J Cell Biol. 1971;49:664–82.
13. Davidorf FH, Mendlovic DB, Bowyer DW, Gresak PM, Foreman BC, Werling KT, Chambers RB. Pathogenesis of retinal dystrophy in the Royal College of Surgeons rat. Ann Ophthalmol. 1991;23(3):87–94.

14. Nandrot EF, Kim Y, Brodie SE, Huang X, Sheppard D, Finnemann SC. Loss of synchronized retinal phagocytosis and age-related blindness in mice lacking αvβ5 integrin. J Exp Med. 2004;200:1539–45.
15. Beatty S, Koh H, Phil M, Henson D, Boulton M. The role of oxidative stress in the pathogenesis of age-related macular degeneration. Surv Ophthalmol. 2000;45(2):115–34.
16. Jeon CJ, Strettoi E, Masland RH. The major cell populations of the mouse retina. J Neurosci. 1998;18(21):8936–46.
17. Bosch E, Horwitz J, Bok D. Phagocytosis of outer segments by retinal pigment epithelium: phagosome-lysosome interaction. J Histochem Cytochem. 1993;41(2):253–63.
18. Young RW. An hypothesis to account for a basic distinction between rods and cones. Vision Res. 1971;11(1):1–5.
19. Hogan MJ. Role of the retinal pigment epithelium in macular disease. Trans Am Acad Ophthalmol Otolaryngol. 1972;76(1):64–80.
20. Steinberg RH, Wood I. Pigment epithelial cell ensheathment of cone outer segments in the retina of the domestic cat. Proc R Soc Lond B Biol Sci. 1974;187(1089):461–78.
21. Steinberg RH, Wood I, Hogan MJ. Pigment epithelial ensheathment and phagocytosis of extrafoveal cones in human retina. Philos Trans R Soc Lond B Biol Sci. 1977;277(958):459–74.
22. Anderson DH, Fisher SK. Disc shedding in rodlike and conelike photoreceptors of tree squirrels. Science. 1975;187(4180):953–5.
23. Anderson DH, Fisher SK. The photoreceptors of diurnal squirrels: outer segment structure, disc shedding, and protein renewal. J Ultrastruct Res. 1976;55(1):119–41.
24. Anderson DH, Fisher SK, Steinberg RH. Mammalian cones: disc shedding, phagocytosis, and renewal. Invest Ophthalmol Vis Sci. 1978;17(2):117–33.
25. Mullen RJ, LaVail MM. Inherited retinal dystrophy: primary defect in pigment epithelium determined with experimental rat chimeras. Science. 1976;192(4241):799–801.
26. LaVail MM. Circadian nature of rod outer segment disc shedding in the rat. Invest Ophthalmol Vis Sci. 1980;19(4):407–11.
27. Goldman AI, Teirstein PS, O'Brien PJ. The role of ambient lighting in circadian disc shedding in the rod outer segment of the rat retina. Invest Ophthalmol Vis Sci. 1980;19(11):1257–67.
28. Fisher SK, Pfeffer BA, Anderson DH. Both rod and cone disc shedding are related to light onset in the cat. Invest Ophthalmol Vis Sci. 1983;24(7):844–56.
29. Basinger S, Hoffman R, Matthes M. Photoreceptor shedding is initiated by light in the frog retina. Science. 1976;194(4269):1074–6.
30. Young RW. The daily rhythm of shedding and degradation of rod and cone outer segment membranes in the chick retina. Invest Ophthalmol Vis Sci. 1978;17(2):105–16.
31. O'Day WT, Young RW. Rhythmic daily shedding of outer-segment membranes by visual cells in the goldfish. J Cell Biol. 1978;76(3):593–604.
32. Mears AJ, Kondo M, Swain PK, Takada Y, Bush RA, Saunders TL, Sieving PA, Swaroop A. Nrl is required for rod photoreceptor development. Nat Genet. 2001;29(4):447–52.
33. Daniele LL, Lillo C, Lyubarsky AL, Nikonov SS, Philp N, Mears AJ, Swaroop A, Williams DS, Pugh Jr EN. Cone-like morphological, molecular, and electrophysiological features of the photoreceptors of the Nrl knockout mouse. Invest Ophthalmol Vis Sci. 2005;46(6):2156–67.
34. Krigel A, Felder-Schmittbuhl MP, Hicks D. Circadian-clock driven cone-like photoreceptor phagocytosis in the neural retina leucine zipper gene knockout mouse. Mol Vis. 2010;16: 2873–81.
35. Bobu C, Craft CM, Masson-Pevet M, Hicks D. Photoreceptor organization and rhythmic phagocytosis in the nile rat Arvicanthis ansorgei: a novel diurnal rodent model for the study of cone pathophysiology. Invest Ophthalmol Vis Sci. 2006;47(7):3109–18.
36. Bobu C, Lahmam M, Vuillez P, Ouarour A, Hicks D. Photoreceptor organisation and phenotypic characterization in retinas of two diurnal rodent species: potential use as experimental animal models for human vision research. Vision Res. 2008;48(3):424–32.
37. Bobu C, Hicks D. Regulation of retinal photoreceptor phagocytosis in a diurnal mammal by circadian clocks and ambient lighting. Invest Ophthalmol Vis Sci. 2009;50(7):3495–502.
38. Immel JH, Fisher SK. Cone photoreceptor shedding in the tree shrew (Tupaia belangerii). Cell Tissue Res. 1985;239(3):667–75.

39. Balkema Jr GW, Bunt-Milam AH. Cone outer segment shedding in the goldfish retina characterized with the 3H-fucose technique. Invest Ophthalmol Vis Sci. 1982;23(3):319–31.
40. Reme CE, Young RW. The effects of hibernation on cone visual cells in the ground squirrel. Invest Ophthalmol Vis Sci. 1977;16(9):815–40.
41. Tabor GA, Fisher SK, Anderson DH. Rod and cone disc shedding in light-entrained tree squirrels. Exp Eye Res. 1980;30(5):545–57.
42. Welsh DK, Logothetis DE, Meister M, Reppert SM. Individual neurons dissociated from rat suprachiasmatic nucleus express independently phased circadian firing rhythms. Neuron. 1995;14(4):697–706.
43. Chen D, Buchanan GF, Ding JM, Hannibal J, Gillette MU. Pituitary adenylyl cyclase-activating peptide: a pivotal modulator of glutamatergic regulation of the suprachiasmatic circadian clock. Proc Natl Acad Sci U S A. 1999;96(23):13468–73.
44. Green CB, Besharse JC. Retinal circadian clocks and control of retinal physiology. J Biol Rhythms. 2004;19(2):91–102.
45. Liu C, Reppert SM. GABA synchronizes clock cells within the suprachiasmatic circadian clock. Neuron. 2000;25(1):123–8.
46. Inouye ST, Kawamura H. Persistence of circadian rhythmicity in a mammalian hypothalamic "island" containing the suprachiasmatic nucleus. Proc Natl Acad Sci U S A. 1979;76(11): 5962–6.
47. Green DJ, Gillette R. Circadian rhythm of firing rate recorded from single cells in the rat suprachiasmatic brain slice. Brain Res. 1982;245(1):198–200.
48. Kalsbeek A, Buijs RM, Engelmann M, Wotjak CT, Landgraf R. *In vivo* measurement of a diurnal variation in vasopressin release in the rat suprachiasmatic nucleus. Brain Res. 1995;682(1–2):75–82.
49. Terman JS, Reme CE, Terman M. Rod outer segment disk shedding in rats with lesions of the suprachiasmatic nucleus. Brain Res. 1993;605(2):256–64.
50. Teirstein PS, Goldman AI, O'Brien PJ. Evidence for both local and central regulation of rat rod outer segment disc shedding. Invest Ophthalmol Vis Sci. 1980;19(11):1268–73.
51. Tosini G, Menaker M. The clock in the mouse retina: melatonin synthesis and photoreceptor degeneration. Brain Res. 1998;789(2):221–8.
52. Tosini G, Kasamatsu M, Sakamoto K. Clock gene expression in the rat retina: effects of lighting conditions and photoreceptor degeneration. Brain Res. 2007;1159:134–40.
53. Tosini G, Pozdeyev N, Sakamoto K, Iuvone PM. The circadian clock system in the mammalian retina. Bioessays. 2008;30(7):624–33.
54. Yoo SH, Yamazaki S, Lowrey PL, Shimomura K, Ko CH, Buhr ED, Siepka SM, Hong HK, Oh WJ, Yoo OJ, et al. PERIOD2::LUCIFERASE real-time reporting of circadian dynamics reveals persistent circadian oscillations in mouse peripheral tissues. Proc Natl Acad Sci U S A. 2004;101(15):5339–46.
55. Ruan GX, Zhang DQ, Zhou T, Yamazaki S, McMahon DG. Circadian organization of the mammalian retina. Proc Natl Acad Sci U S A. 2006;103(25):9703–8.
56. Pavan B, Frigato E, Pozzati S, Prasad PD, Bertolucci C, Biondi C. Circadian clocks regulate adenylyl cyclase activity rhythms in human RPE cells. Biochem Biophys Res Commun. 2006;350(1):169–73.
57. Yoshikawa A, Shimada H, Numazawa K, Sasaki T, Ikeda M, Kawashima M, Kato N, Tokunaga K, Ebisawa T. Establishment of human cell lines showing circadian rhythms of bioluminescence. Neurosci Lett. 2008;446(1):40–4.
58. Ruan GX, Allen GC, Yamazaki S, McMahon DG. An autonomous circadian clock in the inner mouse retina regulated by dopamine and GABA. PLoS Biol. 2008;6(10):e249.
59. Baba K, Sengupta A, Tosini M, Contreras-Alcantara S, Tosini G. Circadian regulation of the PERIOD 2::LUCIFERASE bioluminescence rhythm in the mouse retinal pigment epithelium-choroid. Mol Vis. 2010;16:2605–11.
60. Tosini G, Menaker M. The pineal complex and melatonin affect the expression of the daily rhythm of behavioral thermoregulation in the green iguana. J Comp Physiol. 1996;179(1): 135–42.

61. Nir I, Haque R, Iuvone PM. Diurnal metabolism of dopamine in the mouse retina. Brain Res. 2000;870(1–2):118–25.
62. Tosini G, Dirden JC. Dopamine inhibits melatonin release in the mammalian retina: in vitro evidence. Neurosci Lett. 2000;286(2):119–22.
63. Doyle SE, Grace MS, McIvor W, Menaker M. Circadian rhythms of dopamine in mouse retina: the role of melatonin. Vis Neurosci. 2002;19(5):593–601.
64. Doyle SE, McIvor WE, Menaker M. Circadian rhythmicity in dopamine content of mammalian retina: role of the photoreceptors. J Neurochem. 2002;83(1):211–9.
65. Dubocovich ML. Melatonin is a potent modulator of dopamine release in the retina. Nature. 1983;306(5945):782–4.
66. Cahill GM, Besharse JC. Resetting the circadian clock in cultured Xenopus eyecups: regulation of retinal melatonin rhythms by light and D2 dopamine receptors. J Neurosci. 1991; 11(10):2959–71.
67. Nguyen-Legros J, Chanut E, Versaux-Botteri C, Simon A, Trouvin JH. Dopamine inhibits melatonin synthesis in photoreceptor cells through a D2-like receptor subtype in the rat retina: biochemical and histochemical evidence. J Neurochem. 1996;67(6):2514–20.
68. Besharse JC, Dunis DA. Methoxyindoles and photoreceptor metabolism: activation of rod shedding. Science. 1983;219(4590):1341–3.
69. Grace MS, Chiba A, Menaker M. Circadian control of photoreceptor outer segment membrane turnover in mice genetically incapable of melatonin synthesis. Vis Neurosci. 1999;16(5): 909–18.
70. Greenberger LM, Besharse JC. Stimulation of photoreceptor disc shedding and pigment epithelial phagocytosis by glutamate, aspartate, and other amino acids. J Comp Neurol. 1985; 239(4):361–72.
71. Finnemann SC, Rodriguez-Boulan E. Macrophage and retinal pigment epithelium phagocytosis: apoptotic cells and photoreceptors compete for αvβ3 and αvβ5 integrins, and protein kinase C regulates αvβ5 binding and cytoskeletal linkage. J Exp Med. 1999;190(6):861–74.
72. Mayerson PL, Hall MO. Rat retinal pigment epithelial cells show specificity of phagocytosis in vitro. J Cell Biol. 1986;103(1):299–308.
73. Finnemann SC, Bonilha VL, Marmorstein AD, Rodriguez-Boulan E. Phagocytosis of rod outer segments by retinal pigment epithelial cells requires αvβ5 integrin for binding but not for internalization. Proc Natl Acad Sci U S A. 1997;94(24):12932–7.
74. Chang Y, Finnemann SC. Tetraspanin CD81 is required for the αvβ5 integrin-dependent particle-binding step of RPE phagocytosis. J Cell Sci. 2007;120(17):3053–63.
75. Nandrot EF, Anand M, Almeida D, Atabai K, Sheppard D, Finnemann SC. Essential role for MFG-E8 as ligand for αvβ5 integrin in diurnal retinal phagocytosis. Proc Natl Acad Sci U S A. 2007;104(29):12005–10.
76. Hanayama R, Tanaka M, Miwa K, Shinohara A, Iwamatsu A, Nagata S. Identification of a factor that links apoptotic cells to phagocytes. Nature. 2002;417(6885):182–7.
77. Burgess BL, Abrams TA, Nagata S, Hall MO. MFG-E8 in the retina and retinal pigment epithelium of rat and mouse. Mol Vis. 2006;12:1437–47.
78. Finnemann SC. Focal adhesion kinase signaling promotes phagocytosis of integrin-bound photoreceptors. EMBO J. 2003;22(16):4143–54.
79. Mao Y, Finnemann SC. Essential diurnal Rac1 activation during retinal phagocytosis requires αvβ5 integrin but not tyrosine kinases FAK or MerTK. Mol Biol Cell. 2012;23:1104–14.
80. Chaitin MH, Hall MO. The distribution of actin in cultured normal and dystrophic rat pigment epithelial cells during the phagocytosis of rod outer segments. Invest Ophthalmol Vis Sci. 1983;24(7):821–31.
81. Bulloj A, Duan W, Finnemann SC. PI 3-kinase independent role for AKT in F-actin regulation during outer segment phagocytosis by RPE cells. Exp Eye Res. 2013;113:9–18.
82. Strick DJ, Feng W, Vollrath D. Mertk drives myosin II redistribution during retinal pigment epithelial phagocytosis. Invest Ophthalmol Vis Sci. 2009;50(5):2427–35.
83. Law AL, Ling Q, Hajjar KA, Futter CE, Greenwood J, Adamson P, Wavre-Shapton ST, Moss SE, Hayes MJ. Annexin A2 regulates phagocytosis of photoreceptor outer segments in the mouse retina. Mol Biol Cell. 2009;20(17):3896–904.

84. Rakoczy PE, Lai CM, Baines M, Di Grandi S, Fitton JH, Constable IJ. Modulation of cathepsin D activity in retinal pigment epithelial cells. Biochem J. 1997;324(Pt 3):935–40.
85. Rakoczy PE, Zhang D, Robertson T, Barnett NL, Papadimitriou J, Constable IJ, Lai CM. Progressive age-related changes similar to age-related macular degeneration in a transgenic mouse model. Am J Pathol. 2002;161(4):1515–24.
86. Kim IT, Kwak JS. Degradation of phagosomes and diurnal changes of lysosomes in rabbit retinal pigment epithelium. Korean J Ophthalmol. 1996;10(2):82–91.
87. Deguchi J, Yamamoto A, Yoshimori T, Sugasawa K, Moriyama Y, Futai M, Suzuki T, Kato K, Uyama M, Tashiro Y. Acidification of phagosomes and degradation of rod outer segments in rat retinal pigment epithelium. Invest Ophthalmol Vis Sci. 1994;35(2):568–79.
88. Ruggiero L, Connor MP, Chen J, Langen R, Finnemann SC. Diurnal, localized exposure of phosphatidylserine by rod outer segment tips in wild-type but not Itgb5$^{-/-}$ or Mfge8$^{-/-}$ mouse retina. Proc Natl Acad Sci U S A. 2012;109(21):8145–8.
89. Gal A, Li Y, Thompson DA, Weir J, Orth U, Jacobson SG, Apfelstedt-Sylla E, Vollrath D. Mutations in MERTK, the human orthologue of the RCS rat retinal dystrophy gene, cause retinitis pigmentosa. Nat Genet. 2000;26(3):270–1.
90. Thompson DA, McHenry CL, Li Y, Richards JE, Othman MI, Schwinger E, Vollrath D, Jacobson SG, Gal A. Retinal dystrophy due to paternal isodisomy for chromosome 1 or chromosome 2, with homoallelism for mutations in RPE65 or MERTK, respectively. Am J Hum Genet. 2002;70(1):224–9.
91. McHenry CL, Liu Y, Feng W, Nair AR, Feathers KL, Ding X, Gal A, Vollrath D, Sieving PA, Thompson DA. MERTK arginine-844-cysteine in a patient with severe rod-cone dystrophy: loss of mutant protein function in transfected cells. Invest Ophthalmol Vis Sci. 2004;45(5): 1456–63.
92. Koike M, Shibata M, Ohsawa Y, Nakanishi H, Koga T, Kametaka S, Waguri S, Momoi T, Kominami E, Peters C, et al. Involvement of two different cell death pathways in retinal atrophy of cathepsin D-deficient mice. Mol Cell Neurosci. 2003;22(2):146–61.
93. Zhang D, Brankov M, Makhija MT, Robertson T, Helmerhorst E, Papadimitriou JM, Rakoczy PE. Correlation between inactive cathepsin D expression and retinal changes in mcd2/mcd2 transgenic mice. Invest Ophthalmol Vis Sci. 2005;46(9):3031–8.
94. Yu CC, Nandrot EF, Dun Y, Finnemann SC. Dietary antioxidants prevent age-related retinal pigment epithelium actin damage and blindness in mice lacking αvβ5 integrin. Free Radic Biol Med. 2012;52(3):660–70.

Chapter 7
Retinal Circadian Rhythms in Mammals Revealed Using Electroretinography

Morven A. Cameron, Annette E. Allen, and Robert J. Lucas

Abstract Light levels can change by up to ten orders of magnitude between midday and midnight. As a result, the visual system is faced with a large diurnal variation in functional demands. Two mechanisms exist to allow the retina to function under such varied conditions: adaptation and circadian rhythmicity. Adaptation occurs in response to the presenting light conditions and circadian rhythmicity allows the tissue to anticipate those light conditions. Circadian rhythmicity has been described at many points along the visual projection from its photoreceptive origins to the highest levels of visual processing. Electroretinography has proved a very useful tool in the assessment of retinal rhythms. It offers a noninvasive and quantitative assessment of the activity of first- and second-order cells in the retina and has been used by a number of researchers to describe diurnal and/or circadian rhythms and probe their mechanistic origins in several mammalian species. Here we review the various attempts to investigate these retinal rhythms, predominately by use of the electroretinogram, in several mammalian species.

Keywords Electroretinogram • Circadian • Retina • Mammalian • Adaptation

Light levels can change by up to ten orders of magnitude between midday and midnight. As a result, the visual system is faced with a diurnal variation in functional demands at least as great as any other physiological system. It is perhaps unsurprising then that daily variations in structure, biochemistry, and function have been described at many points along the visual projection from its photoreceptive

M.A. Cameron, B.Sc., Ph.D.
School of Medicine, University of Western Sydney, Campbelltown, NSW 2751, Australia
e-mail: m.cameron@uws.edu.au

A.E. Allen, B.Sc., Ph.D. • R.J. Lucas, B.Sc., Ph.D. (✉)
Faculty of Life Sciences, University of Manchester, Manchester, M13 9PT, UK
e-mail: Annette.allen@manchester.ac.uk; Robert.lucas@manchester.ac.uk

G. Tosini et al. (eds.), *The Retina and Circadian Rhythms*, Springer Series in Vision Research 1, DOI 10.1007/978-1-4614-9613-7_7,

origins to the highest levels of visual processing [1–5]. These diurnal rhythms reflect, in part, direct responses to the changing environmental illuminant (light adaptation), but are also influenced by endogenous circadian clocks. The inclusion of such circadian control has an important functional benefit—by providing a sense of time it allows elements of the visual system to change in advance of predicted alterations in illuminance.

Other chapters in this volume address rhythms in retinal anatomy, biochemistry, and physiology in a variety of vertebrate species. Here we concentrate upon systems level investigations in mammals, with a primary focus on data obtained using the electroretinogram (ERG). Electroretinography has proved a very useful tool in the assessment of retinal rhythms. It offers a noninvasive and quantitative assessment of the activity of first- and second-order cells in the retina and has been used by a number of researchers to describe diurnal and/or circadian rhythms and probe their mechanistic origins in several mammalian species. The ERG, however, cannot reveal the ultimate output of the retina—the pattern of action potentials produced by retinal ganglion cells (RGCs) and sent to the brain. We aim also to review the rather few attempts that have been made to describe rhythmicity in this visual code by recording the activity of either RGCs or retinorecipient nuclei in the brain.

7.1 Electroretinography

The ERG is a graphical representation of field potential changes across the eye elicited by a light stimulus. It is often used clinically as a diagnostic tool to assess the physiological integrity of the retina. However, it is also widely used for research purposes in both human and animal subjects. The process of recording an ERG is relatively straightforward and noninvasive, making it an attractive research tool for recording from laboratory animals. An active electrode is placed on, or near, the cornea and a reference electrode is applied elsewhere on the body. Field potential changes are then recorded that are associated with extracellular currents in the retina elicited by a visual stimulus. The visual stimuli used can vary depending on the response required, and will be discussed later. The most commonly used technique is the full-field ERG, where a brief flash (usually <10 ms) is presented to the retina and the combined response of all cells recorded. However, the development of the multifocal ERG by Erich Sutter 20 years ago allowed for assessment of ERG activity in small, focal, areas of retina in response to the same visual stimuli [6]. This technique was invaluable for detecting local areas of dysfunction in human retina such as macular degeneration or small scotomas. For most assessments of circadian rhythmicity, however, the full-field ERG has been utilized, as circadian regulation is assumed to impact all areas of the retina.

The full-field ERG response to a brief flash of light gives a trace similar to that shown in Fig. 7.1. This trace is a composite waveform to which several defined, and likely some undefined, physiological events contribute [7]. Nevertheless, it is commonly described in terms of three major components, each of which is attributed to a specific event in the transfer of visual information through the retina. These can have one of two notations. Ragnar Granit (1933) designated the different

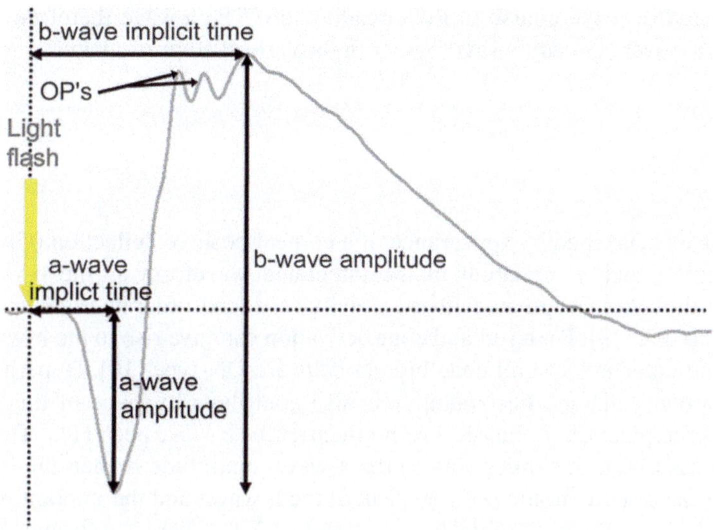

Fig. 7.1 Schematic representation of a flash ERG and the retinal events contributing to it. A representative ERG trace is shown in *gray* (*x*-axis: time; *y*-axis: voltage; arbitrary units), and *yellow arrow* shows time of flash. The earliest element of the ERG is a negative deflection (the a-wave) originating from photoreceptor activation. The next step in signal transfer is activation of ON bipolar cells (the origin of the b-wave, the large slow positive deflection). Subsequently, bipolar cells activate retinal ganglion cells which pass the light signal down the optic nerve. Horizontal and amacrine cells modulate the signal flow through the retina and it is thought that the amacrine cells and/or the retinal ganglion cells contribute to oscillatory potentials (OPs; high frequency wavelets superimposed on the b-wave)

components in the sequence of their disappearance following anesthesia as PI, PII, and PIII. Those experiments emphasized that the ERG waveform is in fact the sum of several separable physiological events. The relative amplitude and latency of these components determines qualitative and quantitative aspects of the ERG waveform. Although the PI–PIII notations are still used today, the more common component names are described below, along with their relationship to Granit's notations.

7.1.1 a-Wave

The earliest component (and most resistant to anesthesia) is a negative deflection (corneal negative potential) termed the a-wave. The a-wave reflects the first physiological event in visual transduction, activation of rod and cone photoreceptors by light, and is a reflection of Granit's PIII component [8]. When assessing the a-wave empirically, the convention is to measure the amplitude from the baseline (prestimulus) to the trough of the transient a-wave (see Fig. 7.1). Additionally, the latency of this wave, the so-called implicit time, can be measured as the time from the onset of the light flash to the trough of the a-wave. These measurements can then be compared over the course of the circadian day. The a-wave therefore gives the potential to reveal circadian rhythmicity in the earliest steps of vision.

7.1.2 b-Wave

The a-wave is curtailed by appearance of a corneal positive deflection (Granit's PII component) which is apparent in the integrated waveform as the b-wave. The b-wave is thought to originate with the activity of second-order neurons, specifically ON bipolar cells [9]. Both rod and cone activation can give rise to the b-wave as all rod bipolar cells and several cone bipolar cells are ON-type [10]. Hyperpolarizing (OFF) bipolar cells and horizontal cells also contribute to shape of this wave by causing repolarization of the ERG response after the b-wave peak [11]. The b-wave can be measured in a similar way to the a-wave: amplitude is measured from the trough of the a-wave, to the positive peak of the b-wave; and the implicit time from the flash onset, to the peak of the b-wave (see Fig. 7.1). There is a strong correlation between the latency and amplitude of a- and b-waves and the magnitude of the light stimulus and, consequently, the visual response. Dim flashes are associated with small amplitudes and long implicit times, and increasing the flash intensity correlates with increasing amplitude and decreasing implicit times. Additionally, light adaptation state of the retina can play a role in the magnitude of these components with adaptation to a background light correlated with an increase in the amplitude and decrease in the implicit time of a- and b-waves [12].

7.1.3 Oscillatory Potentials

High frequency wavelets, termed oscillatory potentials, frequently appear superimposed on the b-wave of the ERG (shown in Fig. 7.1). Although their cellular origin remains controversial, it is generally accepted that they reflect third-order neural events (activation of amacrine and/or ganglion cells; [13, 14]).

7.1.4 Other Components

Other components of the ERG (that, for simplicity, have not been annotated in Fig. 7.1) may appear depending on the stimulus and recording conditions such as a slow corneal positive "c-wave" (corresponding to Granit's PI), and a "d-wave" that is sometimes observed at the cessation of the stimulus. The c-wave is thought to result from an increase in the transepithelial potential of the retinal pigment epithelium caused by electrical separation of the basal and apical membranes by tight junctions [15]. Multiple cellular events are thought to define the d-wave, but the largest contribution is from OFF bipolar cells [9, 16–18].

7.1.5 Separating Rod and Cone Responses

Given the shift in reliance from rod- to cone-based vision and back again over the natural diurnal cycle, it is important that assessments of retinal rhythmicity target visual events from both outer-retinal photoreceptor types. As discussed above, different recording conditions can bias the ERG to reveal responses originating from either rod or cone photoreceptors. However, it is important to remember that rod and cone signals are mixed at all steps in information transfer [19], including at the level of the photoreceptors themselves. As a result, even if an ERG waveform is elicited by selective activation of cones, its characteristics will be impacted also by the physiological state of rods and the retinal pathways downstream of them (and vice versa).

The species under investigation and the type of recording condition can alter the balance of rod and cone input to the a-wave. Nocturnal species, e.g., mice, contain so few cones that only a minimal cone a-wave can be recorded, while species with more cone-rich retinae (including humans) display large cone a-waves. Mammalian rod-dominated retinas usually all show large rod-driven a-wave responses. The easiest way to separate these responses is to take advantage of the fact that, under dark-adapted conditions, rods are roughly 1,000 times more sensitive than cones [20]. Consequently, using dim stimuli, it is possible to isolate rod-dependent (scotopic) responses. At absolute threshold for the ERG response, the waveform can take on a different shape comprising simply a small corneal negative potential (scotopic threshold response), the cellular origin of which is not entirely clear [21, 22]. However, at slightly higher flash intensities a classical rod-dependent b-wave appears. The b-wave can be recorded at lower intensities than the a-wave, likely reflecting substantial amplification of the initial response when passed to the second-order cells.

The difference in absolute sensitivity between rod and cone photoreceptors unfortunately cannot be easily used to isolate cone pathways as brighter flash stimuli under dark-adapted conditions activate both cones and rods. Strategies to isolate a purely cone-dependent (photopic) ERG usually involve eliminating rod influence by saturating them with bright light. A cone ERG may therefore be elicited by a very bright flash presented against a constant background light of sufficient intensity to saturate the rod light response. The rod influence can also be eliminated using the “paired-flash” protocol where a cone ERG can be recorded by a probe flash presented shortly after a very bright flash that saturates rods [23]. Temporal response differences between rods and cones can also be exploited, as cones possess superior temporal resolution compared to rods. The frequency at which a flickering light stimulus is perceived as a constant light (critical fusion frequency) is much higher for cones than rods. Therefore, using a flicker that is above the rod critical flicker fusion can isolate cone responses [24]. Finally, in some species there is a clear difference in spectral sensitivity of rod and cone photoreceptors, allowing judicious use of different wavelengths to bias stimuli towards targeting one or other receptor class. While all these cone-isolating techniques exclude the rod response to the presenting flash stimuli, it must be noted that rods are of course "activated" in all these protocols. Assessing the activity of cones without activating rods is thus only possible using transgenic models that permanently remove or inactivate rod responses [25, 26].

7.2 Rhythmicity of the ERG

There are many advantages of using a technique such as the ERG to examine retinal rhythms. The noninvasive nature of the technique allows many different species to be assessed (including humans), with relative ease and speed. It is also possible to record responses over the course of the circadian day from the same animal, instead of using cross-sectional experimental design. The ERG has therefore been used extensively to study circadian control of the retina at a systems level.

The difference between diurnal and circadian rhythms has likely been extensively reviewed in previous chapters, but we would like to emphasize this distinction before reviewing the ERG literature, particularly because light adaptation is such a ubiquitous feature of visual processes. Even in the absence of endogenous circadian clocks therefore, strong 24 h rhythms in retinal function would be expected under the natural light:dark cycle. In order to determine the extent to which diurnal rhythms reflect circadian control, as opposed to such direct effects of light exposure, experiments should include an assessment under constant lighting conditions (usually constant darkness (DD)).

7.2.1 Local Versus Central Clocks

A second important variable to consider when assessing retinal rhythms, diurnal or circadian, is the influence of the central circadian pacemaker, the suprachiasmatic nucleus (SCN) of the hypothalamus, versus endogenous circadian clocks in the retina itself. The existence of local clocks within the retina was first revealed by Besharse and Iuvone [27] in the retina of the *Xenopus laevis* 30 years ago [27]. They reported a circadian rhythm in activity of the enzyme serotonin *N*-acetyltransferase in isolated eye-cups in vitro and this rhythm was able to be entrained to exactly 24 h by administration of a 12:12 h light/dark (LD) cycle. Since then, autonomous local retinal circadian clocks have been reported in many other vertebrates, including mammals [28]. This provides clear evidence that the retina contains autonomous circadian clocks, but, in vivo, it does receive innervation and hormonal inputs from the brain, and SCN itself, providing an opportunity for central clocks to exert an influence [29–32]. The question of which exerts the predominate influence was elegantly addressed by White and Hock in 1992, who examined a daily suppression in the dark-adapted (scotopic) ERG around the time of rod outer segment disk shedding [33]. Rabbits were entrained to a high amplitude 12:12 h LD cycle for 3 weeks and then one eye was sutured shut and patched to occlude light input. Scotopic ERGs were measured after 12 h in the dark, and the time at which a drop in the amplitude of the b-wave was observed was noted. For the unpatched eye, this always occurred ~30 min after subjective dawn, but for the patched eye the drop in b-wave amplitude occurred earlier as the days progressed. After 4 days of continuous DD for the patched eye, the b-wave reduction, and histological disk shedding correlates, occurred 10 min prior to subjective dawn, ~40 min earlier than the unpatched eye. This corresponded to a "free-running" circadian rhythm of ~23 h 50 min, and provides support for the notion that, even in an intact in vivo system, retinal clocks are the predominate timekeeper in the mammalian retina. Further evidence for this was reported recently in the mouse retina. Storch et al. [34] used a constitutive retinal knockout of *Bmal1* (an essential component of the molecular circadian clock) to display loss of circadian rhythmicity of the light-adapted (photopic) ERG in animals possessing a functional SCN [34]. It is therefore clear that circadian rhythmicity of the retina likely depends largely on intrinsic clocks located within this tissue. This raises a number of supplementary questions including how these local clocks regulate retinal physiology and how they retain synchrony with external time—are they reset by local photoreceptors or do they rely upon a synchronizing signal from the central hypothalamic oscillator (as is the case with other non-SCN clocks)?

7.2.2 ERG Rhythms in Human and Mouse

As previously stated, the noninvasive nature of the ERG technique means it can easily be used to assess rhythms in human retinal function. However, due to the somewhat tedious and impractical nature of circadian experiments that must be conducted in constant conditions (usually DD), most human studies have concentrated on diurnal retinal rhythms. Several studies correlated a change in the scotopic ERG b-wave amplitude (usually a drop) shortly after dawn [35–37], with rod disk shedding and subsequent phagocytosis thought to happen around this time. In fact, disk shedding, and scotopic b-wave amplitude and threshold sensitivity reduction, is highly correlated in both rabbit and rats [33, 38, 39]. However, this reduction in sensitivity could not be entirely accounted for by the shortening of rod outer segments, suggesting the influence of other circadian processes [39]. Studies on photopic ERGs in humans have generally reported alteration of the implicit time of the b-wave, and additionally the d-wave, with longer implicit times exhibited at night [40–42].

More recently, this analysis has been extended to mice, due to the vast opportunity for genetic manipulation in these animals. In C57BL/6 mice we have previously reported a marked rhythm in the amplitude and implicit time of the cone-isolated photopic ERG recorded against a rod saturating background light, with smaller, longer latency responses recorded during the subjective night [43]. Dark-adapted ERGs also showed circadian rhythmicity but only at flash intensities high enough to reflect mixed activation of both rods and cones. This observation was also repeated in sighted C3H$^{f+/+}$ mice that can synthesize the circadian "neurohormone" melatonin, indicating that in wild-type mice, it is primarily retinal responses to bright stimuli that are subject to circadian regulation [44]. Some aspects of these findings may reflect the nocturnal temporal niche, and underground burrowing system utilized by rodents. Under these conditions, rod vision could be useful at all diurnal phases and therefore should be maintained across the circadian cycle. On the other hand, irrespective of such considerations, cone vision should be a feature only of "daytime," and therefore could benefit from being under circadian control.

7.3 Mechanisms Underlying ERG Rhythmicity

A variety of pharmacological and genetic approaches have been used to probe the mechanisms responsible for the ERG rhythmicity.

7.3.1 Rod/Cone Pathway Balance

As rod and cone pathways are intrinsically shared in the mammalian retina [19], the segregation of these two pathways, possibly though modulation of gap-junction coupling, appeared an attractive candidate to explain the enhancement of photopic b-wave amplitude and implicit time in the subjective day. We employed the *Gnat1*$^{-/-}$ mouse that lacks normal rod phototransduction (although morphologically intact rods remain [25]) to assess the influence of rod phototransduction on cone ERG rhythmicity [45]. However, these animals retained large amplitude circadian rhythms in the ERG b-wave, and instead showed specific deficits in light adaptation [45]. Nonetheless, as rods are morphologically still intact in these animals, it is possible that circadian rhythmicity of gap-junctions connecting rod and cone pathways remains intact.

7.3.2 Melanopsin-Containing RGCs

The recently discovered melanopsin-containing retinal ganglion cells (mRGCs) are generally known to be responsible for non-image forming vision, providing light information to the SCN for the entrainment of circadian rhythms, the olivary pretectal nuclei (OPN) for the pupil light reflex (PLR), and several other higher brain areas [46–49]. In addition to these higher inputs we have shown that mRGCs perform a local modulatory function within the retina. Although, mRGCs probably do not make a direct contribution to the ERG waveform, ERG studies in humans and mice have revealed the influence of these photoreceptors on ERG rhythmicity. In the case of humans, mRGCs appear to be responsible for some aspects of long-term light adaptation. The nocturnal increase in cone b-wave implicit time (discussed above) can be reversed by acute bright light stimulation. The action spectrum for this effect (λ_{max} = 483 nm) matches the reported spectral sensitivity of mRGCs [50]. The role of mRGCs in the diurnal and circadian modulation of the ERG was further investigated in melanopsin-deficient mice (*Opn4*$^{-/-}$), which lack mRGC photosensitivity [51]. Circadian rhythmicity of the cone-isolated photopic ERG (described above) was lost in these animals with ERGs resembling a middle ground between "day-time"- and "night-time"-like [52]. This suggests that mRGCs are required to maintain ERG rhythmicity under circadian conditions, perhaps by retaining the synchrony of local clocks.

7.3.3 Melatonin

The so-called circadian neurohormone melatonin plays a widespread role in the temporal regulation of many aspects of physiology [53]. Its influence is no less important in the retina, with several studies assessing the impact of systemic and/or local release of melatonin on retinal function [53]. Retinal melatonin is synthesized in photoreceptors under the direct control of a circadian clock [54, 55], where levels are high during the night and low during the day. Similar to centrally released melatonin from the pineal gland, the synthesis of melatonin can also be suppressed by light stimulation [56]. For this reason, it is thought to provide a night-time cue to influence retinal physiology by playing a role in both diurnal and circadian modulation. Exogenous melatonin administration appears to have a differential effect in humans and rodents, possibly reflecting the difference in ecological niche. In humans, high salivary melatonin was correlated with the low photopic ERG b-wave amplitudes under diurnal conditions [57] and oral dosing with melatonin caused a reduction in the amplitude of the b-wave under both scotopic and photopic conditions [58]. However, other studies in humans have produced varied results [59–61].

In mice, studies involving melatonin have been limited, due to the rarity of mouse strains available that are capable of synthesizing the hormone [62, 63]. The C3H mouse does produce melatonin, but is often coupled with the *rd1* mutation that causes rapid photoreceptor degeneration in these animals. Baba et al. (2009) recently crossed C3Hf$^{+/+}$ mice lacking the *rd1* mutation with mice containing a targeted deletion of the melatonin 1 receptor (MT1$^{-/-}$) to allow further insight into the function of melatonin in the mouse retina [64]. The MT1 receptor is localized on photoreceptors, inner retinal neurons, and ganglion cells providing several loci for melatonin to exert an influence. C3Hf$^{+/+}$MT1$^{-/-}$ mice lack both diurnal and circadian rhythmicity of the photopic ERG with overall suppressed responses, showing a similar phenotype to the *Opn4*$^{-/-}$ animals [44, 64]. Injection of melatonin in the day in C3Hf$^{+/+}$ decreased the scotopic response threshold and increased a- and b-wave amplitudes, but not in C3Hf$^{+/+}$MT1$^{-/-}$ animals. Interestingly, circadian dopamine metabolism, thought to exert a strong influence on ERG rhythmicity (see below), was normal in the retinae of C3Hf$^{+/+}$MT1$^{-/-}$ animals, indicating that the circadian rhythm in dopamine release is not sufficient to drive circadian rhythmicity of the photopic ERG [44].

7.3.4 Dopamine

Dopamine plays an important role in the nervous system where it functions both as a neurotransmitter and neuromodulator acting on two subfamilies of dopamine receptor. D1-like receptors (D1 and D5) couple positively to adenylyl cyclase and D2-like (D2–D4) negatively to produce opposing responses based on cellular receptor expression. Dopamine release in the retina is strongly light-dependent [65–68] and modulated by a circadian clock [69] with greater release in the subjective day. Many aspects of retinal physiology are thought to be affected by light and/or circadian-induced changes in dopamine release, including gap-junction coupling, melatonin synthesis, disk shedding, and growth and development [70]. It is therefore unsurprising that dopamine plays a role in the rhythmicity of the ERG. Dopamine is solely produced in dopaminergic (DA) amacrine cells within the retina that are identifiable due to high expression levels of the rate-limiting enzyme in dopamine metabolism, tyrosine hydroxylase (TH). Jackson et al. have recently published an extensive report on the influence of dopamine on the mouse ERG, using retina-specific dopamine-deficient mice (rTHKO) and selective dopamine receptor subtype knockouts (D1RKO, D4RKO, D5RKO) [71]. In agreement with previous work, a circadian rhythm in the mouse scotopic ERG was not observed; therefore, the authors focused their efforts on the photopic light-adapted ERG using the rod saturating background protocol developed in our laboratory [43]. Mice lacking dopamine in the retina showed abolishment of photopic ERG rhythmicity after 2 days in DD with reduced cone ERG amplitudes in comparison to wild-types. In contrast to the *Gnat1*$^{-/-}$ mice, surprisingly, adaptation (growth of the b-wave in response to background illumination) was unaffected in these animals. Furthermore, circadian rhythmicity of *Per2* (an integral clock gene) was maintained, suggesting the retinal clock remains fully functional in these animals. As these animals display a similar phenotype to the *Opn4*$^{-/-}$ it could be postulated that the loss of melanopsin abolishes the rhythmic release of dopamine in the retina, possibly through a desynchrony of the DA amacrines. We have shown that mRGCs are not sufficient to elicit a global change in dopamine [72], but an excitatory input from mRGCs to DA amacrines has been identified [73] which may play a large role in synchronizing the circadian release of dopamine from these cells.

Jackson et al. expanded their study by examining dopamine receptor-specific knockouts. Their main findings implicated D1 and D4 receptors in the regulation of the ERG (and also higher visual function). D4RKOs displayed similar responses to the rTHKO with the abolishment of circadian rhythmicity after 2 days in DD suggesting this receptor may influence circadian rhythmicity through rhythmical dopamine release. D1RKOs, interestingly, showed deficits in light adaptation of the ERG but intact circadian rhythms. The mechanism of action for this result is unclear, especially due to the lack of adaptation phenotype in the rTHKO animals. However, it could indicate that rapid changes in D1-receptor expression levels underlie the adaptation seen in the wild-type ERG, rather than in dopamine release per se.

7.3.4.1 Importance of Retinal Clocks for General Retinal Function

When considering the prevalence of physiological processes within the retina that are affected by the circadian clock, it has been hypothesized that this circadian modulation is vital to retinal function [5]. This hypothesis has been addressed by studying ERGs in animals lacking integral components of the molecular circadian clock, *Cry1*$^{-/-}$*Cry2*$^{-/-}$ mice [43]. These animals lack behavioral circadian rhythms, showing highly disrupted wheel-running rhythms in constant conditions. They do, however, show "normal" diurnal wheel-running behavior when exposed to 12:12 LD cycles, active in the dark and inactive in the light [74]. Whether this reflects actual diurnal rhythmicity or not is still a matter for discussion, but does indicate that some manner of visual function in these animals remains intact. When analyzing the ERG, we observed that, as may have been expected, circadian rhythmicity was abolished in these animals. However, surprisingly, ERG b-wave amplitudes were constitutively enhanced under both scotopic and photopic conditions indicating that clocks are certainly dispensable for general retinal health, at least under laboratory conditions. In another study, Storch et al. [34] also examined ERGs in animals bearing either a retina-specific or whole animal lesion of the circadian clockwork, namely the core clock gene *Bmal1* [34]. Again, loss of circadian rhythmicity of the photopic ERG was apparent. However, in contrast to our data from the *Cry1*$^{-/-}$*Cry2*$^{-/-}$ mice, modest decreases in b-wave amplitude of cone and mixed rod+cone ERGs were observed. When this data is taken together with our *Cry1*$^{-/-}$*Cry2*$^{-/-}$ findings, it appears likely that this reduction in ERG amplitude is attributed to the loss of circadian regulation of retinal pathways rather than any generalized pathology. Indeed, the difference between the ERG phenotype of *Bmal1*$^{-/-}$ and *Cry1*$^{-/-}$*Cry2*$^{-/-}$ mice is intriguing. One possible explanation is that because *Bmal1* and *Cry1/2* are active at different phases of the molecular oscillation, *Bmal1* is a transcriptional promoter and *Cry1/2* are transcriptional repressors, these two lesions stop the clock in a different state. This could explain why *Bmal1*$^{-/-}$ ERGs are constitutively suppressed (equivalent to the wild-type response at night) and *Cry1*$^{-/-}$*Cry2*$^{-/-}$ ERGs are constitutively enhanced (equivalent to the wild-type response during the day).

7.4 Ganglion Cell Rhythmicity

All light signals elicited in the retina, if they are to be passed to higher visual areas in the brain, must pass through RGCs. As the ERG is generally used to reflect the function of photoreceptors and bipolar cells in the retina, it is important to discuss circadian rhythmicity in the cells that actually carry the ultimate output signals from the retina. Of course, any rhythmicity arising in these cells could come from presynaptic origins so it is difficult to assign the loci of circadian regulation by recording from these cells alone. However, as previously discussed, mRGCs act as photoreceptors and it is possible to identify intrinsic circadian rhythmicity in these cells using the relevant light stimulation techniques. Often referred to as the "circadian photoreceptor," relating to their light input into the central circadian pacemaker, could circadian control mechanisms intrinsic to these cells contribute to the time-of-day dependence observed in the SCN light response [75]?

Core molecular clock components have been localized to the majority of cell types in the mammalian retina [76]; however, a large heterogeneity exists between cell type in the level and rhythmical occurrence of clock protein expression [77]. mRGCs themselves exhibit all six of the "core" clock proteins, making it likely that these cells are capable of autonomous circadian rhythm generation [77]. Certainly, a circadian variation in melanopsin mRNA levels has been observed in these cells [78, 79]; however, this rhythm is abolished after degeneration of rods and cones [80]. Weng et al. [81] addressed circadian rhythmicity of mRGCs using multielectrode array recordings of rat mRGCs pharmacologically isolated from synaptic inputs. They found that although mRGCs did exhibit an increase in response gain of mRGC phototransduction (increased firing rate) in the subjective night, no significant circadian rhythmicity in light sensitivity was observed [81]. In some respects, these results make sense for an input to the circadian clock that must retain sustained irradiance-coding for accurate relay of the presenting light conditions to the central circadian clock.

The circadian rhythmicity of mRGCs could contribute to time-of-day dependence in central light responses. The SCN circadian clock itself shows a strong rhythmicity in photosensitivity, with light adjusting clock phase only when presented during the subjective night [82]. Many individual SCN neurons remain photoresponsive during the day, indicating that retinal afferents are active at all circadian phases. However, electrophysiological recordings indicate that SCN light responses are qualitatively and quantitatively different across the circadian cycle [75, 83], raising the possibility that some of the variation in clock photosensitivity could originate in the retina. Another mRGC-driven response to show strong circadian variation is the PLR. The PLR, although driven by all three types of photoreceptor [84], relies solely on the output of mRGCs [85]. Zele et al. analyzed the post-illumination pupil light response (PIPR), sustained pupilloconstriction after light cessation that is thought to represent activation of melanopsin specifically [86]. They found that the PIPR did display a circadian rhythm with a minimum post-illumination pupil response occurring in the subjective evening ~1.31 h after melatonin onset [87].

This circadian change was effected via altered redilation kinetics observed at this time of day. It is certainly possible that these results could reflect central circadian modulation at the level of the OPN where the PLR is controlled, but also could suggest that the mRGC input could vary over the course of the circadian day.

7.5 Conclusions

The ERG represents an attractive technique for studying circadian control of the retina. It can be applied to both laboratory animals and human subjects, using comparable protocols, facilitating translation of basic physiological insights into an understanding of human biology and pathophysiology. Moreover, because the ERG waveform reflects the integrated response of the retina to a visual stimulus, it reveals circadian regulation at the systems level, in a way that targeted recordings of individual retinal cells cannot. Finally, repeated recordings can be taken from the same individual, raising the possibility of tracking rhythmic changes within a single subject.

The ERG has therefore proved a very good method of describing retinal circadian rhythmicity. Its capacity to determine the physiological origins of such rhythmicity has been less widely exploited. Because the ERG waveform is a composite of multiple physiological events, it is often not possible to assign an alteration in a given ERG parameter to a particular change in the behavior of the retinal circuitry. Thus, for example, reduced a-wave could reflect changes in photoreceptor physiology (e.g., rod:cone coupling), or a reduction in b-wave latency indicative of alterations in ON bipolar cell activity. Combining ERG recordings with intravitreal injections of pharmacological agents can be a good way of bridging this gap, as can the use of genetically modified organisms. Viral gene targeting methods and RNA silencing methods are also increasingly applied to the retina and will surely be incorporated into circadian ERG studies in future.

References

1. Lythgoe JN, Shand J. Endogenous circadian retinomotor movements in the neon tetra (Paracheirodon innesi). Invest Ophthalmol Vis Sci. 1983;24(9):1203–10.
2. Kolbinger W, Wagner D, Wagner HJ. Control of rod retinomotor movements in teleost retinae: the role of dopamine in mediating light-dependent and circadian signals. Cell Tissue Res. 1996;285(3):445–51.
3. Manglapus MK, Uchiyama H, Buelow NF, Barlow RB. Circadian rhythms of rod-cone dominance in the Japanese quail retina. J Neurosci. 1998;18(12):4775–84.
4. Green CB. Molecular control of Xenopus retinal circadian rhythms. J Neuroendocrinol. 2003; 15(4):350–4.
5. Green CB, Besharse JC. Retinal circadian clocks and control of retinal physiology. J Biol Rhythms. 2004;19(2):102.

6. Sutter EE, Tran D. The field topography of ERG components in man—I. The photopic luminance response. Vision Res. 1992;32(3):433–46.
7. Frishman LJ. Origins of the electroretinogram. In: Heckenlively J, Arden GB, editors. Principles and practice of clinical electrophysiology of vision, 2nd edition. MIT Press: Cambridge, MA; 2006. p. 139–185.
8. Kt B. The electroretinogram: its components and their origins. Vision Res. 1968;8:677.
9. Stockton RA, Slaughter MM. B-wave of the electroretinogram. A reflection of ON bipolar cell activity. J Gen Physiol. 1989;93(1):122.
10. Sterling P, Smith RG, Rao R, Vardi N. Functional architecture of mammalian outer retina and bipolar cells. In: Archer S, Djamgoz MBA, Vallerga S, editors. Neurobiology and clinical aspects of the outer retina. London: Chapman & Hall; 1995. p. 325–48.
11. Bush RA, Sieving PA. A proximal retinal component in the primate photopic ERG a-wave. Invest Ophthalmol Vis Sci. 1994;35(2):635–45.
12. Alexander KR, Raghuram A, Rajagopalan AS. Cone phototransduction and growth of the ERG b-wave during light adaptation. Vision Res. 2006;46(22):3941–8.
13. Heynen H, Wachtmeister L, van Norren D. Origin of the oscillatory potentials in the primate retina. Vision Res. 1985;25(10):1365–73.
14. Yu M, Peachey NS. Attenuation of oscillatory potentials in nob2 mice. Doc Ophthalmol. 2007;115(3):173–86.
15. Brindley GS, Hamasaki DI. The properties and nature of the R membrane of the frog's eye. J Physiol. 1963;167:599–606.
16. Gurevich L, Slaughter MM. Comparison of the waveforms of the ON bipolar neuron and the b-wave of the electroretinogram. Vision Res. 1993;33(17):2431–5.
17. Sieving PA, Murayama K, Naarendorp F. Push-pull model of the primate photopic electroretinogram: a role for hyperpolarizing neurons in shaping the b-wave. Vis Neurosci. 1994; 11(3):519–32.
18. Szikra T, Witkovsky P. Contributions of AMPA- and kainate-sensitive receptors to the photopic electroretinogram of the Xenopus retina. Vis Neurosci. 2001;18(2):187–96.
19. Sharpe LT, Stockman A. Rod pathways: the importance of seeing nothing. Trends Neurosci. 1999;22(11):497–504.
20. Fu Y, Yau KW. Phototransduction in mouse rods and cones. Pflugers Arch. 2007;454(5):805–19.
21. Saszik SM, Robson JG, Frishman LJ. The scotopic threshold response of the dark-adapted electroretinogram of the mouse. J Physiol. 2002;543(Pt 3):899–916.
22. Bui BV, Fortune B. Ganglion cell contributions to the rat full-field electroretinogram. J Physiol. 2004;555(Pt 1):153–73.
23. Verdon WA, Schneck ME, Haegerstrom-Portnoy G. A comparison of three techniques to estimate the human dark-adapted cone electroretinogram. Vision Res. 2003;43(19):2089–99.
24. Peachey NS, Alexander KR, Derlacki DJ, Fishman GA. Light adaptation and the luminance-response function of the cone electroretinogram. Doc Ophthalmol. 1992;79(4):363–9.
25. Calvert PD, Krasnoperova NV, Lyubarsky AL, Isayama T, Nicolo M, Kosaras B, et al. Phototransduction in transgenic mice after targeted deletion of the rod transducin alpha-subunit. Proc Natl Acad Sci U S A. 2000;97(25):13913–8.
26. Humphries MM, Rancourt D, Farrar GJ, Kenna P, Hazel M, Bush RA, et al. Retinopathy induced in mice by targeted disruption of the rhodopsin gene. Nat Genet. 1997;15(2):216–9.
27. Besharse JC, Iuvone PM. Circadian clock in Xenopus eye controlling retinal serotonin N-acetyltransferase. Nature. 1983;305(5930):133–5.
28. Tosini G, Pozdeyev N, Sakamoto K, Iuvone PM. The circadian clock system in the mammalian retina. Bioessays. 2008;30(7):624–33.
29. Smeraski CA, Sollars PJ, Ogilvie MD, Enquist LW, Pickard GE. Suprachiasmatic nucleus input to autonomic circuits identified by retrograde transsynaptic transport of pseudorabies virus from the eye. J Comp Neurol. 2004;471(3):298–313.
30. Gastinger M, Bordt AS, Bernal MP, Marshak DW. Serotonergic retinopetal axons in the monkey retina. Curr Eye Res. 2005;30(12):1089–95.

31. Gastinger MJ, Tian N, Horvath T, Marshak DW. Retinopetal axons in mammals: emphasis on histamine and serotonin. Curr Eye Res. 2006;31(7–8):655–67.
32. Korf HW, von Gall C. Mice, melatonin and the circadian system. Mol Cell Endocrinol. 2006;252(1–2):57–68.
33. White MP, Hock PA. Effects of continuous darkness on ERG correlates of disc shedding in rabbit retina. Exp Eye Res. 1992;54(2):173–80.
34. Storch KF, Paz C, Signorovitch J, Raviola E, Pawlyk B, Li T, et al. Intrinsic circadian clock of the mammalian retina: importance for retinal processing of visual information. Cell. 2007; 130(4):730–41.
35. Nozaki S, Wakakura M, Ishikawa S. Circadian rhythm of human electroretinogram. Jpn J Ophthalmol. 1983;27(2):346–52.
36. Birch DG, Berson EL, Sandberg MA. Diurnal rhythm in the human rod ERG. Invest Ophthalmol Vis Sci. 1984;25(2):238.
37. Birch DG, Sandberg MA, Berson EL. Diurnal rhythm in the human rod ERG. Relationship to cyclic lighting. Invest Ophthalmol Vis Sci. 1986;27(2):268–70.
38. White MP, Hock PA, Marmor MF. Electrophysiological correlates of rod outer segment disc shedding in rabbit retina. Vision Res. 1987;27(3):361.
39. Sandberg MA, Pawlyk BS, Berson EL. Electroretinogram (ERG) sensitivity and phagosome frequency in the normal pigmented rat. Exp Eye Res. 1986;43(5):781–9.
40. Hankins MW, Jones RJ, Ruddock KH. Diurnal variation in the b-wave implicit time of the human electroretinogram. Vis Neurosci. 1998;15(1):67.
41. Hankins MW, Jones SR, Jenkins A, Morland AB. Diurnal daylight phase affects the temporal properties of both the b-wave and d-wave of the human electroretinogram. Brain Res. 2001;889(1–2):343.
42. Danilenko KV, Plisov IL, Cooper HM, Wirz-Justice A, Hebert M. Human cone light sensitivity and melatonin rhythms following 24-hour continuous illumination. Chronobiol Int. 2011; 28(5):407–14.
43. Cameron MA, Barnard AR, Hut RA, Bonnefont X, van der Horst GT, Hankins MW, et al. Electroretinography of wild-type and Cry mutant mice reveals circadian tuning of photopic and mesopic retinal responses. J Biol Rhythms. 2008;23(6):489–501.
44. Sengupta A, Baba K, Mazzoni F, Pozdeyev NV, Strettoi E, Iuvone PM, et al. Localization of melatonin receptor 1 in mouse retina and its role in the circadian regulation of the electroretinogram and dopamine levels. PLoS One. 2011;6(9):e24483.
45. Cameron MA, Lucas RJ. Influence of the rod photoresponse on light adaptation and circadian rhythmicity in the cone ERG. Mol Vis. 2009;15:2209–16.
46. Hankins MW, Peirson SN, Foster RG. Melanopsin: an exciting photopigment. Trends Neurosci. 2008;31(1):27–36.
47. Schmidt TM, Do MT, Dacey D, Lucas R, Hattar S, Matynia A. Melanopsin-positive intrinsically photosensitive retinal ganglion cells: from form to function. J Neurosci. 2011;31(45): 16094–101.
48. Do MT, Yau KW. Intrinsically photosensitive retinal ganglion cells. Physiol Rev. 2011; 90(4):1547–81.
49. Bailes HJ, Lucas RJ. Melanopsin and inner retinal photoreception. Cell Mol Life Sci. 2010; 67(1):99–111.
50. Hankins MW, Lucas RJ. The primary visual pathway in humans is regulated according to long-term light exposure through the action of a nonclassical photopigment. Curr Biol. 2002; 12(3):198.
51. Hattar S, Lucas RJ, Mrosovsky N, Thompson S, Douglas RH, Hankins MW, et al. Melanopsin and rod-cone photoreceptive systems account for all major accessory visual functions in mice. Nature. 2003;424(6944):81.
52. Barnard AR, Hattar S, Hankins MW, Lucas RJ. Melanopsin regulates visual processing in the mouse retina. Curr Biol. 2006;16(4):395.
53. Wiechmann AF, Summers JA. Circadian rhythms in the eye: the physiological significance of melatonin receptors in ocular tissues. Prog Retin Eye Res. 2008;27(2):137–60.

54. Cahill GM, Besharse JC. Circadian clock functions localized in xenopus retinal photoreceptors. Neuron. 1993;10(4):573–7.
55. Liu C, Fukuhara C, Wessel 3rd JH, Iuvone PM, Tosini G. Localization of Aa-nat mRNA in the rat retina by fluorescence in situ hybridization and laser capture microdissection. Cell Tissue Res. 2004;315(2):197–201.
56. Iuvone PM, Brown AD, Haque R, Weller J, Zawilska JB, Chaurasia SS, et al. Retinal melatonin production: role of proteasomal proteolysis in circadian and photic control of arylalkylamine N-acetyltransferase. Invest Ophthalmol Vis Sci. 2002;43(2):564–72.
57. Rufiange M, Dumont M, Lachapelle P. Correlating retinal function with melatonin secretion in subjects with an early or late circadian phase. Invest Ophthalmol Vis Sci. 2002;43(7):2491–9.
58. Emser W, Dechoux R, Weiland M, Wirz-Justice A. Melatonin decreases the amplitude of the b-wave of the human electroretinogram. Experientia. 1993;49(8):686–7.
59. Lavoie J, Gagne AM, Lavoie MP, Sasseville A, Charron MC, Hebert M. Circadian variation in the electroretinogram and the presence of central melatonin. Doc Ophthalmol. 2010;120(3): 265–72.
60. Gagne AM, Danilenko KV, Rosolen SG, Hebert M. Impact of oral melatonin on the electroretinogram cone response. J Circadian Rhythms. 2009;7:14.
61. Danilenko KV, Plisov IL, Wirz-Justice A, Hebert M. Human retinal light sensitivity and melatonin rhythms following four days in near darkness. Chronobiol Int. 2009;26(1):93–107.
62. Goto M, Oshima I, Tomita T, Ebihara S. Melatonin content of the pineal gland in different mouse strains. J Pineal Res. 1989;7(2):195–204.
63. Tosini G, Menaker M. The clock in the mouse retina: melatonin synthesis and photoreceptor degeneration. Brain Res. 1998;789(2):221–8.
64. Baba K, Pozdeyev N, Mazzoni F, Contreras-Alcantara S, Liu C, Kasamatsu M, et al. Melatonin modulates visual function and cell viability in the mouse retina via the MT1 melatonin receptor. Proc Natl Acad Sci U S A. 2009;106(35):15043–8.
65. Iuvone PM, Galli CL, Garrison-Gund CK, Neff NH. Light stimulates tyrosine hydroxylase activity and dopamine synthesis in retinal amacrine neurons. Science. 1978;202(4370): 901–2.
66. Parkinson D, Rando RR. Effect of light on dopamine turnover and metabolism in rabbit retina. Invest Ophthalmol Vis Sci. 1983;24(3):384–8.
67. Umino O, Dowling JE. Dopamine release from interplexiform cells in the retina: effects of GnRH, FMRFamide, bicuculline, and enkephalin on horizontal cell activity. J Neurosci. 1991;11(10):3034–46.
68. Witkovsky P, Gabriel R, Haycock JW, Meller E. Influence of light and neural circuitry on tyrosine hydroxylase phosphorylation in the rat retina. J Chem Neuroanat. 2000;19(2): 105–16.
69. Doyle SE, Grace MS, McIvor W, Menaker M. Circadian rhythms of dopamine in mouse retina: the role of melatonin. Vis Neurosci. 2002;19(5):601.
70. Witkovsky P. Dopamine and retinal function. Doc Ophthalmol. 2004;108(1):17–40.
71. Jackson CR, Ruan GX, Aseem F, Abey J, Gamble K, Stanwood G, et al. Retinal dopamine mediates multiple dimensions of light-adapted vision. J Neurosci. 2012;32(27):9359–68.
72. Cameron MA, Pozdeyev N, Vugler AA, Cooper H, Iuvone PM, Lucas RJ. Light regulation of retinal dopamine that is independent of melanopsin phototransduction. Eur J Neurosci. 2009;29(4):761–7.
73. Zhang DQ, Wong KY, Sollars PJ, Berson DM, Pickard GE, McMahon DG. Intraretinal signaling by ganglion cell photoreceptors to dopaminergic amacrine neurons. Proc Natl Acad Sci U S A. 2008;105(37):14181–6.
74. van der Horst GTJ, Muijtjens M, Kobayashi K, Takano R, Kanno S, Takao M, et al. Mammalian Cry1 and Cry2 are essential for maintenance of circadian rhythms. Nature. 1999;398(6728): 627–30.
75. Brown TM, Wynne J, Piggins HD, Lucas RJ. Multiple hypothalamic cell populations encoding distinct visual information. J Physiol. 2011;589(Pt 5):1173–94.

76. Ruan G-X, Zhang D-Q, Zhou T, Yamazaki S, McMahon DG. Circadian organization of the mammalian retina. Proc Natl Acad Sci U S A. 2006;103(25):9703–8.
77. Liu X, Zhang Z, Ribelayga CP. Heterogeneous expression of the core circadian clock proteins among neuronal cell types in mouse retina. PLoS One. 2012;7(11):e50602.
78. Hannibal L, Georg B, Hindersson P, Fahrenkrug J. Light and darkness regulate melanopsin in the retinal ganglion cells of the albino Wistar rat. J Mol Neurosci. 2005;27(2):147–55.
79. Sakamoto K, Liu C, Kasamatsu M, Pozdeyev NV, Iuvone PM, Tosini G. Dopamine regulates melanopsin mRNA expression in intrinsically photosensitive retinal ganglion cells. Eur J Neurosci. 2005;22(12):3129–36.
80. Sakamoto K, Liu C, Tosini G. Circadian rhythms in the retina of rats with photoreceptor degeneration. J Neurochem. 2004;90(4):1019–24.
81. Weng S, Wong KY, Berson DM. Circadian modulation of melanopsin-driven light response in rat ganglion-cell photoreceptors. J Biol Rhythms. 2009;24(5):391–402.
82. Daan S, Pittendrigh CS. A functional analysis of circadian pacemakers in nocturnal rodents. II The variability of phase response curves. J Comp Physiol. 1976;106:253–66.
83. Meijer JH, Watanabe K, Detari L, Schaap J. Circadian rhythm in light response in suprachiasmatic nucleus neurons of freely moving rats. Brain Res. 1996;741(1–2):352–5.
84. Lucas RJ, Hattar S, Takao M, Berson DM, Foster RG, Yau KW. Diminished pupillary light reflex at high irradiances in melanopsin-knockout mice. Science. 2003;299(5604):245–7.
85. Guler AD, Ecker JL, Lall GS, Haq S, Altimus CM, Liao HW, et al. Melanopsin cells are the principal conduits for rod-cone input to non-image-forming vision. Nature. 2008; 453(7191):102–5.
86. Gamlin PD, McDougal DH, Pokorny J, Smith VC, Yau KW, Dacey DM. Human and macaque pupil responses driven by melanopsin-containing retinal ganglion cells. Vision Res. 2007; 47(7):946–54.
87. Zele AJ, Feigl B, Smith SS, Markwell EL. The circadian response of intrinsically photosensitive retinal ganglion cells. PLoS One. 2011;6(3):e17860.

Chapter 8
Circadian Effects on Retinal Light Damage

Paul Wong, Daniel T. Organisciak, Alison Ziesel, M.A. Chrenek, and M.L. Patterson

Abstract Light-induced retinal damage has long served as a model of retinal dysfunction and visual cell loss arising from inherited disease or caused by oxidative stress. Its utility resides in the fact that nearly the entire complement of retinal photoreceptors is simultaneously involved in a now well-defined progression of cellular degeneration and active cell death. Numerous extrinsic factors are known to influence the extent of visual cell loss, including previous light-rearing history, light intensity and duration, and diet. However, visual cell damage is also impacted by intrinsic factors such as circadian rhythms. These endogenous rhythms are known to affect the cellular machinery involved with light reception and metabolism, receptor function and signaling, and the cascade of apoptotic cell death. Herein we describe the progression of light-induced oxidative damage and visual cell death in retina and how circadian-dependent gene expression affects the process. The time course of retinal gene expression and light damage susceptibility has been compared at various times of the day or night. In addition to genes known to exhibit a circadian profile of regulation, we also describe a number of genes previously not recognized as affecting the progression of visual cell damage and loss.

P. Wong, Ph.D. (✉) • A.C. Ziesel, B.Sc. • M.A. Chrenek, B.Sc.
Department of Ophthalmology, Emory University,
B5512 1365B Clifton Rd NE, Atlanta, GA 30322, USA
e-mail: pwong@emory.edu; aziesel@emory.edu; micah.chrenek@emory.edu

D.T. Organisciak, Ph.D.
Petticrew Research Laboratory, Department of Biochemistry and Molecular Biology,
Boonshoft School of Medicine, Wright State University,
3640 Colonel Glenn Highway, Dayton, OH 45435, USA
e-mail: dto@wright.edu

M.L. Patterson, B.Sc., Ph.D.
Department of Biological Sciences, University of Alberta,
CW405 Biological Sciences Bldg, Edmonton, AB, Canada, T6G 2E9
e-mail: canadianlostintexas@hotmail.com

G. Tosini et al. (eds.), *The Retina and Circadian Rhythms*, Springer Series in Vision Research 1, DOI 10.1007/978-1-4614-9613-7_8,

Keywords Retinal light damage • Gene array • Circadian rhythm • Oxidation • Rat retina

8.1 Introduction

Photoreceptor cells are differentiated postmitotic retinal neurons that function over a wide range of ambient light conditions for the efficient capture of photons and initiation of visual transduction. Because of these functions and their anatomical location between the inner nuclear layers in retina and the retinal pigment epithelium (RPE), photoreceptors are normally subjected to high levels of incident light while being maintained in an oxygen-rich environment. Given the high photon flux and oxygen tension in the outer retina and the high levels of polyunsaturated fatty acids present in rod outer segment (ROS) membranes, the potential for visual cell injury is great. In most species prolonged intense visible light exposure can lead to photoreceptor cell damage, while in nocturnal animals the light intensity required need only be 2–3 times above normal room lighting. The extent of retinal light damage is modulated by light intensity and the duration of exposure [1, 2], the wavelengths of light used [2–4], diet, age, and genetic factors (see [5–7] for reviews). Long-term adaptive processes regulated by different light-rearing intensities [8, 9], or by dark rearing [10, 11], are additional extrinsic factors that alter the susceptibility of visual cells to light-mediated damage. The current discussion focuses on the fact that susceptibility to retinal damage is also a circadian-dependent process that relies on endogenous factors [12–15]. Intense light or even physiological levels of light can establish and regulate the near 24 h cycles of synthetic and catabolic processes, which in turn can impact both visual cell responses and visual cell damage. Herein, we review the current light damage literature; we then take a systems biology approach to describe circadian factors that can affect retinal gene expression and that may lead to photoreceptor cell resistance or susceptibility to the toxic actions of light.

8.2 Retinal Morphology and Apoptotic Cell Death Following Light Exposure

Albino rats exposed to bright light or to continuous illumination exhibit photoreceptor damage and loss that is characterized by morphological changes. Histologically, the outer retinal layers appear swollen and pyknotic following intense light exposure [16–19], ROS become distorted with the formation of sacs or whorls containing elongated tubules [20, 21], and rod inner segments (RIS), nuclei, and synaptic terminals become vacuolated [20]. ROS also become increasingly disorganized, until they are phagocytized by the RPE or ingested by invading macrophages. The morphology of the retina after exposure to high intensity light is consistent with a tissue

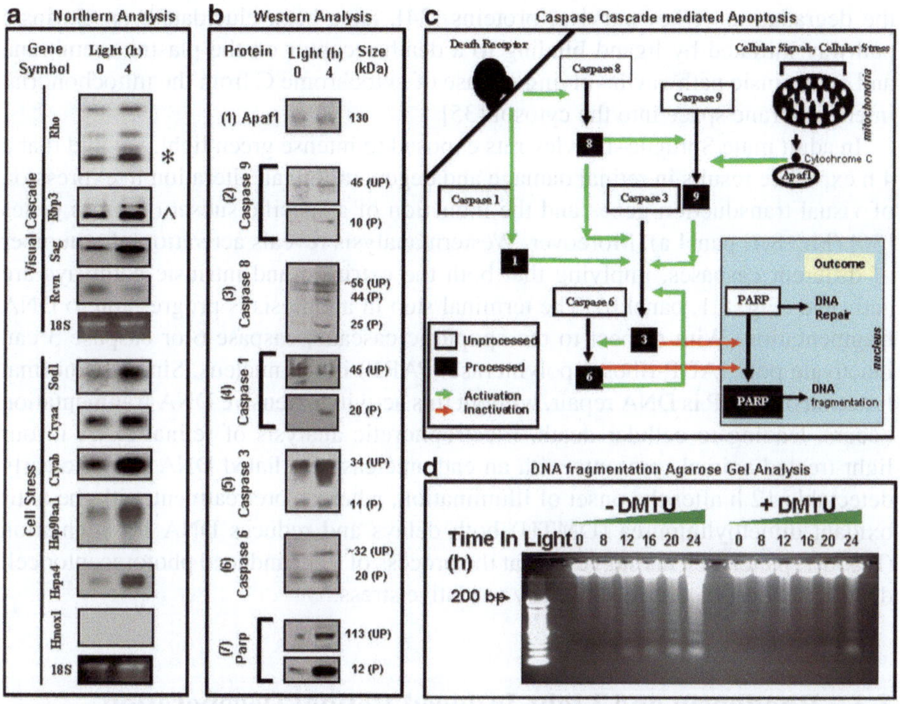

Fig. 8.1 *Panel a*: Dark-reared adult male albino rats (p60), treated or untreated with DMTU, were exposed to intense light for 0–24 h. The rats were killed in a saturated CO_2 chamber in dim red light and retinas excised and then used for either Northern (panel a), Western (panel b), or DNA analysis (panel d) as described [36]. In panel a Northern analysis for gene markers of the visual cascade (*Rho* rhodopsin, *Rbp3* retinol-binding protein 3, *Sag* rod arrestin, *Rcvn* recoverin) and cell stress (*Sod-1* superoxide dismutase 1, *Cryaa* crystallin alpha A, *Cryab* crystalline alpha B, *Hsp90aa* heat shock protein 90 kDa alpha (cytosolic) class A member 1, *Hspaa* heat shock 70 kDa protein 4, *Hmox-1* heme oxygenase-1) is shown. *Panel b*: Western analyses to detect processed (P) and unprocessed forms (UP) of various caspases and other components of the apoptotic cell death cascade. *Panel c*: Schematic showing known caspase-mediated pathways. *Panel d*: Analysis of retinal DNA after intense light exposure reveals an electrophoretic pattern of fragmentation containing a 200 base pair DNA ladder. Pretreatment with the antioxidant DMTU both delays and reduces DNA fragmentation, suggesting that light-induced photoreceptor cell death is in part mediated by oxidative stress

undergoing classical programmed cell death [22, 23]. Analysis of retinal DNA after intense light exposure reveals an electrophoretic pattern of fragmentation containing a 200 base pair DNA ladder, along with TUNEL reactivity localized within the ONL [24–26]. Furthermore, after some debate, it has become clear that light-induced retinal degeneration in rats involves a caspase-mediated apoptotic pathway [27, 28]. Caspases define a family of cysteine aspartate-specific proteases that coordinate and execute the process of programmed cell death [29–31]. These enzymes are initially synthesized as inactive zymogens and become activated by proteolytic cleavage [32, 33]; see Fig. 8.1. Two major caspase-mediated pathways, leading to the degradation of key survival proteins [34], have been elucidated: an extrinsic pathway initiated by ligand binding to a death receptor on the plasma membrane and an intrinsic pathway involving release of cytochrome C from the mitochondrial intermembrane space into the cytosol [35].

In adult male Sprague–Dawley rats exposed to intense green light, we find that a 4 h exposure results in retinal damage and degeneration, an alteration in expression of visual transduction genes and the induction of a specific subset of stress genes [36] (Fig. 8.1, panel a). Moreover, Western analysis reveals activation of a number of different caspases, implying that both the extrinsic and intrinsic pathways are activated (Fig. 8.1, panel b). The terminal step in apoptosis is progression to DNA fragmentation. With respect to the apoptotic cascade, caspase 6 or caspase 3 can inactivate poly (ADP-ribose) polymerase (PARP) in the nucleus. Since the normal function of PARP is DNA repair, without this activity extensive DNA fragmentation occurs, leading to cellular death. Electrophoretic analysis of retinal DNA in our light-treated animals presents with an endonuclease-mediated DNA ladder clearly detectable 12 h after the onset of illumination, whereas pretreatment with the anti-oxidant dimethylthiourea (DMTU) both delays and reduces DNA fragmentation (Fig. 8.1, panel d). This suggests that the process of light-induced photoreceptor cell death is at least in part mediated by oxidative stress.

8.3 Rhodopsin and Light-Induced Retinal Degeneration

Central to all light damage hypotheses is the importance of rhodopsin as the trigger for photoreceptor cell damage [2, 37, 38]. First described almost 50 years ago [1, 2], retinal light damage has long served as a model of retinal degenerations arising from genetic inheritance and environmental or toxic insult. Its utility as a model resides in the near synchronous involvement of the entire complement of rod photo-receptors and, depending on exposure conditions, damage in the adjacent RPE. By careful experimental manipulation of prior light-rearing conditions, light intensity and duration, or diet, we know that retinal light damage begins within ROS with the bleaching of rhodopsin. Genetic manipulation of retinal proteins also points to a central role for rhodopsin bleaching, as rhodopsin knockout (KO) and RPE-65 KO mice are protected against retinal light damage [37–40]. Because opsin, but not rhodopsin, is present in RPE-65 KO mice and rhodopsin KO mice lack the protein altogether, this implicates rhodopsin regeneration in addition to its bleaching. In arrestin or rhodopsin kinase KO mice, retinal light damage occurs under ordinary room lighting, while dark rearing prevents damage [41, 42]. As these proteins act to attenuate the light-driven photo-response, this implicates visual transduction as well in the mechanism of damage. Other genetic evidence suggests that two distinct pathways of light damage may exist, one involving bright light and activator protein-1 (AP-1) and a low light apoptotic pathway primarily involving activated transducin [43].

8.4 Light-Induced Visual Cell Damage is Species-Dependent

Arrestin and transducin each exhibit light-driven translocation into or out of ROS [44], and their relative expression [11] and protein levels [10] have been found to correlate with retinal light damage susceptibility. Conversely, RPE-65 gene expression and its protein levels do not correlate with retinal light damage in rats [45], or with their enhanced susceptibility to damage at night [46]. Inhibition of the visual cycle with 13-*cis* retinoic acid, however, does lead to a reduction in the extent of light-induced photoreceptor cell loss [47]. Accordingly, although species differences do exist, rhodopsin regeneration can impact retinal light damage in both mice and rats. Similarly, an action spectrum for retinal light damage that has been reported for rats is nearly identical to the rhodopsin absorption spectrum [2, 3]. Other evidence points to a peak for photodamage in the near UV, possibly due to free all-trans retinal (reviewed in [48]). Unexplained at this time is how extensive bleaching of rhodopsin in rod photoreceptors during the bright light conditions that exist throughout the day can occur without extensive damage in most species. Likewise, we do not yet understand how identical levels of rhodopsin bleaching and regeneration during the day or at night [14, 46] can lead to such remarkable differences in photoreceptor damage. We sought to address some of these questions by profiling retinal gene expression as a function of time of day to identify genes as potential endogenous effectors of visual cell damage.

8.5 Intense Light Induces Oxidative Stress in the Retina

Traditional antioxidants prevent retinal light damage and photoreceptor cell loss. These include the natural L-stereoisomer of ascorbic acid as well as D-ascorbate, which is also an antioxidant but not an enzymatic cofactor for collagen synthesis and wound repair [49, 50]. Synthetic antioxidants, including the radioprotective dye WR 77913, which quenches singlet oxygen [51], *N*-acetyl cysteine [52], and phenyl-N-*tert*-butylnitrone [53], are also known to prevent retinal light damage. Similarly, DMTU, a quencher of hydroxyl radical formation [54, 55], effectively protects against the light-induced loss of visual cells [56]. It also prevents the loss of retinol dehydrogenase activity, which has been shown to decrease during intense light exposure [57]. Irrespective of their natural or synthetic origins, most antioxidants are only effective when given before the onset of light, suggesting that oxidation is a relatively early event in the damage process. Evidence for the rapid formation of light-induced retinal hydroperoxides has been reported for an oxidatively susceptible rat model of Smith–Lemli–Opitz syndrome [58]. In vitro evidence also supports the early onset of oxidation. In isolated photoreceptors, Demontis et al. [59] found that light-induced oxidation within the ROS and RIS correlated with the time course of rhodopsin photo-bleaching and metabolic changes in mitochondria, respectively. Similarly, Yang et al. [60] found rapid light-induced changes in fluorescent dyes indicative of increased reactive oxygen species in the photoreceptor cell-mitochondria-rich ellipsoid region.

A unifying factor in photoreceptor/RPE damage appears to be light-induced reactive oxygen species generated by rhodopsin bleaching or by photosensitive compounds such as A2E [61]. Because of their intimate metabolic and morphologic relationship, their high oxygen tension, and the high rate and duration of photon flux, damage in either the retina or RPE can lead to degeneration in the other [62]. One outcome of intense light exposure is its effects on the omega-3 polyunsaturated fatty acid docosahexaenoic acid (DHA). Intermediates of DHA oxidation have been found in lipid extracts of retina after only 1–4 h of light exposure [63], and carboxyethylpyrrole (CEP) protein adducts, involving a DHA oxidation fragment, have been detected after light treatment [64]. Other evidence indicates that neuroprotectin D1 (NPD1), a lipoxygenase-mediated epoxide of DHA, is formed in RPE and that it protects against apoptotic cell death [65]. Accordingly, both physiological and pathological products of DHA oxidation can be formed in ocular tissues and a balance between the two may normally exist. Whether the formation of CEP lipid–protein adducts and NPD1 synthesis are primarily tissue specific or whether their formation also occurs in a circadian fashion is currently unknown.

8.6 Crystallins in the Retina and RPE Are Protective

Postmitotic tissues like the retina contain α-, β-, and γ-crystallins [66–68]. Among these, α-crystallins are members of a superfamily of small heat shock proteins (HSP) that share considerable amino acid homology and molecular chaperone properties with other HSPs [67–70]; for a recent review see [71]. They also undergo posttranslational modifications [67], many of which are present in aging retina [72] or in genetically induced retinal degenerations [73]. Alpha B-crystallin is known to aggregate upon cellular stress and to redistribute from the cytoplasm to the perinuclear region, a process reversed during recovery [74, 75]. Intense light-induced oxidative stress leads to higher αB-crystallin levels in ROS and to increases in all three classes of crystallins in retina [76]. Alpha B-crystallin is also constitutively expressed in RPE, and its expression increases during oxidative stress [77]. RPE cells transfected with an αB-crystallin gene are resistant to apoptosis [77], possibly from binding with proapoptotic members of the Bcl family [78, 79] or by binding with and preventing caspase activation [80, 81]. The β- and γ-crystallins contain amino acid homologies to enzymes such as lactate dehydrogenase and aldose reductase [82] and have about 50 % sequence homology with a γN-crystallin ancestor [83]. Six γ-crystallins are expressed in mouse retina [84, 85] where they may serve a protective role analogous to that of αB-crystallin [84]. Gamma-crystallin mRNA levels can be upregulated by intense light beginning at 1 am [86] and are normally downregulated by the high melatonin levels at night [87]. We now know that both α- and γ-crystallins can translocate into and out of ROS in a light-independent circadian manner [86], a process that differs from the strictly light-driven process affecting visual cell arrestin and transducin migration [44].

8.7 Circadian Rhythms and Retinal Light Damage

Compelling evidence exists that the retina possesses a circadian oscillator independent of the master pacemaker located in the suprachiasmatic nucleus (SCN). Synthesis of a number of visual cell transduction proteins and phagocytosis by RPE are either light entrained or expressed in a circadian fashion [88–93]. Likewise, expression of the retinal transcription factors c-fos and c-jun is light entrained [94, 95], and cone cell survival appears to depend on the expression of a circadian-driven voltage-gated potassium channel [96]. Intense light exposure causes an increase in retinal heme oxygenase-1 (Hmox-1) message and a simultaneous reduction in interphotoreceptor cell retinol-binding protein (IRBP) mRNA [14], changes which are otherwise circadian-dependent Hmox-1 is a 32 kDa inducible stress protein that catalyzes conversion of the prooxidant heme to biliverdin, which is subsequently converted to the antioxidant bilirubin [97]. IRBP is synthesized in photoreceptor cells and is a useful marker of visual cell function following damage [14]. Retinal glutathione peroxidase is also induced by photic injury [98] or by bright light-rearing conditions [99], and neuroprotective peptides are known to be upregulated in bright light [100, 101]. Thus, intense light can change the expression of potentially protective proteins or processes that appear to be enhanced or repressed by the time of day of light exposure. More recent evidence indicates that melanopsin, a photopigment located in a subset of retinal ganglion cells, serves as a retinal circadian irradiance detector [102–104]. Cryptochromes are also known to be present in the retina as well as in peripheral tissues [105]. In melanopsin KO mice, circadian-related behavior and phase resetting were attenuated but still present [106, 107]. Although melanopsin may function as a photoreceptive pigment, it is not thought to mediate photic damage because the inner retinal layers are not normally destroyed by intense visible light [2].

Repeated nighttime exposures of rats to intense visible light or repeated treatments with melatonin enhance retinal light damage [12, 13, 108–110], whereas the melatonin receptor antagonist luzindole prevents light damage [111]. This implicates melatonin as an effector, not a photosensitizer, in light-induced photoreceptor cell damage. Because melatonin levels are elevated at night and because it enhances susceptibility to retinal light damage ([108–110]; reviewed in [112]), we studied the effects of a single, relatively brief period, of intense visible light starting at various times in the circadian cycle. Visual cell loss was enhanced 2–3-fold by light exposure at 1 am compared to identical light treatment beginning at 9 am or 5 pm [14]. Figure 8.2 panel a shows that both cyclic light- and dark-reared rats are susceptible to light damage at 1 am and least vulnerable at 5 pm. Light exposure starting at 9 am, however, results in a major difference; cyclic light-reared rats are resistant to visual cell loss, while dark-reared animals remain susceptible to light damage. This difference is nicely demonstrated by the appearance of retinal histology 2 weeks after light exposure [15].

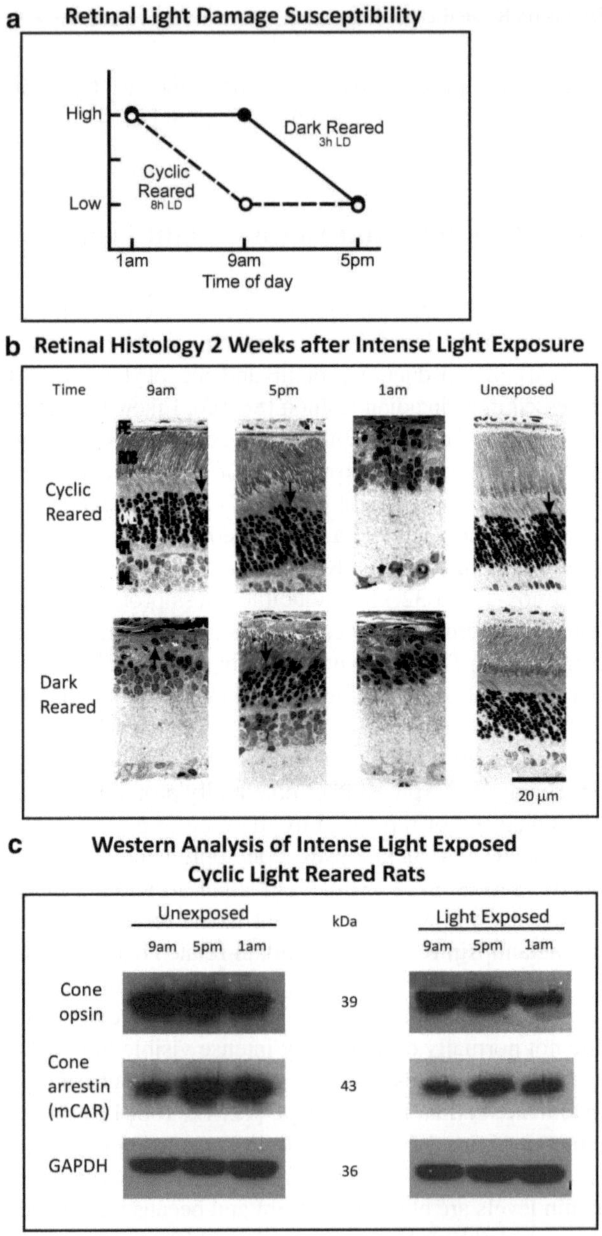

Fig. 8.2 *Panel a*: Schematic showing the relative retinal susceptibility of dim cyclic light-reared rats (20 lx, 12 h per day; lights on at 8 am) and littermates dark maintained for 40 days before intense light exposure beginning at 1 am, 9 am, or 5 pm. All rats were dark adapted for 16 h before intense light treatment. As determined by retinal rhodopsin measurements and retinal DNA, both cyclic light- and dark-reared animals are highly susceptible to light damage at 1 am and both are resistant to exposure beginning at 5 pm. Dark-reared rats retain their retinal light damage susceptibility for light treatment beginning at 9 am, whereas cyclic light rats are largely resistant to damage at that time. *Panel b*: Retinal histology 2 weeks after intense light exposure of male albino rats (P60) exposed to light for either 8 h (cyclic light-reared) or 3 h (dark-reared) starting at 9 am, 5 pm,

As shown in Fig. 8.2 panel b, intense light exposure starting at 9 am resulted in little visible effect in the ONL, ROS, and RIS of cyclic light-reared animals but massive photoreceptor cell damage and loss along with the loss of nuclei from the ONL in dark-reared rats. Light damage was also seen in dark-reared rats exposed at 5 pm, but it was much reduced from the pattern seen for the 9 am light treatment. Intense light exposure at 1 am, however, resulted in substantial losses of photoreceptor cells in both the cyclic light- and dark-reared animals. Few visual cell nuclei remain in the ONL, with nuclei from the INL now present in the area normally occupied by photoreceptor cells. Some nuclei appear to be those of remnant cones (arrows) that have also been retained in the severely damaged ONL. To determine if cones actually survived in these light-damaged retinas, we used Western analysis to detect cone cell opsin and arrestin (mCAR). Figure 8.2 panel c shows that cone cell opsin and mCAR are present in rat retina at all times of the day and night. Their levels are reduced by light treatment, compared with those found in retinal extracts from unexposed animals, and light exposure at 1 am appears to induce the greatest loss of both cone opsin and mCAR. As expected, 5 pm retinas exhibit the highest levels of cone cell markers in rats treated with intense light.

Circadian, light-induced visual cell death proceeds by an apoptotic process that is enhanced by light exposure at 1 am and preventable by antioxidant treatment [14, 15]. Circadian rhythms are also normally retained when animals from a diurnal light cycle are dark adapted for several days and reversed when the normal light–dark cycles are reversed [113]. In this regard, light damage susceptibility was also retained in cyclic light rats that had been dark adapted for as long as 3 days and reversed in those phase-shifted to reset their circadian clock [14]. Because rhodopsin bleaching and/or regeneration triggers light-induced retinal damage and rhodopsin mutations can lead to an accelerated rate of photoreceptor cell loss [114], we studied the profiles of retinal degeneration in transgenic animals expressing mutations at either the n-terminal (P23H) or c-terminal (S334ter) region of rhodopsin. Both transgenic

Fig. 8.2 (continued) or 1 am, as described [15]. In these representative sections, dramatic differences in the ONL can be seen in dark-reared animals exposed to light at 9 am versus their cyclic light counterparts. Photoreceptor cell damage was much reduced or absent when light exposures began at 5 pm, with extensive damage in both types of rats induced by intense light starting at 1 am. Arrows point to remnant cone nuclei in the ONL, magnification bar = 20 μm (Figure adapted from Vaughan et al. [15] with permission). *Panel c*: Cyclic light-reared male albino rats ($n = 4$) were acclimated to hyperthermia at 97 °C for 2 h in darkness and then exposed to intense visible light or maintained in darkness for 2 additional hours [2]. Following light treatment the animals were placed into darkness at room temperature for 2 days. The rats were killed in a saturated CO_2 chamber in dim red light and retinas excised and used for Western analysis, as described [64]. Cone opsin staining was detected with an antibody from Millipore (Temecula, CA), diluted 1:1,000. Anti-cone arrestin antibody (mCAR) was generously supplied by Dr. Cheryl Craft, USC, Los Angeles, CA, and used at a 1:2,000 dilution. The levels of immuno-reactive cone opsin and mCAR were decreased by light exposure, especially for light exposure beginning at 1 am. The cone cell marker proteins were highest in the unexposed and light exposed rats at 5 pm, with intermediate levels at 9 am. SDS gel electrophoresis was run with 20 μg of retinal protein per lane

animal models were found to exhibit circadian-dependent retinal light damage, with greatest susceptibility at 1 am, and to undergo apoptotic visual cell loss [115, 116]. The extent and rate of photoreceptor cell degeneration was also influenced by environmental light-rearing conditions and transgene expression levels. Although the P23H and S334ter rhodopsin mutations resulted in greater susceptibility to light damage than normal, and rhodopsin kinase and opsin have both been detected in the pineal gland [117, 118], their photoreceptor cell degenerations do not appear to arise from a uniformly enhanced circadian susceptibility to light damage.

8.8 Retinal Gene Expression Profiles and Analysis

Toward an understanding of light-induced oxidation and damage in retina and RPE are questions about the relationships between the time course of susceptibility and the nature of any time-dependent endogenous factors that can initiate or maintain visual cell damage. With this in mind, we undertook a study of gene expression profiles in the retinas of rats sacrificed at various times of the day and night (9 am, 5 pm, and 1 am). These studies have dovetailed nicely with the pursuit of putative circadian-regulated retinal expressed genes and circadian effects on susceptibility and resistance to light-induced retinal degeneration.

The molecular basis of the circadian clock consists of transcriptional–translational feedback loops involving a group of genes known collectively as "clock genes" (reviewed in [119, 120]). Results of mining the NCBI Gene Database for specific terms suggest that there are at least 38 genes that can be considered under the category of rat circadian clock genes and as many as 260 genes that may be circadian related. In our current discussion we will confine our descriptions to 14 specific clock genes (Fig. 8.2).

Core genes within this network include Clock, Arntl (Bmal1), and Npas2. In the first major feedback loop, these three genes form heterodimers that bind to the E-box elements in promoters of period genes (Per 1, Per 2), cryptochromes (Cry1, Cry2), and Nr1d1 to induce transcriptional activation. Eventually built up Per proteins will complex with Cry proteins (as well as others) and the complex will repress their own expression by blocking Clock- and Bmal1-mediated transactivation. A second functional loop involves the transactivation of Nr1d1 and Rora genes by Clock/Arntl. Nr1d1 and Rora proteins in turn competitively bind to RRE (also known as RORE) elements in the promoters of Arntl, thus driving a daily rhythm of Arntl expression. The third gene loop that defines the core clock network involves the transactivation of gene expression through the D-box element by hepatic leukemia factor (HLF), thyrotroph embryonic factor (TEF), and D-site-binding protein (DBP) (Fig. 8.3). Dbp, Tef, and Hlf are under the direct control of the core circulatory loop and form the clock-controlled output genes [121]. Posttranslational regulation (acetylation, degradation phosphorylation, ubiquitination, and sumoylation) is also important in regulating circadian oscillation. In addition to controlling the expression of components within the immediate clock network, the core clock network also sets into motion a host of circadian clock-regulated genes that may

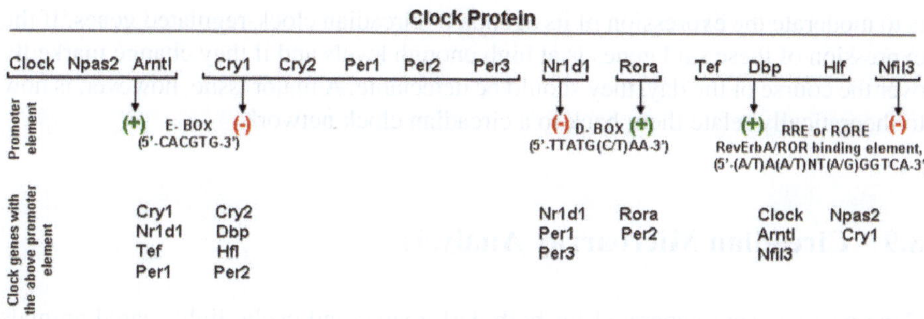

Fig. 8.3 Core clock genes and their mechanism to mediate gene expression

underlie the changes in biological processes that occur on a daily basis. This core system is entrained to the light–dark cycle with Clock: Bmal1 mRNA being high during the light period and Per:Cry mRNA being high during the dark period in the SCN regardless of whether the animal is diurnal or nocturnal (reviewed in [122]).

The central "master" clock is located in the SCN (reviewed in [123]) and regulates the timing of most circadian rhythms in mammals. Many organs and tissues of the body, including the retina, can however generate circadian rhythms independently of the SCN. Tosini et al. [124] first demonstrated that cultured mammalian retinal cells release melatonin in a circadian pattern. The circadian profile of melatonin release free-ran in constant darkness was entrained by light in vitro and was temperature-compensated, thus fulfilling the three fundamental properties that define a circadian rhythm [124, 125]. It is widely agreed that retinal photoreceptors play a key role in the circadian organization of the whole organism because they are the only source of photic input to the SCN [126]. Injection of Royal College of Surgeons (RCS) rats with kainic acid (which destroys the inner nuclear layer but spares photoreceptor cells) was found to repress the circadian rhythms of Aanat mRNA in RCS rats with no photoreceptors but not in the congenic wild-type animals. The conclusion from this observation is that there were at least two independent circadian clocks within the retina, one in the photoreceptor cell and one in the inner nuclear layer [127, 128]. In the mouse, multiple core circadian clock proteins have been observed in cone photoreceptors; horizontal, bipolar, dopaminergic amacrine cells; catecholaminergic amacrines; and melanopsin expressing intrinsically photosensitive ganglion cells [129, 130]. Core clock proteins are expressed in most mouse retinal neurons with a large degree of homogeneity within the same cell type and a heterogeneous expression profile (expression levels and circadian profile) between cell types, inferring a unique circadian clock in each cell population [130]. This disparity in clock protein profiles suggests that the perceived circadian profiles observed are built upon distinct subpopulations of neuronal cellular clocks. It is therefore unlikely that measuring mRNA levels from the whole retina will provide useful information about the functional characteristics of retinal circadian clocks [119]. We agree with this; however, in our opinion we see the system as a model to identify additional circadian-regulated genes. There is already sufficient evidence that there is more than one retinal clock to consider and that the goal of each clock is to moderate the expression of its agenda of circadian clock-regulated genes. If the expression of these said genes is at high enough levels and if they change markedly over the course of the day, they should be detectable. A major issue, however, is how to theoretically relate them back to a circadian clock network.

8.9 Circadian Microarray Analysis

Gene profiles were generated for both dark-reared and cyclic light-reared animals using standard methodologies (Fig. 8.4). In each case retinal tissues taken at 1 am, 9 am, and 5 pm were analyzed. For dark-reared animals 1 am and 9 am define a light-induced damage-susceptible period, while 5 pm defines a light-induced damage-resistant period. In the case of cyclic light-reared animals, 1 am defines a

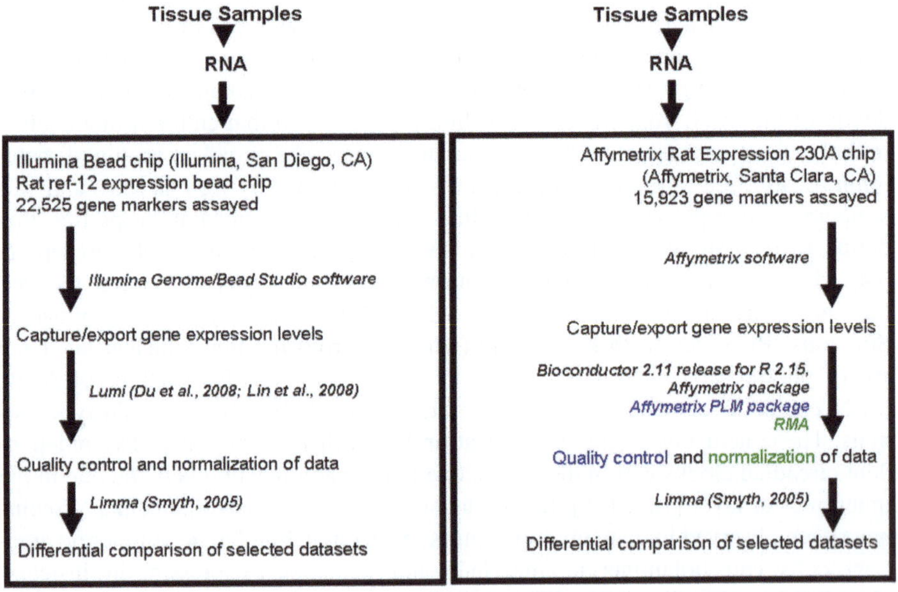

Fig. 8.4 Summary of gene profiling analyses: Retinas from dark- and cyclic light-reared animals (P60) were obtained at 1 am, 9 am, and 5 pm. Tissue was obtained from three different animals to define a single tissue sample, and three independent tissue samples were collected for each time point. Total RNA was extracted with Trizol reagent using the supplier's protocol (Invitrogen, Carlsbad, CA) and further purified through an RNA easy column (Qiagen, Germantown, MD). The integrity of the RNA was assayed using an Agilent 2100 BioAnalyzer NanoChip (Agilent Technologies, Santa Clara, CA). Each RNA sample was run three times on three different arrays. For a single gene profiling experiment (1 am, 9 am, 5 pm), a total of 27 microarrays were run. In the case of dark-reared animals, gene profiles were run on Illumina rat arrays [140–142], as well as on Affymetrix rat arrays. In each case a different set of samples was run on each array type. The computer software used to process the array data is as shown

light-induced damage-susceptible period, while 9 am and 5 pm define light damage-resistant periods.

Differential comparisons were run for the collective arrays defining 9 am versus 1 am (to define the progression of 1 am to 9 am), 5 pm versus 9 am (to define the progression of 9 am to 5 pm), and 1 am versus 5 pm (to define the progression of 5 pm to 1 am). Fold changes (FC) in gene expression and the adjusted p-value (adjusted for multiple testing) for the FC were estimated; an adjusted p-value of less than 0.05 was used as the cut-off to define the threshold at which a transcript was considered statistically significantly differentially expressed.

In each gene profiling study (dark-reared retinal profiles on Illumina arrays, dark-reared retinal profiles on Affymetrix arrays, or cyclic light-reared retinal profiles on Affymetrix arrays), if the FC for a given gene (gene marker) had an adjusted p-value of <0.05 for any successive pair of time points during the day, the FC at the other comparisons (regardless of the adjusted p-value) were also pulled and annotated so that the progression in changes in gene expression throughout the day could be viewed. Likewise the data for a gene with a significant FC in one study were also pulled and annotated for the other studies so that a match set of data was generated. In total 410 gene expression FC profiles were generated for each profiling study, and using the progressive FC values, relative expression levels could be estimated for each time point and an expression trend could be established (Fig. 8.5). Goodness of fit testing was applied and those profiles with insignificant variation were discarded from further analysis, leaving a final list of 137 gene markers defining 120 genes and their "circadian" expression profile trends (Tables 8.1, 8.2, and 8.3). In total we screened 38,448 unique gene markers to identify 137 (roughly 0.35 %) gene markers of interest. Expression profiles for 20 of these genes were confirmed by qRT-PCR using independent RNA sample sets. Moreover, 55 of our selected genes had the same 1 am, 9 am, and 5 pm FC trends on both Illumina and Affymetrix arrays (roughly 67 % of genes detected on both arrays).

8.10 Relating Selected Retinal Genes to the Circadian Clock

In the list of genes identified as putative circadian clock-regulated retinal genes, we reidentified several known circadian clock-regulated retinal genes (Fos, Jun, Kcnv2, Irbp, and seven different crystallin genes), validation for the approach. Moreover, given that the core circadian clock network is an intricate network of transcription regulators and transcription modifiers, we examined our 120 genes (defined by 137 gene markers) for E-box, D-box, and RRE promoter elements, the three promoter elements characteristic of core clock protein genes [131]. We were able to retrieve 20 kb of promoter sequence for 99 of the 120 genes. Fuzznuc (Emboss) was used to analyze promoter sequences [131]. Eighty-eight of the genes had at least one E-box (89 %), 25 genes had at least one RRE site (25 %), and 13 had at least one D-box element within their promoter regions (10 %). In the 14 core circadian clock genes (Fig. 8.3), 8 have at least one E-box site in the promoter (57 %), 5 had at least one

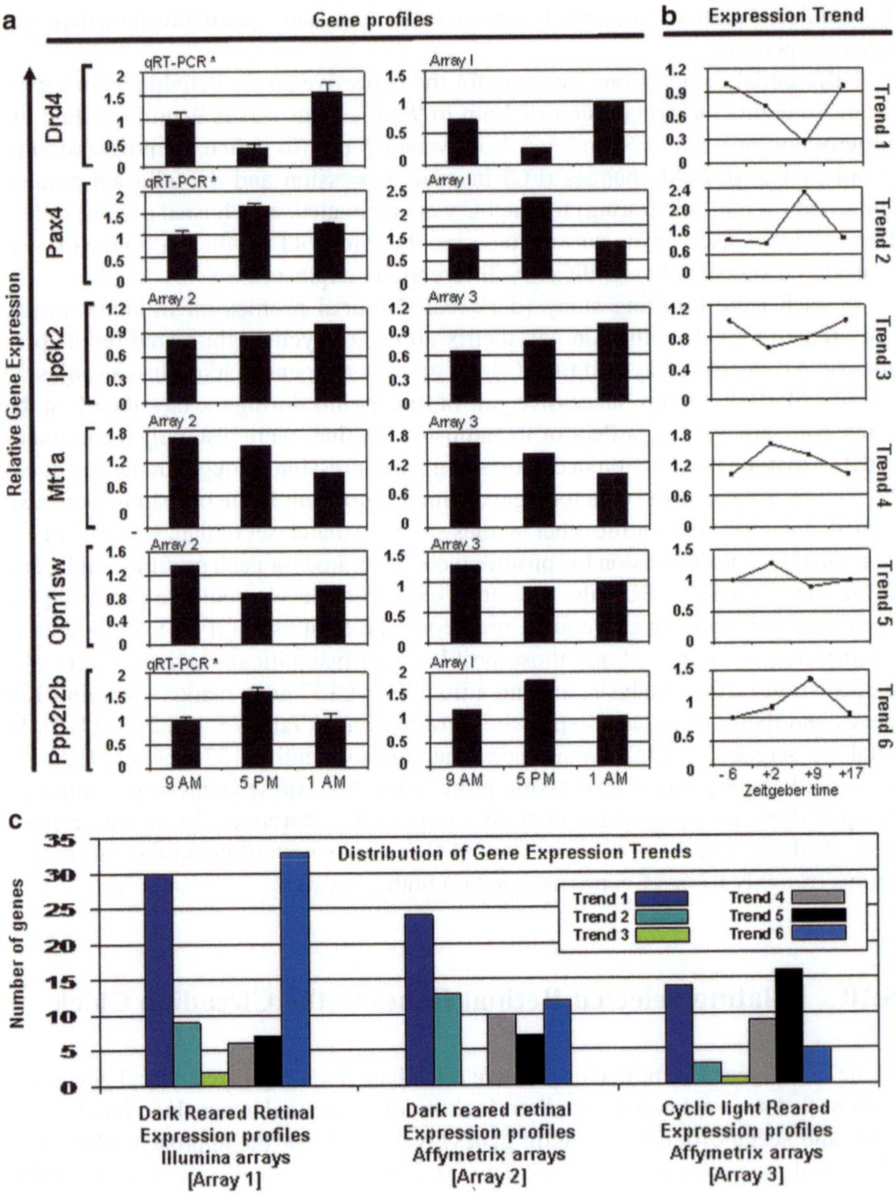

Fig. 8.5 *Panel a*: Retinal gene expression profiles established from microarray studies or qRT-PCR. *Panel b*: Extrapolation of gene profiling data to generate a circadian expression profile. In total six different expression trends were observed. *Panel c*: Distribution of expression trends in each of the three gene profiling studies (dark-reared retinal profiles on Illumina arrays, dark-reared retinal profiles on Affymetrix arrays, or cyclic light-reared retinal profiles on Affymetrix arrays)

Table 8.1 Dark Reared Retinal gene profiles-Illumina arrays

Gene Symbol	Gene ID	9AM/1AM Fold Change		5PM/9AM Fold Change		1AM/5PM Fold Change		Daily Expression Profile Trend	
Acbd7	361277	1.12		1.29		-1.39	*	Trend 6	**
Actb	81822	1.36	*	1.19		-1.60		Trend 6	**
Actb	81822	1.29		1.24		-1.57		Trend 6	**
Actb	81822	1.33		1.13		-1.51		Trend 6	**
Adi1	298934	-1.22		-1.29	*	1.47	*	Trend 1	**
Adra1b	24173	-1.19		-1.81	*	1.94	*	Trend 1	**
Ak4	29223	-1.34	*	-1.42	*	1.80	*	Trend 1	**
Ak4	29223	X		X		X		X	
Anxa1	25380	1.03		1.60	*	-1.56	*	Trend 6	**
Aplf	500247	1.29		-1.35	*	1.04		Trend 5	**
Aqp1	25240	-1.10		-1.77	*	1.85	*	Trend 1	**
Aqp1	25240	X		X		X		X	
Arfgef1	312915	1.01		1.33	*	-1.33	*	Trend 6	**
Bbs4	300754	1.14		-1.42		1.23		Trend 5	
Bcl2l10	114552	-1.33	*	-1.12		1.38	*	Trend 1	**
Casp7	64026	1.11		1.53	*	-1.56	*	Trend 6	**
Ccdc136	362331	1.30		1.19		-1.45	*	Trend 6	**
Ccng1	25405	1.19		1.35	*	-1.57	*	Trend 6	**
Cd24	25145	1.20		1.19		-1.37	*	Trend 6	**
Cds1	81925	1.06		1.32	*	-1.40	*	Trend 6	**
Cerk	300129	-1.15		-1.64	*	1.85		Trend 1	**
Col1a2	84352	-1.22	*	1.15		1.04		Trend 3	
Col1a2	84352	X		X		X		X	
Coq10b	301416	-1.37	*	-1.17		1.55	*	Trend 1	**
Cpt1a	25757	1.08		2.23	*	-2.25	*	Trend 6	**
Cpt1a	25757	X		X		X		X	
Cryab	25420	2.03	*	-1.36	*	-1.46	*	Trend 4	**
Cryba1	25583	2.66	*	-1.12		-2.28	*	Trend 4	**
Cryba2	286925	2.59	*	-1.26	*	-1.95	*	Trend 4	**
Crygc	24277	2.30	*	1.16		-2.45	*	Trend 6	**
Crygd	24278	2.43	*	-1.11		-1.97	*	Trend 4	**
Cryge	24279	2.56	*	-1.29		-1.76	*	Trend 4	**
Csnk1e	58822	X		X		X		X	
Csta	288067	-1.01		1.43	*	-1.31	*	Trend 2	**
Dio1	25430	-1.12		1.54	*	-1.30	*	Trend 2	**
Dlg4	29495	X		X		X		X	
Dlg4	29495	X		X		X		X	
Dnah11	117253	1.08		1.30	*	-1.30	*	Trend 6	**

(continued)

Table 8.1 (continued)

Gene Symbol	Gene ID	9AM/1AM Fold Change		5PM/9AM Fold Change		1AM/5PM Fold Change		Daily Expression Profile Trend	
Drd4	25432	-1.39		-2.81	*	3.79	*	Trend 1	**
Drd4	25432	X		X		X		X	
Drd4	25432	X		X		X		X	
Dusp1	114856	-1.21		-1.34	*	1.59	*	Trend 1	**
Dusp1	114856	X		X		X		X	
Dyrk2	314862	1.18		1.54	*	-1.67	*	Trend 6	**
Dyrk2	314862	X		X		X		X	
Egr1	24330	5.46	*	-24.60	*	3.72	*	Trend 5	**
Ell2	309918	-1.48	*	-1.10		1.55	*	Trend 1	**
Emp1	25314	-1.04		1.90	*	-1.67	*	Trend 2	**
Emp1	25314	X		X		X		X	
Fos	314322	X		X		X		X	
Frmpd1	313244	-1.06		-1.52	*	1.56	*	Trend 1	**
G0s2	289388	-1.28		-1.79	*	2.12	*	Trend 1	**
Glmn	289437	1.03		1.31		-1.33		Trend 6	**
Gnaz	25740	-1.26		-1.73	*	2.09	*	Trend 1	**
Gnaz	25740	X		X		X		X	
Gpx4	29328	1.00		-1.47	*	1.46	*	Trend 5	**
Hook1	313370	1.06		1.40	*	-1.43		Trend 6	**
Htra1	65164	-1.27		-1.12		1.44		Trend 1	**
Ier2	494344	1.14		-1.72	*	1 412	*	Trend 5	**
Ifitm3	361673	1.06		1.23	*	-1.24	*	Trend 6	
Ip6k2	59268	X		X		X		X	
Jun	24516	X		X		X		X	
Junb	24517	1.12		-1.48	*	1.26		Trend 5	**
Kcnv2	294065	-1.04		-1.17	*	1.16	*	Trend 1	
Kdm4c	298144	1.18		1.32	*	-1.49	*	Trend 6	**
Krt8	25626	-1.05		1.22	*	-1.13		Trend 2	
Lpl	24539	-1.17		1.18	*	-1.01		Trend 2	
Lypd2	300017	1.01		1.10		-1.08	*	Trend 6	
Mafg	64188	-1.06		-1.50	*	1.55		Trend 1	**
Mast2	313819	1.25		2.11	*	-2.50	*	Trend 6	**
Mpp4	58808	-1.12		-1.49	*	1.66	*	Trend 1	**
Mt1a	24567	-1.26		1.69	*	-1.30		Trend 2	**
Nr2c1	252924	-1.06		-1.42	*	1.42	*	Trend 1	**
Nr2c1	252924	X		X		X		X	
Nr4a1	79240	-1.06		1.01		1.04		Trend 3	
Opn1sw	81644	1.02		-1.43	*	1.34	*	Trend 5	

(continued)

Table 8.1 (continued)

Gene Symbol	Gene ID	9AM/1AM Fold Change		5PM/9AM Fold Change		1AM/5PM Fold Change		Daily Expression Profile Trend	
P4ha1	64475	-1.20		-1.42		1.63	*	Trend 1	**
Pax4	83630	-1.02		2.34	*	-2.19	*	Trend 2	**
Pbrm1	306254	1.07		1.50		-1.51		Trend 6	**
Pbx3	311876	1.22		1.38		-1.66		Trend 6	**
Pcp2	304195	-1.43		-1.35		1.88	*	Trend 1	**
Pdk1	116551	-1.29	*	-1.11		1.38	*	Trend 1	
Perp	292949	X		X		X		X	
Pla2r1	295631	1.27	*	1.24	*	-1.48	*	Trend 6	**
Plaa	116645	1.04		1.43	*	-1.42		Trend 6	**
Plac8	360914	-1.01		1.38	*	-1.30	*	Trend 2	**
Plekhb1	64471	-1.11		-1.35	*	1.49	*	Trend 1	**
Pmm1	300089	-1.30		-1.67	*	2.10		Trend 1	**
Ppp1r10	65045	X		X		X		X	
Ppp2r2b	60660	1.21		1.49	*	-1.69		Trend 6	**
Ptprn	116660	1.10		1.15		-1.23		Trend 6	
Raly	296301	X		X		X		X	
Rasl11a	304268	-1.28	*	-1.31	*	1.57	*	Trend 1	**
Rbm3	114488	1.01		1.37	*	-1.34		Trend 6	**
Rbp1	25056	-1.24	*	1.32	*	-1.06		Trend 2	
Rcvrn	140936	1.34		-1.28		-1.05		Trend 4	**
RGD1304963	313891	1.18		1.34	*	-1.48	*	Trend 6	**
RGD1305045	296050	1.12		1.11		-1.23	*	Trend 6	
RGD1307355	311825	-1.07		1.69	*	-1.52	*	Trend 2	**
RGD1560513	500271	-1.03		-1.45	*	1.46		Trend 1	**
RGD1560601	317432	1.05		1.04		-1.07		Trend 6	
RGD1560925	501196	X		X		X		X	
RGD1563351	294350	-1.57	*	1.58	*	-1.01		Trend 2	**
RGD1564296	502490	-1.88	*	1.23		1.43		Trend 3	**
RGD1564980	500019	-1.76		-1.46		2.28	*	Trend 1	**
RGD1566136	300278	-1.30		-1.21		1.47	*	Trend 1	**
Rorb	309288	-1.01		-1.14		1.12		Trend 1	
Rpe65	89826	-1.24		1.39	*	-1.11		Trend 2	
Rps9	81772	-1.39		-1.13		1.55	*	Trend 1	**
Rps9	81772	-1.38		-1.33		1.78	*	Trend 1	**
S100a6	85247	1.15		1.32	*	-1.43		Trend 6	**
Scgn	306942	1.01		-1.47	*	1.41		Trend 5	**
Sebox	58922	X		X		X		X	
Sft2d3	364835	1.14	*	1.56	*	-1.62	*	Trend 6	**

(continued)

Table 8.1 (continued)

Gene Symbol	Gene ID	9AM/1AM Fold Change		5PM/9AM Fold Change		1AM/5PM Fold Change		Daily Expression Profile Trend	
Sgip1	313413	1.04		1.72	*	-1.75	*	Trend 6	**
Sik1	59329	-1.58	*	1.03		1.45		Trend 3	**
Slc16a3	80878	-1.43		-1.49		2.04		Trend 1	**
Slc31a2	298091	X		X		X		X	
Slc3a1	29484	-1.07		-1.46		1.50	*	Trend 1	**
Slc7a5	50719	-1.02		-1.24	*	1.24		Trend 1	
Sostdc1	266803	-1.03		1.17	*	-1.11		Trend 2	
Stard7	296128	1.15		1.20		-1.36	*	Trend 6	**
Stxbp1	25558	1.37		1.44		-1.97		Trend 6	**
Stxbp1	25558	X		X		X		X	
Svil	361256	1.14		1.26	*	-1.38	*	Trend 6	**
Sys1	685079	X		X		X		X	
Tfrc	64678	-1.00		-1.06		1.05		Trend 1	
Tfrc	64678	X		X		X		X	
Tmem108	300967	1.30		-1.54	*	1.16		Trend 5	**
Tmem116	690442	X		X		X		X	
Tmem237	316412	X		X		X		X	
Tmem98	303356	-1.02		1.01		1.00		Trend 3	
Uckl1	499956	1.25		1.16		-1.44	*	Trend 6	**
Uri1	308537	X		X		X		X	
Vip	117064	-1.32		1.45	*	-1.08		Trend 2	**
Wdr89	314243	X		X		X		X	
Zmynd10	363139	-1.23		-1.51	*	1.76		Trend 1	**

* Fold Change with an adjusted p value <0.05; ** trend levels are above background

Note: Multiple entrees for the same gene symbol define different gene regions assayed for expression

D-Box element (36 %), and 5 genes had at least one RRE site in their promoter regions (36 %) (reviewed in 120).

Moreover, Ak4, Csnkie, JunB, Nr2c1, Rcvrn, Pcp2, Mt1e, ActB, Jun, Rorb, and Fos, from our circadian clock-regulated retinal expressed gene list (~10 %), have known direct interactions with one or more of the 14 core clock proteins shown in Fig. 8.3. Because of the nature of the circadian clock to coordinate and control complex biological events, systematic control of gene expression should manifest in the eventual generation and modification of functional protein networks. An unknown variable is the degree at which a set of circadian core genes can mediate an effect and of course how to get a measure of this. Most known clock-controlled genes lack the E-box response elements which are necessary for a direct regulation by the circadian pacemaker (reviewed in [132]). This suggests that a good part of circadian control is indirect. In the genes that we isolated, the presence of E-box response

Table 8.2 Dark Reared Retinal gene profiles-Affymetrix arrays

Gene Symbol	Gene ID	9AM/1AM Fold Change		5PM/9AM Fold Change		1AM/5PM Fold Change		Daily Expression Profile Trend	
Acbd7	361277	X		X		X		X	
Actb	81822	1.04		1.09		-1.14		Trend 6	
Actb	81822	1.04		1.01		-1.05		Trend 6	
Actb	81822	X		X		X		X	
Adi1	298934	-1.09		-1.40	*	1.56	*	Trend 1	**
Adra1b	24173	-1.12		-1.80	*	1.98	*	Trend 1	**
Ak4	29223	-1.06		-1.60	*	1.75	*	Trend 1	**
Ak4	29223	-1.09		-1.00		1.09		Trend 1	
Anxa1	25380	1.11		-1.20		1.04		Trend 5	
Aplf	500247	1.15		-1.00		-1.11		Trend 4	
Aqp1	25240	1.09		-1.90	*	1.74	*	Trend 5	**
Aqp1	25240	-1.03		-1.50	*	1.59	*	Trend 1	**
Arfgef1	312915	1.03		1.29	*	-1.33	*	Trend 6	
Bbs4	300754	1.04		-1.40	*	1.31	*	Trend 5	
Bcl2l10	114552	1.06		-1.30		1.20		Trend 5	
Casp7	64026	1.07		1.17	*	-1.13		Trend 6	
Ccdc136	362331	X		X		X		X	
Ccng1	25405	1.01		1.17	*	-1.19		Trend 6	
Cd24	25145	1.12		1.23		-1.38		Trend 6	**
Cds1	81925	-1.06		1.25	*	-1.18	*	Trend 2	
Cerk	300129	-1.07		-1.10		1.20		Trend 1	
Col1a2	84352	1.44		-1.30		-1.09		Trend 4	**
Col1a2	84352	1.40		-1.30		-1.05		Trend 4	**
Coq10b	301416	-1.16		-1.20		1.35	*	Trend 1	
Cpt1a	25757	-1.09		1.76	*	-1.76	*	Trend 2	**
Cpt1a	25757	-1.00		1.53	*	-1.41	*	Trend 2	**
Cryab	25420	1.09		1.33		-1.44		Trend 6	**
Cryba1	25583	1.06		1.66		-1.75		Trend 6	**
Cryba2	286925	1.14		1.69		-1.92		Trend 6	**
Crygc	24277	-1.10		3.35		-3.06		Trend 2	**
Crygd	24278	-1.03		2.54		-2.48		Trend 2	**
Cryge	24279	1.07		2.24		-2.39		Trend 6	**
Csnk1e	58822	-1.09		-1.30	*	1.45	*	Trend 1	**
Csta	288067	X		X		X		X	
Dio1	25430	-1.02		1.03		-1.01		Trend 2	
Dlg4	29495	1.04		-1.00		1.01		Trend 5	
Dlg4	29495	-1.11		-1.10		1.22		Trend 1	
Dnah11	117253	X		X		X		X	

(continued)

Table 8.2 (continued)

Gene Symbol	Gene ID	9AM/1AM Fold Change		5PM/9AM Fold Change		1AM/5PM Fold Change		Daily Expression Profile Trend	
Drd4	25432	-1.13		-2.10	*	2.29	*	Trend 1	**
Drd4	25432	-1.33		-2.30	*	2.59	*	Trend 1	**
Drd4	25432	-1.08		-2.70	*	3.55	*	Trend 1	**
Dusp1	114856	-1.14		-1.20		1.58	*	Trend 1	**
Dusp1	114856	-1.35		-1.20	*	1.37	*	Trend 1	**
Dyrk2	314862	1.18		1.56	*	-1.85	*	Trend 6	**
Dyrk2	314862	1.16		1.59	*	-1.86	*	Trend 6	**
Egr1	24330	2.52		-3.60	*	1.44		Trend 5	**
Ell2	309918	-1.16		-1.10		1.30	*	Trend 1	
Emp1	25314	1.26		-1.20		-1.18		Trend 4	**
Emp1	25314	1.72		-1.50		-1.02		Trend 4	**
Fos	314322	1.66		-2.90	*	1.72		Trend 5	**
Frmpd1	313244	X		X		X		X	
G0s2	289388	X		X		X		X	
Glmn	289437	-1.07		1.28	*	-1.19		Trend 2	
Gnaz	25740	-1.12		-1.60	*	1.80	*	Trend 1	**
Gnaz	25740	-1.16		-1.70	*	1.99	*	Trend 1	**
Gpx4	29328	1.00		-1.30	*	1.26	*	Trend 5	
Hook1	313370	-1.02		1.33	*	-1.30	*	Trend 2	
Htra1	65164	-1.03		-1.20	*	1.23	*	Trent 1	
Ier2	494344	1.03		-1.40	*	1.32		Trend 5	
Ifitm3	361673	1.38		-1.50		1.07		Trend 5	**
Ip6k2	59268	-1.23	*	1.05		1.17		Trend 3	
Jun	24516	1.02		-1.10		1.04		Trend 5	
Junb	24517	-1.02		-1.20	*	1.19		Trend 1	
Kcnv2	294065	-1.14		-1.50	*	1.69	*	Trend 1	**
Kdm4c	298144	X		X		X		X	
Krt8	25626	1.37		-1.20		-1.11		Trend 4	**
Lpl	24539	1.59		-1.60		-1.01		Trend 4	**
Lypd2	300017	2.76		-2.50	*	-1.10		Trend 4	**
Mafg	64188	1.01		-1.00		1.03		Trend 5	
Mast2	313819	-1.04		1.94	*	-1.90	*	Trend 2	**
Mpp4	58808	1.08		-1.40	*	1.29	*	Trend 5	
Mt1a	24567	1.59		-1.20		-1.38		Trend 4	**
Nr2c1	252924	-1.03		-1.50	*	1.53	*	Trend 1	**
Nr2c1	252924	-1.01		-1.00		1.02		Trend 1	
Nr4a1	79240	-1.17		-1.40	*	1.61	*	Trend 1	**
Opn1sw	81644	1.27		-1.40	*	1.11		Trend 5	**

(continued)

Table 8.2 (continued)

Gene Symbol	Gene ID	9AM/1AM Fold Change		5PM/9AM Fold Change		1AM/5PM Fold Change		Daily Expression Profile Trend	
P4ha1	64475	-1.08		-1.50	*	1.57	*	Trend 1	**
Pax4	83630	-1.32	*	2.15	*	-1.63	*	Trend 2	**
Pbrm1	306254	-1.13		1.45	*	-1.29		Trend 2	**
Pbx3	311876	1.03		1.24	*	-1.28	*	Trend 6	
Pcp2	304195	-1.00		1.00		1.00		Trend 3	
Pdk1	116551	-1.27		-1.20		1.46	*	Trend 1	**
Perp	292949	1.40		-1.40		1.03		Trend 5	**
Pla2r1	295631	X		X		X		X	
Plaa	116645	-1.03		1.33	*	-1.28	*	Trend 2	
Plac8	360914	1.10		-1.10		-1.05		Trend 4	
Plekhb1	64471	-1.09		-1.40	*	1.48	*	Trend 1	**
Pmm1	300089	-1.15		-1.50	*	1.72	*	Trend 1	**
Ppp1r10	65045	1.14		1.16		-1.32	*	Trend 6	
Ppp2r2b	60660	1.10		1.58	*	-1.73	*	Trend 6	**
Ptprn	116660	1.09		-1.00		-1.04		Trend 4	
Raly	296301	1.03		1.36	*	-1.40	*	Trend 6	**
Rasl11a	304268	X		X		X		X	
Rbm3	114488	-1.04		1.28	*	-1.23		Trend 2	
Rbp1	25056	1.28		-1.20		-1.11		Trend 4	
Rcvrn	140936	-1.03		-1.10	*	1.15	*	Trend 1	
RGD1304963	313891	X		X		X		X	
RGD1305045	296050	1.00		1.33	*	-1.33	*	Trend 6	**
RGD1307355	311825	X		X		X		X	
RGD1560513	500271	X		X		X		X	
RGD1560601	317432	-1.02		1.35	*	-1.33	*	Trend 2	**
RGD1560925	501196	-1.07		2.43	*	-2.26	*	Trend 2	**
RGD1563351	294350	X		X		X		X	
RGD1564296	502490	X		X		X		X	
RGD1564980	500019	X		X		X		X	
RGD1566136	300278	X		X		X		X	
Rorb	309288	-1.15		-1.40	*	1.64	*	Trend 1	**
Rpe65	89826	1.78		-1.70		-1.07		Trend 4	**
Rps9	81772	X		X		X		X	
Rps9	81772	-1.04		1.01		1.03		Trend 3	
S100a6	85247	1.06		1.12		-1.19		Trend 6	
Scgn	306942	X		X		X		X	
Sebox	58922	-1.03		1.40	*	-1.36	*	Trend 2	**
Sft2d3	364835	X		X		X		X	

(continued)

Table 8.2 (continued)

Gene Symbol	Gene ID	9AM/1AM Fold Change		5PM/9AM Fold Change		1AM/5PM Fold Change		Daily Expression Profile Trend	
Sgip1	313413	X		X		X		X	
Sik1	59329	-1.11		-1.10		1.23		Trend 1	
Slc16a3	80878	-1.04		-1.50	*	1.57	*	Trend 1	**
Slc31a2	298091	-1.06		1.42	*	-1.34	*	Trend 2	**
Slc3a1	29484	-1.34		1.04		1.29		Trend 3	
Slc7a5	50719	-1.09		-1.10		1.15		Trend 1	
Sostdc1	266803	1.47		-1.30		-1.11		Trend 4	**
Stard7	296128	1.01		1.26	*	-1.28	*	Trend 6	
Stxbp1	25558	-1.08		1.29	*	-1.19	*	Trend 2	
Stxbp1	25558	-1.23		1.49	*	-1.19		Trend 2	**
Svil	361256	1.03		-1.00		1.01		Trend 5	
Sys1	685079	-1.10		-1.40	*	1.50	*	Trend 1	**
Tfrc	64678	-1.12		-1.20	*	1.62	*	Trend 1	**
Tfrc	64678	-1.31		-1.30	*	1.46	*	Trend 1	**
Tmem108	300967	X		X		X		X	
Tmem116	690442	1.08		1.41	*	-1.53	*	Trend 6	**
Tmem237	316412	1.03		-1.40	*	1.45	*	Trend 5	**
Tmem98	303356	X		X		X		X	
Uckl1	499956	X		X		X		X	
Uri1	308537	1.07		1.39	*	-1.46	*	Trend 6	**
Vip	117064	-1.22		1.21		1.01		Trend 3	
Wdr89	314243	-1.03		1.41	*	-1.37	*	Trend 2	**
Zmynd10	363139	X		X		X		X	

*** Fold Change with an adjusted p value <0.05; ** trend levels are above background**

Note: Multiple entrees for the same gene symbol define different gene regions assayed for expression

elements is very high, suggesting that we may have isolated a group of genes fairly close functionally to elements of the pacemaker. In order to examine this possibility, we generated a functional walk from the 14 core circadian clock genes (Fig. 8.3) toward our putative circadian clock-regulated retinal expressed genes (120 genes). Our approach was to interrogate the Ingenuity gene database to find proteins with known direct functional interactions (protein to protein interactions or protein to DNA interactions) with any one of the core clock genes in question. We subsequently then defined which of these second generation clock-related genes/proteins had known direct functional interactions with the genes/proteins in the retinal circadian clock-regulated gene list. In total, 71 of the retinal circadian clock-regulated genes/proteins (~60 %) were found to have known direct interactions with 158 of the proteins defining generation 2 core clock-related genes. The analysis relating the circadian core clock to our circadian core clock-regulated retinal expressed genes is

Table 8.3 Cylic Light Reared Retinal gene profiles-Affymetrix arrays

Gene Symbol	Gene ID	9AM/1AM Fold Change		5PM/9AM Fold Change		1AM/5PM Fold Change		Daily Expression Profile Trend	
Acbd7	361277	X		X		X		X	
Actb	81822	1.26		-1.10		-1.19		Trend 4	
Actb	81822	1.41		-1.10		-1.24		Trend 4	**
Actb	81822	X		X		X		X	
Adi1	298934	-1.01		-1.40	*	1.38	*	Trend 1	
Adra1b	24173	-1.01		-1.80	*	1.87	*	Trend 1	**
Ak4	29223	-1.06		-1.00		1.07		Trend 1	
Ak4	29223	-1.02		-1.50	*	1.55		Trend 1	**
Anxa1	25380	-1.21		1.08		1.12		Trend 3	
Aplf	500247	1.20		-1.20		-1.03		Trend 4	
Aqp1	25240	1.06		-1.40	*	1.30		Trend 5	
Aqp1	25240	1.03		-1.50	*	1.42	*	Trend 5	**
Arfgef1	312915	1.15		1.10		-1.27		Trend 6	
Bbs4	300754	1.12		-1.50	*	1.37	*	Trend 5	**
Bcl2l10	114552	-1.04		-1.00		1.07		Trend 1	
Casp7	64026	-1.05		1.10		-1.05		Trend 2	
Ccdc136	362331	X		X		X		X	
Ccng1	25405	1.02		1.07		-1.10		Trend 6	
Cd24	25145	1.00		1.19		-1.19		Trend 6	
Cds1	81925	1.02		1.16		-1.18	*	Trend 6	
Cerk	300129	1.06		-1.10		1.07		Trend 5	
Col1a2	84352	-1.04		-1.10		1.11		Trend 1	
Col1a2	84352	-1.18		-1.00		1.21		Trend 1	
Coq10b	301416	-1.00		-1.30		1.33		Trend 1	
Cpt1a	25757	-1.04		1.55	*	-1.49	*	Trend 2	**
Cpt1a	25757	1.02		1.21		-1.23		Trend 6	
Cryab	25420	1.05		1.12		-1.18		Trend 6	
Cryba1	25583	1.08		-1.20		1.09		Trend 5	
Cryba2	286925	1.06		-1.00		-1.06		Trend 4	
Crygc	24277	-1.15		-1.30		1.51		Trend 1	**
Crygd	24278	-1.32		-1.20		1.54		Trend 1	**
Cryge	24279	-1.04		-1.00		1.07		Trend 1	
Csnk1e	58822	-1.01		-1.40		1.39		Trend 1	
Csta	288067	X		X		X		X	
Dio1	25430	-1.10		1.10		1.00		Trend 3	
Dlg4	29495	1.34		-1.60	*	1.16		Trend 5	**
Dlg4	29495	1.40		-1.50	*	1.11		Trend 5	**
Dnah11	117253	X		X		X		X	

(continued)

Table 8.3 (continued)

Gene Symbol	Gene ID	9AM/1AM Fold Change		5PM/9AM Fold Change		1AM/5PM Fold Change		Daily Expression Profile Trend	
Drd4	25432	-1.08		-1.60	*	1.72	*	Trend 1	**
Drd4	25432	-1.06		-1.40	*	1.52		Trend 1	**
Drd4	25432	1.10		-2.10	*	1.86	*	Trend 5	**
Dusp1	114856	-1.01		-1.30		1.31		Trend 1	
Dusp1	114856	-1.03		-1.20	*	1.28		Trend 1	
Dyrk2	314862	1.32		1.20		-1.58	*	Trend 6	**
Dyrk2	314862	1.36		1.11		-1.51		Trend 6	**
Egr1	24330	3.60		-5.70	*	1.59		Trend 5	**
Ell2	309918	-1.33	*	-1.10		1.48	*	Trend 1	**
Emp1	25314	-1.12		1.02		1.09		Trend 3	
Emp1	25314	-1.23		-1.10		1.31		Trend 1	
Fos	314322	3.38		-6.50	*	1.92		Trend 5	**
Frmpd1	313244	X		X		X		X	
G0s2	289388	X		X		X		X	
Glmn	289437	1.11		1.01		-1.12		Trend 6	
Gnaz	25740	-1.09		-1.40	*	1.54	*	Trend 1	**
Gnaz	25740	1.21		-2.00	*	1.66	*	Trend 5	**
Gpx4	29328	1.05		-1.20		1.19		Trend 5	
Hook1	313370	1.10		1.13		-1.24		Trend 6	
Htra1	65164	1.01		-1.20		1.18		Trend 5	
Ier2	494344	1.53		-1.90	*	1.23		Trend 5	**
Ifitm3	361673	-1.01		1.06		-1.06		Trend 2	
Ip6k2	59268	-1.54	*	1.20		1.29		Trend 3	**
Jun	24516	1.35		-1.60	*	1.16		Trend 5	**
Junb	24517	1.16		-1.40	*	1.21		Trend 5	
Kcnv2	294065	-1.07		-1.40		1.45	*	Trend 1	**
Kdm4c	298144	X		X		X		X	
Krt8	25626	1.25		-1.20		-1.07		Trend 4	
Lpl	24539	1.41		-1.40		1.01		Trend 5	**
Lypd2	300017	-1.25		1.09		1.15		Trend 3	
Mafg	64188	1.05		-1.10		1.08		Trend 5	
Mast2	313819	-1.14		1.66	*	-1.45	*	Trend 2	**
Mpp4	58808	1.02		-1.40	*	1.36	*	Trend 5	
Mt1a	24567	1.66		-1.10		-1.51		Trend 4	**
Nr2c1	252924	-1.06		1.02		1.04		Trend 3	
Nr2c1	252924	-1.15		-1.40	*	1.61	*	Trend 1	**
Nr4a1	79240	-1.12		-1.50	*	1.69	*	Trend 1	**
Opn1sw	81644	1.37		-1.50	*	1.12		Trend 5	**

(continued)

Table 8.3 (continued)

Gene Symbol	Gene ID	9AM/1AM Fold Change		5PM/9AM Fold Change		1AM/5PM Fold Change		Daily Expression Profile Trend	
P4ha1	64475	1.04		-1.50	*	1.42	*	Trend 5	**
Pax4	83630	-1.12		1.79	*	-1.60	*	Trend 2	**
Pbrm1	306254	1.30		1.06		-1.39		Trend 6	**
Pbx3	311876	1.04		1.10		-1.15		Trend 6	
Pcp2	304195	1.09		-1.10		1.04		Trend 5	
Pdk1	116551	1.08		-1.40		1.26		Trend 5	
Perp	292949	-1.23		1.10		1.12		Trend 3	
Pla2r1	295631	X		X		X		X	
Plaa	116645	1.16		1.10		-1.28		Trend 6	
Plac8	360914	-1.14		1.09		1.04		Trend 3	
Plekhb1	64471	1.02		-1.40	*	1.35	*	Trend 5	
Pmm1	300089	-1.12		-1.30	*	1.47	*	Trend 1	**
Ppp1r10	65045	1.34		-1.00		-1.31	*	Trend 4	**
Ppp2r2b	60660	1.06		1.36	*	-1.44	*	Trend 6	**
Ptprn	116660	1.39	*	-1.30	*	-1.06		Trend 4	**
Raly	296301	1.09		1.14		-1.24		Trend 6	
Rasl11a	304268	X		X		X		X	
Rbm3	114488	1.16		-1.00		-1.16		Trend 4	
Rbp1	25056	1.54		-1.30		-1.19		Trend 4	**
Rcvrn	140936	1.07		-1.10		1.00		Trend 5	
RGD1304963	313891	X		X		X		X	
RGD1305045	296050	1.08		1.17		-1.27	*	Trend 6	
RGD1307355	311825	X		X		X		X	
RGD1560513	500271	X		X		X		X	
RGD1560601	317432	1.09		1.16		-1.26		Trend 6	
RGD1560925	501196	1.00		1.99	*	-1.99	*	Trend 6	**
RGD1563351	294350	X		X		X		X	
RGD1564296	502490	X		X		X		X	
RGD1564980	500019	X		X		X		X	
RGD1566136	300278	X		X		X		X	
Rorb	309288	-1.22		-1.20		1.52	*	Trend 1	**
Rpe65	89826	2.64		-1.90		-1.36		Trend 4	**
Rps9	81772	X		X		X		X	
Rps9	81772	1.08		1.06		-1.14		Trend 6	
S100a6	85247	-1.11		1.15		-1.04		Trend 2	
Scgn	306942	X		X		X		X	
Sebox	58922	1.10		1.10		-1.21		Trend 6	
Sft2d3	364835	X		X		X		X	

(continued)

Table 8.3 (continued)

Gene Symbol	Gene ID	9AM/1AM Fold Change		5PM/9AM Fold Change		1AM/5PM Fold Change		Daily Expression Profile Trend	
Sgip1	313413	X		X		X		X	
Sik1	59329	-1.04		-1.00		1.05		Trend 1	
Slc16a3	80878	1.06		-1.40	*	1.36	*	Trend 5	**
Slc31a2	298091	1.03		1.14		-1.17		Trend 6	
Slc3a1	29484	-1.03		-1.00		1.05		Trend 1	
Slc7a5	50719	1.36		-1.30		-1.01		Trend 4	**
Sostdc1	266803	1.42		-1.20		-1.17		Trend 4	**
Stard7	296128	1.04		1.14		-1.18		Trend 6	
Stxbp1	25558	1.10		1.20		-1.32		Trend 6	
Stxbp1	25558	1.03		1.19		-1.23		Trend 6	
Svil	361256	-1.09		1.03		1.05		Trend 3	
Sys1	685079	1.09		-1.50	*	1.37	*	Trend 5	**
Tfrc	64678	1.02		-1.50	*	1.52		Trend 5	**
Tfrc	64678	-1.11		-1.40	*	1.51	*	Trend 1	**
Tmem108	300967	X		X		X		X	
Tmem116	690442	1.14		1.13		-1.28		Trend 6	
Tmem237	316412	-1.06		1.07		-1.01		Trend 2	
Tmem98	303356	1.52		-1.40		-1.08		Trend 4	**
Uckl1	499956	X		X		X		X	
Uri1	308537	1.09		1.21		-1.32		Trend 6	
Vip	117064	-1.04		1.01		1.03		Trend 3	
Wdr89	314243	1.25		1.05		-1.31	*	Trend 6	
Zmynd10	363139	X		X		X		X	

*** Fold Change with an adjusted p value <0.05; ** trend levels are above background**

Note: Multiple entrees for the same gene symbol define different gene regions assayed for expression

summarized in Table 8.4. Arntl seems to be the circadian clock protein that has the most influence with our circadian clock-regulated retinal expressed genes/proteins. With respect to the retinal expressed genes/proteins, COL1A2 has the most core clock interactions of extracellular proteins. ACTB has the most core clock interactions of cytoplasmic proteins, ANXA1 has the most core clock interactions of plasma membrane proteins, and FOS and JUN have the most interactions with respect to nuclear proteins. Overall most interactions mediated by the core clock candidates target proteins within the nucleus.

Table 8.4 Functional interactions between circadian core components and circadian clock regulated retinal expressed gene products

Core Clock genes	CLOCK	NPAS2	ARNTL	CRY1	CRY2	PER1	PER2	PER3	NR1D1	RORA	TEF	DBP	HLF	NFIL3	Total Interactions
A) Extracellular proteins															
Col1a2	1	3	3	2			1		3	1	3	1	1	1	20
EMC7			1												1
Htra1	1									1					2
Rbp1															0
Sostdc1										1					1
Vip						1									1
	2	3	4	2	0	1	1	0	3	3	3	1	1	1	25
B) Cytoplasmic proteins															
Dnah11			1												1
Cpt1a		2	4							2	1				9
Rpe65		1													1
P4ha1			3				1								4
Pmm1			3				1								4
Gpx4			1												1
Cds1			1												1
Aplf	2								1						3
Dio1		1	2						3	1					7
Lpl		2	2						2	2	1				9
RPL29															0
Arfgef1			1												1
Ccdc136										1					1
Ppp2r2b			2					1							3

Core Clock genes	CLOCK	NPAS2	ARNTL	CRY1	CRY2	PER1	PER2	PER3	NR1D1	RORA	TEF	DBP	HLF	NFIL3	Total Interactions
C) Nuclear proteins															
Adi1			1						1						2
Rasl11a							1								1
Ppp1r10			2												2
Junb		2	4						1	6	2	2	1	3	21
Rorb		1								2					3
Mafg	2	1	4			1			1	2		2		1	14
Jun	11	7	10	1	2	1	1	1	6	14	2		3	6	65
Egr1	1	3	5			1			2	4		2	2	3	30
Sebox															0
Fos	10	6	12			1	3		4	11	1	3	1	5	57
Nr4a1	5	6	5					1	2	9		1	1	1	31
Ell2	3		1						1						5
Nr2c1		2	2	1		1	3		1	1				1	12
Pax4															0
Pbx3															0
Ccng1			2	2					1						5
Kdm4c						1			1						2
Cryab	1	2	2						1	3		1			10
Pbrm1			2						1	1					4
Dusp1	2	2	2				1		1	1		1		1	11
Plac8			1	1			1			1					4
Rbm3		1	2							1					4
Raly			2				1								3
Uri1			1						1						2
	42	33	60	5	2	6	11	2	25	56	5	12	8	21	288

(continued)

Table 8.4 (continued)

Plekhb1															0
Bcl2l10			1												1
Frmpd1															0
Coq10b			1				1								2
Casp7	2		2			1		1	1						7
Crygc															0
Crygd	1														1
Hook1			1												1
Rcvrn															0
S100a6		1	3		1										5
Dyrk2		2	2					1	1						6
Mast2		1	1	1					1						4
Ifitm3			2							1					3
Sgip1		1													1
Glmn			1												1
Scgn															0
Mpp4															0
Ip6k2		2	1	2					1						6
Csta			1									1	1		3
Mt1a			1						1	1					3
Actb	3	4	6	1		1	1	1	1						18
Rps9	1		2					1	1						5
Pcp2															0
Pdk1			3						1						4
Plaa		1	2												3
Krt8	1	1	5		1					1					9
G0s2			1								1				2
Bbs4			2							2					4
Ier2			2				1								3
Ak4							1		1						2
Csnk1e	2	2		1	1	1	4	1	1						13
Sik1	1		1			1	1		1					1	6
Stxbp1	1	1	1						1	1					5
Sys1			1												1
Tmem98			1												1
Uckl1															0
Zmynd10															0
	14	22	64	5	3	4	11	6	18	12	3	1	1	1	165

D) Plasma membrane

Opn1sw															0
Adra1b	2														2
Pla2r1															0
Kcnv2															0
Gnaz			1				1		1						3
Drd4			1												1
Ptprn	1											1			2
Anxa1	1		3					1		3	1	1			10
Lypd2															0
Cd24			1												1
Slc3a1															0
Slc31a2															0
Slc7a5	1		1												3
Slc16a3							1								0
Svil		1	1			1						1			4
Tfrc	1		3						1						5
Cerk			1						2						3
Perp		1	1												2
Emp1															0
Aqp2			2	1					1					1	5
Dlg4	3		1				1		1	5					11
	9	2	16	1	0	1	3	1	6	8	1	3	0	1	52

8.11 Susceptibility and Resistance to Light-Induced Damage

Biological factors that influence the progression of light-induced damage have already been discussed. There is currently limited information at the molecular level that defines light damage-susceptible and light damage-resistant factors. However, light damage studies using genetically modified mice have demonstrated that c-Fos ablation and erythropoietin overexpression protect against light-induced retinal damage [133–135]. Similar studies using other KO mouse models have shown that ablation of p53 and JunD does not prevent light damage [136, 137] and rd mice still undergo retinal degeneration in the absence of c-Fos [138]. Although the mouse KO studies are compelling, comparable genetic studies in rat models remain elusive. We have made a start by characterizing circadian changes in gene expression in the rat model of light-induced retinal degeneration. Integration of these gene expression results in different rodent models will require additional time and experimentation.

In this study two different biological paradigms were examined, retina that is more susceptible to light-induced damage (dark-reared retina at 1 am and 9 am, cyclic light-reared retina at 1 am) and retina that is more resistant to light-induced damage (dark-reared retina at 5 pm and cyclic light-reared retina at 9 am and 5 pm). We took a comparative analysis to screen for changes at the level of gene expression to see if there were any conserved events between light damage resistance and susceptibility. For each profiling study (dark-reared retinal profiles on Illumina arrays, dark-reared retinal profiles on Affymetrix arrays, or cyclic light-reared retinal profiles on Affymetrix arrays), two comparisons could be made. These results in turn were compared from one profiling study to the other to examine for consistencies in FC changes (Table 8.5). Thirteen genes (Table 8.5A) were found to be consistently repressed during the light-induced damage-susceptible state as compared to the resistant state. Seventeen genes (Table 8.5B) were found to be consistently elevated during the light-induced damage-susceptible state as compared to the resistant state. We regard these 30 genes as those that putatively define retinal susceptibility/resistance to light-induced damage (LID susceptibility/resistance genes). The remainder of the genes had contradictions either within a specific profiling study (Table 8.5C, Cryab), between different profiling studies (Table 8.5C, Nr4a1 and Anaxa1), or genes that were not tested in up to two of the three profiling studies (examples are shown in Table 8.5C).

In order to interrogate the collection of 30 LID susceptibility/resistance genes and their gene products for similarities and associations, we submitted the list of genes to Ingenuity Pathway Analysis. Of the LID susceptibility/resistance genes listed, nine genes (Adi1, Dusp1, Ell2, Nr2c1, Rorb, Ccng1, Pbra1, Pbx3, Rbm3) code for nuclear proteins. Of the examples of suggestive genes, Cryab, Nr4a1, and Csnk1e all code for nuclear proteins. In the context of the core circadian clock network, Rorb is a paralog of Nr1d1. Rors (retinoic acid-related orphan receptors) are transcriptionally controlled by the same mechanism as that of Per and Cry genes. While NR1D1 negatively regulates the gene expression of Bmal1, RORs positively regulate the expression. Although these transcription factors are not involved in

Table 8.5 30 genes that putatively define retinal susceptibility/resistance to light-induced damage

a

Entrez Gene Name	Location	Type(s)	Gene		Dark Reared Retinal gene profiles-Illumina arrays		Dark reared Retinal gene profiles-Affymetrix arrays		Cylic Light Reared Retinal gene profiles-Affymetrix arrays	
			Gene Symbol	Gene ID	5PM/9AM Fold Change	5PM/1AM Fold Change	5PM/9AM Fold Change	5PM/1AM Fold Change	9AM/1AM Fold Change	5PM/1AM Fold Change
acireductone dioxygenase 1	Nucleus	enzyme	**Adi1**	**298934**	-1.29 *	-1.47 *	-1.43 *	-1.56 *	-1.01	-1.38 *
adrenoceptor alpha 1B	Plasma Membrane	G-protein coupled receptor	**Adra1b**	**24173**	-1.81 *	-1.94 *	-1.77 *	-1.98 *	-1.01	-1.87 *
adenylate kinase 4	Cytoplasm	kinase	**Ak4**	**29223**	-1.42 *	-1.80 *	-1.60 *	-1.75 *	-1.06	-1.07
BCL2-like 10	Cytoplasm	other	**Bcl2l10**	**114552**	-1.12	-1.38 *	-1.26	-1.20	-1.04	-1.07
coenzyme Q10 homolog B	Cytoplasm	other	**Coq10b**	**301416**	-1.17	-1.55 *	-1.17	-1.35 *	-1.00	-1.33
dopamine receptor D4	Plasma Membrane	G-protein coupled receptor	**Drd4**	**25432**	-2.81 *	-3.79 *	-2.12 *	-2.29 *	-1.08	-1.72 *
dual specificity phosphatase 1	Nucleus	phosphatase	**Dusp1**	**114856**	-1.34 *	-1.59 *	-1.17	-1.58 *	-1.01	-1.31
elongation factor, RNA polymerase II	Nucleus	transcription regulator	**Ell2**	**309918**	-1.10	-1.55 *	-1.13	-1.30 *	-1.33 *	-1.48
guanine nucleotide binding protein, alpha z polypeptide	Plasma Membrane	enzyme	**Gnaz**	**25740**	-1.73 *	-2.09 *	-1.61 *	1.80 *	-1.09	-1.54 *
potassium channel, subfamily V, member 2	Plasma Membrane	ion channel	**Kcnv2**	**294065**	-1.17 *	-1.16 *	-1.49 *	-1.69 *	-1.07	-1.45 *
nuclear receptor subfamily 2, group C, member 1	Nucleus	transcription regulator	**Nr2c1**	**252924**	-1.42 *	-1.42 *	-1.48 *	-1.53 *	-1.06	-1.04
phosphomannomutase 1	Cytoplasm	enzyme	**Pmm1**	**300089**	-1.67 *	-2.10	-1.49 *	-1.72 *	-1.12	-1.47 *
RAR-related orphan receptor B	Nucleus	ligand-dependent nuclear receptor	**Rorb**	**309288**	-1.14	-1.12	-1.43 *	-1.64 *	-1.22	-1.52 *

b

actin, beta	Cytoplasm	other	**Actb**	**81822**	1.19		1.60		1.09		1.14		1.26		1.19	
ADP-ribosylation factor guanine nucleotide-exchange factor 1 (brefeldin A-inhibited)	Cytoplasm	other	**Arfgef1**	**312915**	1.33	*	1.33	*	1.29	*	1.33	*	1.15		1.27	
cyclin G1	Nucleus	other	**Ccng1**	**25405**	1.35	*	1.57	*	1.17	*	1.19		1.02		1.10	
CD24a antigen	Plasma Membrane	other	**Cd24**	**25145**	1.19		1.37	*	1.23		1.38		1.00		1.19	
CDP-diacylglycerol synthase 1	Cytoplasm	enzyme	**Cds1**	**81925**	1.32	*	1.40	*	1.25	*	1.18	*	1.02		1.18	*
dual-specificity tyrosine-(Y)-phosphorylation regulated kinase 2	Cytoplasm	kinase	**Dyrk2**	**314862**	1.54	*	1.67	*	1.56	*	1.85	*	1.32		1.58	*
glomulin, FKBP associated protein	Cytoplasm	other	**Glmn**	**289437**	1.31		1.33		1.28	*	1.19		1.11		1.12	
hook homolog 1 (Drosophila)	Cytoplasm	other	**Hook1**	**313370**	1.40	*	1.43		1.33	*	1.30	*	1.10		1.24	
polybromo 1	Nucleus	other	**Pbrm1**	**306254**	1.50		1.51		1.45	*	1.29		1.30		1.39	
pre-B-cell leukemia homeobox 3	Nucleus	transcription regulator	**Pbx3**	**311876**	1.38		1.66		1.24	*	1.28	*	1.04		1.15	
phospholipase A2-activating protein	Cytoplasm	other	**Plaa**	**116645**	1.43	*	1.42		1.33	*	1.28	*	1.16		1.28	
protein phosphatase 2, regulatory subunit B, beta	Cytoplasm	phosphatase	**Ppp2r2b**	**60660**	1.49	*	1.69		1.58	*	1.73	*	1.06		1.44	*
RNA binding motif (RNP1, RRM) protein 3	Nucleus	other	**Rbm3**	**114488**	1.37	*	1.34		1.28	*	1.23		1.16		1.16	
ER membrane protein complex subunit 7	Extracellular Space	other	RGD1305045	**296050**	1.11		1.23	*	1.33	*	1.33	*	1.08		1.27	*
similar to Jumonji/ARID domain-containing protein 1C (SmcX protein)	unknown	other	RGD1560601	**317432**	1.04		1.07		1.35	*	1.33	*	1.09		1.26	

(continued)

Table 8.5 (continued)

C	StAR-related lipid transfer (START) domain containing 7	unknown	other	**Stard7**	**296128**	1.20		1.37	*	1.26	*	1.28	*	1.04		1.18	
	syntaxin binding protein 1	Cytoplasm	transporter	**Stxbp1**	**25558**	1.44		1.97		1.29	*	1.19	*	1.10		1.32	
	crystallin, alpha B	Nucleus	other	**Cryab**	**25420**	-1.36	*	1.46	*	1.33		1.44		1.05		1.18	
	nuclear receptor subfamily 4, group A, member 1	Nucleus	ligand-dependent nuclear receptor	**Nr4a1**	**79240**	1.01		-1.04		-1.38	*	-1.61	*	-1.12		-1.69	*
	annexin A1	Plasma Membrane	other	**Anxa1**	**25380**	1.60	*	1.56	*	-1.16		-1.04		-1.21		-1.12	
	casein kinase 1, epsilon	Cytoplasm	kinase	**Csnk1e**	**58822**	X		X		-1.33	*	-1.45	*	-1.01		-1.39	

X: Not tested, *: Adjusted p value for FC is <0.05

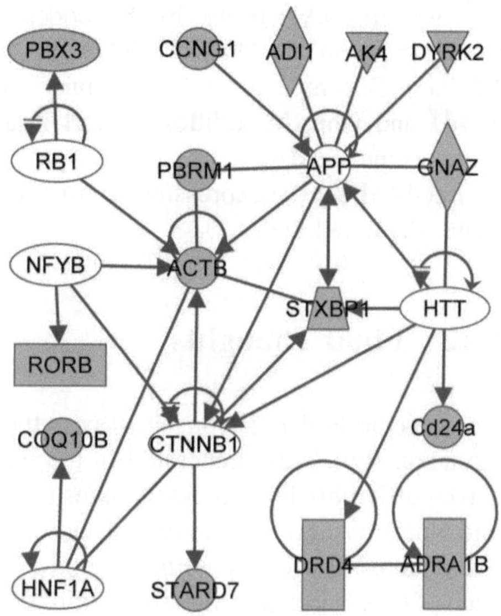

Fig. 8.6 LID susceptibility/resistance protein network. Full names for LID susceptibility/resistance factors are given in Table 8.5 and highlighted in grey. Proteins incorporated into the network but not identified as LID susceptibility/resistance factors are as follows: amyloid beta (A4), precursor protein (APP), catenin (cadherin-associated protein), beta 1 (CTNNB1), HNF1 homeobox A (HNF1A), huntingtin (HTT), nuclear transcription factor Y, beta (NFYB), and retinoblastoma 1 (RB1). *Solid lines* indicate direct protein to protein or protein to DNA interactions

rhythm generation, they are important in controlling the phase and the amplitude of gene expression [139]. Moreover, casein kinase 1 epsilon (CSNK1E) phosphorylates PER proteins and tags them for degradation and can positively regulate BMAL1 activity by phosphorylation (reviewed in [122]).

With respect to gene ontology, 12 genes are associated with the process of cell death and survival (Actb, Adi1, Adra1b, Bcl2l10, Ccng1, Cd24a, Dusp1, Nr2c1, Pbx3, Ppp2r2b, Rbm3, Stxbp1), and 11 genes are associated with the process of cellular development (Actb, Adra1b, Bcl2l10, Ccng1, Cd24a, Dusp1, Glmn, Hook, Pbrn1, Pbx3, and Rbm3). Because most gene products are multifunctional, eight of the genes listed above are associated with both the processes of cell death and survival as well as cellular development.

Gene products for 15 of the 30 LID susceptibility/resistance genes fall into a single functional protein network (Fig. 8.6). Interactions were limited to direct protein to protein or protein to DNA interactions. Each protein in the network was limited to being a maximum of one protein away from another protein stemming from the list of LID susceptibility/resistance genes. Associated with the HTT (the huntingtin gene product) node is DRD4, the functional product of the dopamine receptor D4 gene (Drd4). In mice with a disruption of the *Drd4 gene*, cyclic AMP levels in the dark-adapted retina are significantly lower compared to wild-type retina, unresponsive to light, and therefore resistant to light-induced damage (Dr. M. Iuvone, Emory University, personal communication, 2013). In our hands Drd4 was selected as a circadian-regulated gene; further network analysis suggests that Drd4 can in turn mediate a feedback loop that affects the core clock circuit. DRD4 can directly bind with the G-protein-coupled receptor adrenoceptor alpha 1B (ADRA1B). ADRA1B, in turn, has an indirect influence on the expression of Dusp1 mRNA, a transcript that codes for a nuclear dual specificity phosphatase. Overexpression of Dusp1 mRNA is associated with decreased activation of NF-kappaB complexes, which in turn can affect the biology of core Clock genes Nr1d1 and Dbp. In addition, Dusp1 is a negative regulator of mitogen-activated protein kinase (MAPK)-mediated signal transduction; increased Dusp1 levels can indirectly affect the expression or function of core clock genes/gene products Dbp, Arntl, Per1, and Per2.

8.12 Final Thoughts

Our gene array data is consistent with the thought that numerous genetically based circadian signals are generated in the retina. In total we identified 120 circadian-driven differentially expressed genes (as defined by gene expression at 1 am, 9 am, and 5 pm). With respect to retinal susceptibility/resistance to light damage, we have found consistent changes in 30 genes. The specific retinal cell types affected by each of these circadian signals remain to be sorted out. Moreover, how these signals are normally integrated into a unified retinal response and how interruption of this time-dependent symphony of signals impacts light-induced retinal degeneration remains an open question which will require additional research.

Acknowledgements This work was supported in part with funding from NSERC, RPB, NIH (NEI) P30 EY006360 and the Knights Templar of Georgia (PW) and NIH (NEI) grant EY01959, the International Retina Research Foundation, Ohio Lions Eye Research Foundation, and the Petticrew Research Laboratory (DTO). Thanks to Dr. Cheryl Craft for providing the mCAR antibody and to Linda Barsalou, Ruth Darrow, and Christine Rapp for their technical assistance.

References

1. Noell WK. Aspects of experimental and hereditary retinal degeneration. In: Graymore CN, editor. Biochemistry of the retina. New York: Academic Press; 1965. p. 51–72.
2. Noell WK, Walker VS, Kang BS, Berman S. Retinal damage by light in rats. Invest Ophthalmol. 1966;5:450–73.
3. Williams TP, Howell WL. Action spectrum of retinal light-damage in albino rats. Invest Ophthalmol Vis Sci. 1983;24:285–7.
4. Grimm C, Reme CE, Rol PO, Williams TP. Blue light's effects on rhodopsin: photoreversal of bleaching in living rat eyes. Invest Ophthalmol Vis Sci. 2000;41:3984–90.
5. Organisciak DT, Winkler BS. Retinal light damage: practical and theoretical considerations. In: Osborne NN, Chader GJ, editors. Progress in retinal and eye research, vol. 13. New York: Pergamon Press; 1994. p. 1–29.
6. Rapp LM. Retinal phototoxicity. In: Chang LW, Dyer RS, editors. Handbook of neurotoxicology. New York: Marcel Dekker; 1995. p. 963–1003.
7. Reme CE, Grimm C, Hafezi F, Marti A, Wenzel A. Apoptotic cell death in retinal degenerations. In: Osborne NN, Chader GJ, editors. Progress in retinal and eye research, vol. 17. New York: Pergamon Press; 1998. p. 443–64.
8. Penn JS, Williams TP. Photostasis. Regulation of daily photon-catch by rat retinas in response to various cyclic illuminances. Exp Eye Res. 1986;43:915–28.

9. Penn JS, Naash ML, Anderson RE. Effect of light history on retinal antioxidants and light damage susceptibility in the rat. Exp Eye Res. 1987;44:779–88.
10. Organisciak DT, Xie A, Wang H-M, Jiang Y-L, Darrow RM, Donoso LA. Adaptive changes in visual cell transduction protein levels: effect of light. Exp Eye Res. 1991;53:773–9.
11. Farber DB, Seager-Danciger J, Organisciak DT. Levels of mRNA encoding proteins of the cGMP cascade as a function of light environment. Exp Eye Res. 1991;53:781–6.
12. White MP, Fisher LV. Degree of light damage to the retina varies with time of day of bright light exposure. Physiol Behav. 1987;39:607–13.
13. Duncan TE, O'Steen WK. The diurnal susceptibility of rat retinal photoreceptors to light-induced damage. Exp Eye Res. 1985;41:497–507.
14. Organisciak DT, Darrow RM, Barsalou L, Kutty RK, Wiggert B. Circadian-dependent retinal light damage in rats. Invest Ophthalmol Vis Sci. 2000;41:3694–701.
15. Vaughan DK, Nemke JL, Fliesler SJ, Darrow RM, Organisciak DT. Evidence for a circadian rhythm of susceptibility to retinal light damage. Photochem Photobiol. 2002;75:547–53.
16. Grignolo A, Orzalesi N, Castellazzo R, Vittone P. Retinal damage by visible light in albino rats, an electron microscope study. Ophthalmologica. 1969;157:43–59.
17. O'Steen WK, Lytle RB. Cellular disruption and phagocytosis in photically induced retinal degeneration. Am J Anat. 1971;130:227–33.
18. O'Steen WK, Shear CR, Anderson KV. Retinal damage after prolonged exposure to visible light. A light and electron microscopy study. Am J Anat. 1972;134:5–22.
19. Rapp LM, Williams TP. A parametric study of retinal light damage in albino and pigmented rats. In: Williams TP, Baker BN, editors. The effects of constant light on visual processes. New York: Plenum Press; 1980. p. 133–59.
20. Kuwabara T, Gorn RA. Retinal damage by visible light: an electron microscopic study. Arch Ophthalmol. 1968;79:69–78.
21. Shear CR, O'Steen WK, Anderson KV. Effects of short-term low intensity light on albino rat retina. An electron microscopic study. Am J Anat. 1973;138:127–32.
22. Kerr JFR, Wyllie AH, Currie AR. Apoptosis: a basic biological phenomenon with wide ranging implications in tissue kinetics. Br J Cancer. 1972;26:230–57.
23. Wyllie AH, Beattie GJ, Hargreaves AD. Chromatin changes in apoptosis. Histochem J. 1981;13:681–92.
24. Abler AS, Chang CJ, Ful J, Tso MO, Lam TT. Photic injury triggers apoptosis of photoreceptor cells. Res Commun Mol Pathol Pharmacol. 1996;92:177–89.
25. Li S, Chang CJ, Abler AS, Fu J, Tso MO, Lam TT. A comparison of continuous versus intermittent light exposure on apoptosis. Curr Eye Res. 1996;15:914–22.
26. Organisciak DT, Darrow RA, Barsalou L, Darrow RM, Lininger LA. Light-induced damage to the retina: differential effects of dimethylthiourea on photoreceptor survival, apoptosis and DNA. Photochem Photobiol. 1999;70:261–8.
27. Chambers ML, Organisciak D, Darrow R, Barsalou L, Stepczynski J, McDonald B, Lagali P, et al. Activation of the caspase cascade appears to be involved in oxidative stress mediated light induced retinal degeneration in rats. Invest Ophthalmol Vis Sci. 2000;41:S20.
28. Perche O, Doly M, Ranchon-Cole I. Caspase-dependent apoptosis in light-induced retinal degeneration. Invest Ophthalmol Vis Sci. 2007;48:2753–9.
29. Cohen GM. Caspases: the executioners of apoptosis. Biochem J. 1997;326:1–16.
30. Thornberry NA, Lazebnik Y. Caspases: enemies within. Science. 1998;281:1312–6.
31. Chang HY, Yang X. Proteases for cell suicide: functions and regulation of caspases. Microbiol Mol Biol Rev. 2000;64:821–46.
32. Slee EA, Adrain C, Martin SJ. Serial killers: ordering caspase activation events in apoptosis. Cell Death Differ. 1999;6:1067–74.
33. Earnshaw WC, Martins LM, Kaufmann SH. Mammalian caspases: structure, activation, substrates, and functions during apoptosis. Annu Rev Biochem. 1999;68:383–424.
34. Wolf BB, Green DR. Suicidal tendencies: apoptotic cell death by caspase family proteinases. J Biol Chem. 1999;274:20049–52.

35. Saleh A, Srinivasula SM, Acharya S, Fishel R, Alnemri ES. Cytochrome c and dATP-mediated oligomerization of Apaf-1 is a prerequisite for procaspase-9 activation. J Biol Chem. 1999;274:17941–5.
36. Wong P, Kutty RK, Darrow R, Shivaram S, Kutty G, Fletcher RT, Wiggert B, et al. Changes in clusterin expression associated with light-induced retinal damage in rats. Biochem Cell Biol. 1994;72:499–503.
37. Humphries MM, Rancourt D, Farrar GJ, Kenna P, Hazel M, Bush RA, Sieving PA, et al. Retinopathy induced in mice by targeted disruption of the rhodopsin gene. Nat Genet. 1997;15:216–9.
38. Grimm C, Wenzel A, Hafezi F, Yu S, Redmond TM, Reme C. Protection of Rpe 65 deficient mice identifies rhodopsin as a mediator of light-induced retinal degeneration. Nat Genet. 2000;25:63–6.
39. Wenzel A, Grimm C, Samardzija M, Reme C. The genetic modifier RPE 65 leu 450: effect on light damage susceptibility in c-fos deficient mice. Invest Ophthalmol Vis Sci. 2003; 44:2798–802.
40. Redmond TM, Yu S, Lee E, Bok D, Hamasaki D, Chen N, Goletz P, et al. Rpe65 is necessary for production of 11-*cis* vitamin A in the retinal visual cycle. Nat Genet. 1998;20:344–51.
41. Chen J, Simon MI, Matthes MT, Yasumura D, LaVail MM. Increased susceptibility to light damage in an arrestin knockout mouse model of Oguchi disease (stationary night blindness). Invest Ophthalmol Vis Sci. 1999;40:2978–82.
42. Chen CK, Burns ME, Spencer M, Niemi GA, Chen J, Hurley JB, Baylor DA, Simon MI. Abnormal photoresponses and light-induced apoptosis in rods lacking rhodopsin kinase. Proc Natl Acad Sci U S A. 1999;96:3718–22.
43. Hao W, Wenzel A, Obin MS, Chen C-K, Brill E, Krasnoperova NV, Eversol-Cire P, et al. Evidence for two apoptotic pathways in light-induced retinal degeneration. Nat Genet. 2002;32:254–60.
44. Sokolov M, Lyubarsky AL, Strissel KJ, Savchenko AB, Govardovskii VI, Pugh EN, Arshavsky VY. Massive light-driven translocation of transducin between the two major compartments of rod cells: a novel mechanism of light adaptation. Neuron. 2002;35:95–106.
45. Iseli H-P, Wenzel A, Hafezi F, Reme CE, Grimm C. Light damage susceptibility and RPE65 in rats. Exp Eye Res. 2002;75:407–13.
46. Beatrice J, Wenzel A, Reme CE, Grimm C. Increased light damage susceptibility at night does not correlate with RPE65 levels and rhodopsin regeneration in rats. Exp Eye Res. 2003;76:695–700.
47. Sieving PA, Chaudry P, Kondo M, Provenzano M, Wu D, Carlson TJ, Bush RA, Thompson DA. Inhibition of the visual cycle in vivo by 13-cis retinoic acid protects from light damage and provides a mechanism for night blindness in isotretinoin therapy. Proc Natl Acad Sci U S A. 2001;98:1835–40.
48. Rozanowska MB. Light-induced damage to the retina: current understanding of the mechanisms and unresolved questions; a symposium in print. Photochem Photobiol. 2012;88: 1303–8.
49. Organisciak DT, Darrow RM, Bicknell IR, Jiang Y-L, Pickford M, Blanks JC. Protection against retinal light damage by natural and synthetic antioxidants. In: Anderson RE, Hollyfield JG, LaVail MM, editors. Retinal degenerations. Boca Raton, FL: CRC Press; 1989. p. 189–201.
50. Organisciak DT, Bicknell IR, Darrow RM. The effects of L- and D-ascorbic acid administration on retinal tissue levels and light damage in rats. Curr Eye Res. 1992;11:231–41.
51. Reme CE, Braschler UF, Roberts J, Dillon J. Light damage in the rat retina: effects of a radioprotective agent (WR-77913) on acute rod outer segment disk disruptions. Photochem Photobiol. 1991;54:137–42.
52. Tanito M, Nishiyama A, Tanaka T, Masutani H, Nakamura H, Yodoi J, Ohira A. Change of redox status and modulation by thiol replenishment in retinal photooxidative damage. Invest Ophthalmol Vis Sci. 2002;43:2392–400.
53. Ranchon I, Chen S, Alvarez K, Anderson RE. Systemic administration of phenyl-N-*tert-butylnitrone* protects the retina from light damage. Invest Ophthalmol Vis Sci. 2001;42:1375–9.

54. Vaughan DK, Peachey NS, Richards MJ, Buchan B, Fliesler SJ. Light-induced exacerbation of retinal degeneration in a rat model of Smith-Lemli-Opitz syndrome. Exp Eye Res. 2006;82:496–504.
55. Lam S, Tso MOM, Gurne DH. Amelioration of retinal photic injury in albino rats by dimethylthiourea. Arch Ophthalmol. 1990;108:1751–7.
56. Organisciak DT, Darrow RM, Jiang Y-L, Marak GE, Blanks JC. Protection by dimethylthiourea against retinal light damage in rats. Invest Ophthalmol Vis Sci. 1992;33:1599–609.
57. Darrow RA, Darrow RM, Organisciak DT. Biochemical characterization of cell specific enzymes in light-exposed rat retinas: oxidative loss of all-*trans* retinol dehydrogenase activity. Curr Eye Res. 1997;16:144–51.
58. Richards MJ, Nagel BA, Fliesler SJ. Lipid hydroperoxide formation in the retina: correlation with retinal degeneration and light damage in a rat model of Smith-Lemli-Opitz syndrome. Exp Eye Res. 2006;82:538–41.
59. Demontis GC, Longoni B, Marchiafava PL. Molecular steps involved in light-induced oxidative damage to retinal rods. Invest Ophthalmol Vis Sci. 2002;43:2421–7.
60. Yang J-H, Basinger SF, Gross RL, Wu SM. Blue light-induced generation of reactive oxygen species in photoreceptor ellipsoids requires mitochondrial electron transport. Invest Ophthalmol Vis Sci. 2003;44:1312–9.
61. Sparrow JR, Nakanishi K. Parish CA The lipofuscin fluorophore A2E mediates blue light induced damage to retinal pigmented epithelial cells. Invest Ophthalmol Vis Sci. 2000;41:1981–9.
62. Noell WK. There are different kinds of retinal light damage in the rat. In: Williams TP, Baker BN, editors. The effects of constant light on visual processes. New York: Plenum Press; 1980. p. 3–8.
63. Sun M, Finnemann SC, Febbraio M, Shan L, Annangudi SP, Podrez EA, Hoppe G, et al. Light-induced oxidation of photoreceptor outer segment phospholipids generates ligands for CD-36-mediated phagocytosis by retinal pigment epithelium. J Biol Chem. 2006;281:4222–30.
64. Organisciak D, Wong P, Rapp C, Darrow R, Ziesel A, Rangarajan R, Lang J. Light-induced retinal degeneration is prevented by zinc, a component in the age-related eye disease study formulation. Photochem Photobiol. 2012;88:1396–407.
65. Mukherjee PK, Marcheselli VL, Bazan NG. Neuroprotectin D1: a docosahexaenoic acid-derived docosatriene protects human retinal pigment epithelial cells from oxidative stress. Proc Natl Acad Sci. 2004;101:8491–6.
66. Horwitz J. The function of alpha-crystallin in vision. Semin Cell Dev Biol. 2000;11:53–60.
67. Groenen PJTA, Merck KB, DeJong WW, Bloemendal H. Structure and modifications of the junior chaperone α-crystallin: from lens transparency to molecular pathology. Eur J Biochem. 1994;225:1–19.
68. Deretic D, Aebersold RH, Morrison HD, Papermaster DS. αA-and αB-crystallin in the retina. J Biol Chem. 1994;269:16853–61.
69. Horwitz J. Alpha-crystallin can function as a molecular chaperone. Proc Natl Acad Sci U S A. 1992;89:10449–53.
70. Jakob U, Gaestel M, Engel K, Buchner J. Small heat shock proteins are molecular chaperones. J Biol Chem. 1993;268:1517–20.
71. Kannan R, Sreekumar PG, Hinton DR. Novel roles for α-crystallins in retinal function and disease. Prog Retin Eye Res. 2012;31:576–604.
72. Kapphahn RJ, Ethen CM, Peters EA, Higgens L, Ferrington DA. Modified αA crystallin in the retina: altered expression and truncation with aging. Biochemistry. 2003;42:15310–25.
73. Organisciak D, Darrow R, Gu X, Barsalou L, Crabb JW. Genetic, age and light mediated effects on crystallin protein expression in the retina. Photochem Photobiol. 2006;82:1088–96.
74. Klemenz R, Frohli E, Steiger RH, Shafer R, Aoyama A. αB crystallin is a small heat shock protein. Proc Natl Acad Sci U S A. 1991;88:3652–6.
75. Voorter CEM, Wintjes L, Bloemendal H, DeJong WW. Relocalization of αB crystallin by heat-shock in ovarian carcinoma cells. FEBS Lett. 1992;309:111–4.
76. Sakaguchi H, Miyagi M, Darrow RM, Crabb JS, Hollyfield JG, Organisciak DT, Crabb JW. Intense light exposure changes the crystallin content in retina. Exp Eye Res. 2003; 76:131–3.

77. Alge CS, Priglinger SG, Neubauer AS, Kampik A, Zillig M, Bloemendal H, Welge-Lussen U. Retinal pigment epithelium is protected against apoptosis by αB-crystallin. Invest Opthalmol Vis Sci. 2002;43:3575–82.
78. Mao Y-M, Liu J-P, Li DW-C. Human αA-and αB-crystallins bind to Bax and Bcl-Xs to sequester their translocation during staurosporine-induced apoptosis. Cell Death Diff. 2004; 11:512–26.
79. Brady JP, Garland DL, Green DE, Tamm ER, Giblin FJ, Wawrousek EF. αB-crystallin in lens development and muscle integrity: a gene knockout approach. Invest Ophthalmol Vis Sci. 2001;42:2924–34.
80. Kamradt MC, Chen F, Cryns VL. The small heat shock protein αB-crystallin negatively regulates cytochrome c-and caspase-8-dependent activation of caspase-3 by inhibiting its autoproteolytic maturation. J Biol Chem. 2001;276:16059–63.
81. Sreekumar PG, Chothe P, Sharma KK, Baid R, Kompella U, Spee C, Kannan N, et al. Antiapoptotic properties of α–crystallin-derived peptide chaperones and characterization of their uptake transporters in human RPE cells. Invest Ophthalmol Vis Sci. 2013;54:2787–98.
82. Wistow GJ, Piatigorsky J. Lens crystallins: the evolution and expression of proteins for a highly specialized tissue. Annu Rev Biochem. 1988;57:479–504.
83. Wistow G, Wyatt K, David L, Gao C, Bateman O, Bernstein S, Tomarev S, et al. γN-crystallin and the evolution of the βγ-crystallin superfamily in vertebrates. FEBS J. 2005;272:2276–91.
84. Jones SE, Jomary C, Grist J, Makawana J, Neal MJ. Retinal expression of γ-crystallins in the mouse. Invest Opthalmol Vis Sci. 1999;40:3017–20.
85. Xi J, Farjo R, Yoshida S, Kern TS, Swaroop A, Andley UP. A comprehensive analysis of the expression of crystallins in mouse retina. Mol Vis. 2003;9:410–9.
86. Organisciak D, Darrow R, Barsalou L, Rapp C, McDonald B, Wong P. Light-induced and circadian effects on retinal photoreceptor cell crystallins. Photochem Photobiol. 2011;87:151–9.
87. Wiechmann AF. Regulation of gene expression by melatonin: a microarray survey of the rat retina. J Pineal Res. 2002;33:178–85.
88. Korenbrot JI, Fernald RD. Circadian rhythm and light regulate opsin mRNA in rod photoreceptors. Nature. 1998;337:454–7.
89. Brann MR, Cohen LV. Diurnal expression of transducin mRNA and translocation of transducin in rods of rat retina. Science. 1987;235:585–7.
90. Bowes G, van Veen T, Farber DB. Opsin, G-protein and 48-kDa protein in normal and rd mouse retinas: developmental expression of mRNAs and proteins and light/dark cycling of mRNAs. Exp Eye Res. 1988;47:369–90.
91. McGinnis JF, Whelan JP, Donoso LA. Transient cyclic changes in mouse visual cell gene products during the light–dark cycle. J Neurosci Res. 1992;31:584–90.
92. Wiechmann AF, Sinacola MK. Diurnal expression of recoverin in the rat retina. Mol Brain Res. 1997;45:321–4.
93. Goldman AI, Teirstein PS, O'Brien PJ. The role of ambient lighting in circadian disc shedding in the rod outer segment of the rat retina. Invest Ophthalmol Vis Sci. 1980;19:1257–67.
94. Yoshida K, Kawamura K, Imaki J. Differential expression of c-fos mRNA in rat retinal cells: regulation by light/dark cycle. Neuron. 1993;10:1049–54.
95. Imaki J, Yamashita K, Yamakawa A, Yoshida K. Expression of jun family genes in rat retinal cells: regulation by light/dark cycle. Mol Brain Res. 1995;30:48–52.
96. Holter P, Kunst S, Wolloscheck T, Keller DK, Sticht C, Wolfrum U, Spessert R. The retinal clock drives the expression of Kcnv2, a channel essential for visual function and cone survival. Invest Ophthalmol Vis Sci. 2012;53:6947–54.
97. Kutty RK, Kutty G, Wiggert B, Chader GJ, Darrow RM, Organisciak DT. Induction of heme oxygenase −1 in the retina by intense light: suppression by the antioxidant dimethylthiourea. Proc Natl Acad Sci. 1995;92:1177–81.
98. Ohira A, Tanito M, Kaidzu S, Kondo T. Glutathione peroxidase induced in rat retinas to counteract photic injury. Invest Ophthalmol Vis Sci. 2003;44:1230–6.
99. Penn JS, Anderson RE. Effects of light history on the rat retina. In: Osborne NN, Chader GJ, editors. Progress in retinal research, vol. 11. New York: Pergamon Press; 1991. p. 75–98.

100. Liu C, Peng M, Laties M, Wen R. Preconditioning with bright light evokes a protective response against light damage in the rat retina. J Neurosci. 1998;18:1337–44.
101. LaVail MM, Unoki K, Yasumura D, Matthes MT, Yancopoulos GD, Steinberg RH. Multiple growth factors, cytokines and neurotrophins rescue photoreceptors from the damaging effects of constant light. Proc Natl Acad Sci U S A. 1992;89:11249–53.
102. Provencio I, Rodriguez IR, Jiang G, Hayes WP, Moreira EF, Rollag MD. A novel human opsin in the inner retina. J Neurosci. 2000;20:600–5.
103. Hattar S, Liao H-W, Takao M, Berson DM, Yau K-W. Melanopsin-containing retinal ganglion cells: architecture, projections, and intrinsic photosensitivity. Science. 2002;295:1065–70.
104. Berson DM, Dunn FA, Takao M. Phototransduction by retinal ganglion cells that set the circadian clock. Science. 2002;295:1070–3.
105. Yagita K, Tamanini F, van Der Horst GTJ, Okamura H. Molecular mechanisms of the biological clock in cultured fibroblasts. Science. 2001;292:278–81.
106. Ruby NF, Brennan TJ, Xie X, Cao V, Franken P, Heller HC, O'Hara BF. Role of melanopsin in circadian responses to light. Science. 2002;298:2211–3.
107. Panda S, Sato TK, Castrucci AM, Rollag MD, DeGrip WJ, Hogenesch JB, Provencio I, Kay SA. Melanopsin (Opn4) requirement for normal light-induced circadian phase shifting. Science. 2002;298:2213–6.
108. Bubenik GA, Purtill RA. The role of melatonin and dopamine in retinal physiology. Can J Physiol Pharmacol. 1980;58:1457–62.
109. Leino M, Ilkka-Matti A, Kari E, Gynther J, Markkanen S. Effects of melatonin and 6-methoxy-tetrahydro-α-carboline in light induced retinal damage: a computer morphometric method. Life Sci. 1984;35:1997–2001.
110. Wiechmann AF, O'Steen WK. Melatonin increases photoreceptor susceptibility to light-induced damage. Invest Ophthalmol Vis Sci. 1992;33:1894–902.
111. Bush RA, Suguwara T, Iuvone PM, Sieving PA. The melatonin receptor antagonist luzindole protects photoreceptors from light damage in rat. Invest Ophthalmol Vis Sci. 1998;39:2458–65.
112. Weichmann AF, Summers JA. Circadian rhythms in the eye: the physiological significance of melatonin receptors in ocular tissues. Prog Retinal Eye Res. 2008;27:137–60.
113. Besharse JC, Iuvone PM. Circadian clock in Xenopus eye controlling retinal serotonin N-acetyltransferase. Nature. 1983;305:133–5.
114. Green ES, Menz MD, LaVail MM, Flannery JG. Characterization of rhodopsin mis-sorting and constitutive activation in a transgenic rat model of retinitis pigmentosa. Invest Ophthalmol Vis Sci. 2000;41:1546–53.
115. Organisciak DT, Darrow RM, Barsalou L, Kutty RK, Wiggert B. Susceptibility to light damage in transgenic rats having rhodopsin mutations. Invest Ophthalmol Vis Sci. 2003;44: 486–92.
116. Vaughan D, Coulibaly S, Darrow RM, Organisciak DT. A morphometric study of light-induced damage in transgenic rat models of retinitis pigmentosa. Invest Ophthalmol Vis Sci. 2003;44:848–55.
117. Zimmerman BL, Tso MOM. Morphological evidence of photoreceptor differentiation of pinealocytes in the neonatal rat. J Cell Biol. 1975;66:60–75.
118. Zhao X, Haeseleer F, Fariss RN, Huang J, Baehr W, Milam AH, Palczewski K. Molecular cloning and localization of rhodopsin kinase in the mammalian pineal. Vis Neurosci. 1997; 14:225–32.
119. Tosini G, Pozeyev N, Sakamoto K, Iuvone M. The circadian clock system in the mammalian retina. Bioessays. 2008;30:624–33.
120. Ukai H, Ueda HR. Systems biology of mammalian circadian clocks. Annu Rev Physiol. 2010;72:579–603.
121. Gachon F. Physiological function of PARb Zip circadian clock-controlled transcription factors. Ann Med. 2007;39:562–71.
122. Sukumaran S, Almon RR, Dubois DC, Jusko WJ. Circadian rhythms in gene expression: relationship to physiology, disease, drug disposition and drug action. Adv Drug Deliv Rev. 2010;62:904–17.

123. Li J-D, Hu W-P, Zhou Q-Y. The circadian output signals from the suprachiasmatic nuclei. Prog Brain Res. 2012;199:119–27.
124. Tosini G, Menaker M. Circadian rhythms in cultured mammalian retina. Science. 1996; 272:419–21.
125. Tosini G, Menaker M. Temperature compensation of retinal circadian oscillation in wild-type and tau mutant hamsters. Neuroreport. 1998;9:1001–5.
126. Paul KN, Saafir TB, Tosini G. The role of retinal photoreceptors in the regulation of circadian rhythms. Rev Endocr Metab Disord. 2009;10:271–8.
127. Sakamoto K, Liu C, Tosini G. Circadian rhythms in the retina of rats with photoreceptor degeneration. J Neurochem. 2004;90:1019–24.
128. Sakamoto K, Lieu C, Kasamatsu M, Iuvone PM, Tosini G. Intraocular injection of kainic acid does not abolish the circadian rhythm of Aanat mRNA in the rat photoreceptors. Mol Vis. 2006;12:117–24.
129. Ruan G-X, Zhang D-Q, Zhou T, Yamazaki S, McMahon G. Circadian organization of the mammalian retina. Proc Natl Acad Sci. 2006;103:9703–8.
130. Liu X, Zhang Z, Ribelayga CP. Heterogeneous expression of the core circadian clock proteins among neuronal cell types in mouse retina. PLoS One. 2012;7:e50602.
131. Yamamoto T, Nakahata Y, Soma H, Akashi M, Mamine T, Takumi T. Transcriptional oscillation of canonical clock genes in mouse peripheral tissues. BMC Mol Biol. 2004;5:18.
132. Delaunay F, Laudet V. Circadian clock and microarrays: mammalian genome gets rhythm. Trends Genet. 2002;18:595–7.
133. Hafezi F, Steinbach JP, Marti A, Wang ZQ, Wagner EF, Aguzzi A, Reme CE. The absence of c-fos prevents light-induced apoptotic cell death of photoreceptors in retinal degeneration in vivo. Nat Med. 1997;3:346–9.
134. Wenzel A, Grimm C, Marti A, Kueng-Hitz N, Hafezi F, Niemeyer G, Reme CE. C-fos controls the "private pathway" of light induced apoptosis of retinal photoreceptors. J Neurosci. 2000;20:81–8.
135. Grimm C, Wenzel A, Stanescu D, Samardzija M, Hotop S, Groszer M, Naash M, et al. Constitutive overexpression of human erythropoietin protects the mouse retina against induced but not inherited retinal degeneration. J Neurosci. 2004;24:5651–8.
136. Hafezi F, Grimm C, Wenzel A, Abegg M, Yaniv M, Reme CE. Retinal photoreceptors are apoptosis-competent in the absence of Jun D/AP-1. Cell Death Differ. 1999;6:934–6.
137. Marti A, Hafezi F, Lansel N, Hegi ME, Wenzel A, Grimm C, Niemeyer G, Reme C. Light-induced cell death of retinal photoreceptors in the absence of p53. Invest Ophthalmol Vis Sci. 1998;39:846–9.
138. Hafezi F, Abegg M, Grimm C, Wenzel A, Munz K, Sturmer J, Farber DB, Reme CE. Retinal degeneration in the rd mouse in the absence of c-fos. Invest Ophthalmol Vis Sci. 1998; 39:2239–44.
139. Mirsky HP, Liu AC, Welsh DK, Kay SA, Doyle FJ. A model of the cell-autonomous mammalian circadian clock. Proc Natl Acad Sci U S A. 2009;106:11107–12.
140. Du P, Kibbe WA, Lin SM. Lumi: a pipeline for processing Illumina microarrays. Bioinformatics. 2008;24:1547–8.
141. Smyth GK. Limma: linear models for microarray data. In: Gentleman R, Carey V, Dudoit S, Irizarry R, Huber W, editors. Bioinformatics and computational biology solutions using R and bioconductor. New York: Springer; 2005. p. 397–420.
142. Lin SM, Du P, Huber W, Kibbe WA. Model-based variance-stabilizing transformation for Illumina microarray data. Nucleic Acids Res. 2008;36:e11.

Chapter 9
Circadian Rhythms and Vision in Zebrafish

Farida Emran and John E. Dowling

Abstract Zebrafish (Danio rerio) offer many advantages for studying vertebrate vision and circadian rhythmicity. They are amenable to high-throughput genetic and behavioral experiments, and the zebrafish retina is similar in structure and function to other vertebrates including mammals. Our goals here are to review some of the advances in our understanding of retinal circadian rhythm phenomena in zebrafish. We emphasize morphological and physiological rhythms, as well as circadian-regulated gene expression of the retina.

Keywords Zebrafish • Circadian rhythms • Vision • Retina • CLOCK genes • Photoreceptors • Horizontal cells • Electrical coupling • Dopamine • Synaptic ribbons

Abbreviations

6-OHDA	6-hydroxydopamine
ATP	Adenosine triphosphate
cAMP	Cyclic adenosine monophosphate
clk^{dg3}	*clock1*
Cry	Cryptochrome
Cx35	Connexin 35

F. Emran, Ph.D. (✉)
Center for Research in Neuroscience, Montreal General Hospital, McGill University, 1650 Cedar Avenue, Montreal, Canada H35-1A4
e-mail: emran@gmail.com

J.E. Dowling, Ph.D.
Department of Molecular and Cellular Biology, Harvard University, Cambridge, MA 02138, USA
e-mail: dowling@mcb.harvard.edu

G. Tosini et al. (eds.), *The Retina and Circadian Rhythms*, Springer Series in Vision Research 1, DOI 10.1007/978-1-4614-9613-7_9,

dpf	Days postfertilization
ERG	Electroretinogram
hpf	Hours postfertilization
IRBP	Interphotoreceptor retinoid binding protein
lag^{dg2}	*lager and lime*
LC opsin	Long-wavelength opsin
MAPK	Mitogen-activated kinases
mPer	Mouse Period
mCry	Mouse Cryptochrome
nbb	*night blindness b*
OKR	Optokinetic reflex
Per	Period
PKA	Protein kinase A
RPE	Retinal pigment epithelium
VMR	Visual-motor response

9.1 Introduction

Zebrafish (Danio rerio) offer many advantages for studying vertebrate vision and circadian rhythmicity. They are amenable to high-throughput genetic and behavioral experiments, and the zebrafish retina is similar in structure and function to other vertebrates including mammals. Zebrafish have excellent color vision with red-, green-, blue-, and ultraviolet-sensitive cones as well as abundant rods. Embryonic development is rapid and in many respects synchronous within populations of embryos [1]: eggs are laid and fertilized at dawn, and by 24-h postfertilization (hpf), all the major organ systems have formed. Hatching takes place on the third day, and larvae begin coordinated swimming and feeding by 4–5 days of age. Key events in retinal differentiation occur at 48–50 hpf; photoreceptors first begin to express photoreceptor-specific markers including the expression of visual pigments. By 60 hpf outer segments and the adult cone mosaic have formed [2]. At this stage, zebrafish also exhibit light responses, and by day 4, the retina is functional [3].

One of the pioneers in promoting zebrafish as a useful model system for the genetic analysis of vertebrate circadian clock mechanisms was Gregory Cahill. He initially investigated the development of melatonin rhythmicity, one of the most studied manifestations of the circadian clock, in the embryonic zebrafish [4]. These studies revealed that a circadian oscillator that regulates melatonin synthesis becomes functional and responsive to light between 20 and 26 hpf [5]. Although both the retina and pineal contain circadian oscillators that drive rhythms in melatonin synthesis, as well as phototransduction mechanisms for entrainment of the oscillator, only the pineal photoreceptors have begun to differentiate at 20–26 hpf, suggesting that the first circadian melatonin rhythms are of pineal origin [4, 6]. Studies on cultured zebrafish pineals and retinas have shown that the pineal photoreceptors are capable of maintaining a robust circadian rhythm of melatonin release with precise timing in constant darkness for 7 days. However, the retinal melatonin rhythm dampens out after a few cycles in constant darkness, indicating that retinal melatonin synthesis is regulated by a damped circadian oscillator [4]. This suggests that cellular signals from the pineal gland, which has been shown to drive rhythms of melatonin independent of any neuronal input or other master clock structure [7], may be essential for maintaining circadian rhythmicity in the retina [8].

9.2 Morphological Rhythms

9.2.1 Photoreceptor Disc Shedding

A fundamental characteristic of photoreceptor cells in all vertebrates is the continuous renewal of their outer segments. The basic mechanism of this process consists of formation of new membrane discs at the base of the outer segments and the shedding of old discs from their tips [9, 10]. The adjacent retinal pigment epithelial cells, which envelop the ends of the outer segments, ingest and degrade the detached membranes [9]. The shedding component of this renewal process follows a daily rhythm [11–13]. Groups of older rod membranes, assayed by an increase in the number of phagosomes in the retinal pigment epithelium (RPE), are shed early in the morning, shortly after light onset. Cone shedding also occurs in the morning in some species, but in others, cone shedding occurs after light offset. Disc shedding has not been studied in zebrafish, but studies in goldfish suggest that rods shed their membranes early in the light period, whereas cones shed membranes early in the dark period [14].

9.2.2 Retinomotor Movements

Since most fish lack strong pupillary reflexes to control the amount of light entering the retina, they rely on other mechanisms to adapt to changes in lighting conditions. Retinomotor movement is one such a mechanism, and it evolved early in the evolution of vertebrates. In fish and some lower vertebrates, photoreceptor inner segments elongate and contract according to changes in ambient lighting and the circadian cycle. Pigment granules in the RPE, found in fingerlike processes that surround photoreceptor outer segments, also disperse and aggregate in response to light and dark. These morphological movements in the photoreceptor layer are collectively known as retinomotor movements (see Fig. 9.1). Cone and rods display a reverse pattern of movements [15–18]. During the day or light adaptation, cones contract, rods elongate, and the pigment granules migrate to the apical part of the RPE and into the processes that surround the cone outer segments. This process

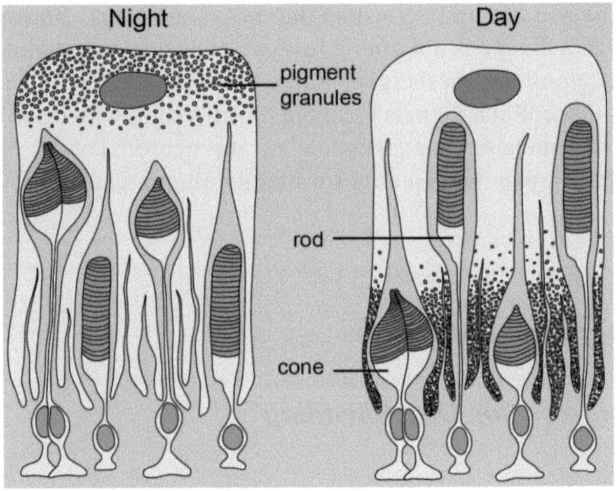

Fig. 9.1 Schematic drawing of retinomotor movements of cone and rod photoreceptors and pigment granules in the retinal pigment epithelium (RPE). Cone and rods display a reverse pattern of movements. During the day, cones contract, rods elongate, and the pigment granules migrate to the apical part of the RPE. This process allows the cone outer segments to be first in line for light absorption and shades the rod outer segments from excessive light exposure. In contrast, during the night or dark adaptation, cones elongate, rods contract, and pigment granules migrate to condense in the basal part of the RPE. This arrangement, in turn, leads to optimal light exposure for rod outer segments

allows the cone outer segments to be first in line for light absorption and shades the rod outer segments from excessive light exposure. In contrast, during the night or dark adaptation, cones elongate, rods contract, and pigment granules migrate to condense in the basal part of the RPE. This arrangement, in turn, leads to optimal light exposure for rod outer segments [19]. Elongation and contraction are mediated in the myoid region of the inner segments between the nucleus and the inner–outer segment boundary of the photoreceptor and are mediated by contractile mechanisms of microtubules and actin.

How these different types of movements (rods, cones, and pigment granules) are regulated by endogenous circadian oscillators and by light-dependent mechanisms vary among teleost species [16–18, 20]. In zebrafish, retinomotor movements follow a clear diurnal rhythm under normal light–dark cycles [21]. Rod inner segments elongate fully within 1 h after light onset, and all cones elongate maximally within the first hour after light offset. Both long single and double cones (blue- and red-, green-sensitive cones, respectively) remain elongated throughout the night until light onset, whereas short single (UV-sensitive) cones begin to contract during the last half of the night. Under continuous light conditions, retinomotor movements generally do not differ from those observed during the objective day, suggesting that light exposure overrides the endogenous circadian cycle. Under constant darkness, retinomotor movements of RPE pigment granules, long single and double cones persist in the absence of light, suggesting that these types of movements are regulated by an internal circadian oscillator [21]. Also, double cones migrate quickly with the migration fully completed in 10–20 min in continuous darkness, whereas RPE pigment granules migrate relatively slowly (1 h) [22].

The circadian morphological changes in the retina are induced by the neuromodulators melatonin and dopamine. Both neuromodulators are synthesized and released rhythmically, and the interplay between them results in two rhythms in counterphase: melatonin release is high during the night, and dopamine release is high during the day [23–26]. Melatonin acts to suppress dopamine release, whereas dopamine acts to inhibit melatonin production in photoreceptor cells [27]. In the fish retina, dopamine is released by interplexiform cells, which receive synaptic input from amacrine and bipolar cells in the inner plexiform layer and make synaptic contacts onto horizontal and bipolar cells in the outer plexiform layer [28, 29]. Dopamine receptors are G-protein-coupled receptors and are grouped into two families, D_1- and D_2-like receptors. The main functional difference between the two groups is based on their effects on cyclic AMP (cAMP) production [27]. Upon activation, the D_1 receptor family stimulates adenylate cyclase leading to an increase in cAMP, whereas the D_2 family inhibits adenylate cyclase leading to a decrease in cAMP production.

Myoid motility of the inner segment of photoreceptors is triggered by cAMP or calcium [30]. Earlier studies on teleost fish, such as the green sunfish, suggested that an increase of cAMP in exposed retinas or isolated cones induces cones to elongate [31, 32], whereas an increase in dopamine induces cones to contract [33, 34]. Thus, dopamine was proposed to mimic the effect of light. Studies on the blue acara, however, indicate that dopamine-depleted retinas (treated with 6-hydroxydopamine to selectively destroy dopaminergic cells) show no significant changes in rod retinomotor movements [35]. Similarly, studies on another fish, Aequidens pulcher, indicate that intraocular injections of dopamine have no significant effect on triggering retinomotor movements of the cones or RPE [17]. In *Xenopus laevis*, melatonin has been shown to mimic darkness by causing cone elongation under photic conditions, but this effect is blocked by an increase in exogenous dopamine concentration [36]. Thus, dopaminergic mechanisms appear to be involved in controlling the light-induced retinomotor migration in many animals; however, it is likely that additional control mechanisms are also involved.

9.2.3 Photoreceptor Synaptic Ribbons

After the discovery of retinomotor responses in fish, it was speculated that other structural changes seen in rods and cones might correlate with changes in their activity. Ultrastructural studies of cone photoreceptors in the cichlid fish revealed that the length of synaptic ribbons correlated with different degrees of light and dark adaptation [37]. Synaptic ribbons are electron-dense bars that are anchored at the

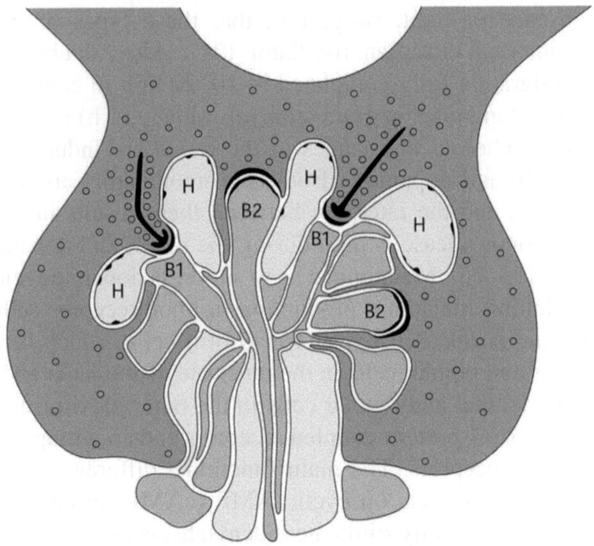

Fig. 9.2 Diagram of photoreceptor terminals in zebrafish larvae at 5 dpf. Bipolar and horizontal cell processes invaginate into the pedicle in a tight bundle to make two types of junctions: ribbon synapses (B1) and flat contacts (B2). Ribbon synapses are made onto presumed ON bipolar (B1) and horizontal (H) cell dendrites, whereas flat contacts are made onto OFF bipolar cells (B2). Synaptic vesicles surround the synaptic ribbon. Flat contacts are found between the ribbon synapses and have dense cytoplasmic material on both sides of the junction. This drawing was modified from [123] Emran et al. 2007. *PNAS*, 104(48):19126–31

synaptic terminal base by a structure called the arciform density to form a special type of synapse called the ribbon synapse [38, 39]. In the fish retina, bipolar and horizontal cell processes extend into a single invagination in rod and cone terminals in a tight bundle to make two types of junctions: ribbon synapses and flat contacts, also known as basal junctions (see Fig. 9.2). Ribbon synapses are made onto ON bipolar cell dendrites, whereas flat contacts are made onto OFF bipolar cell dendrites and are found between the ribbon synapses in fish. Cone pedicles typically contain several synaptic ribbons, whereas rod terminals usually contain just one or two. In many fish, synaptic ribbons completely disassemble at night and reform in the morning [37, 40]. Indeed, electron microscopic studies on larval zebrafish retinas demonstrated that the plasticity of synaptic ribbons in cone pedicles is a circadian phenomenon [41]. That is, whereas synaptic ribbons assemble and are more numerous during the day and disaggregate at night, this dynamic process of assembly and disassembly persists in constant darkness [41, 42]. How the ribbons are degraded is not well understood, but small round densities observed in the terminals at night suggest that ribbons are formed from an aggregation of subunits [41].

9.2.4 *Spinule Formation in Horizontal Cells*

In fish, such as the Aequidens pulcher, structural changes of synaptic ribbons on the presynaptic side in cone pedicles are correlated with changes on the postsynaptic side in horizontal cells. As mentioned above, horizontal and bipolar cells send their dendritic processes into the photoreceptor terminals to form synapses. In the fish retina, two distinct types of horizontal cells are observed, either cone or rod specific [43–46]. Dendrites of cone horizontal cells display numerous fingerlike projections, called spinules in the day. These spinules undergo rhythmic morphological changes: they extend during the day, but retract at dusk. Formation and disappearance of spinules takes about an hour and is circadian related [40]. Studies in goldfish revealed that both light-evoked and circadian-regulated cone spinule formation is dependent on dopamine [47]. When fish were treated with 6-hydroxydopamine (6-OHDA) to selectively destroy the dopaminergic neurons in the retina, formation of spinules was severely suppressed in the light. Thus, spinule formation in cone horizontal cells appears to be regulated by rhythmic exogenous dopamine levels in the retina [17]. As yet, spinule formation and disassembly have not been studied in zebrafish.

9.3 Physiological Rhythms

9.3.1 *Behavioral Rhythms in Zebrafish*

Another reason why zebrafish is a good model system to study not only vision but also circadian rhythms is their amenability for behavioral assays. For example, Cahill and colleagues examined circadian regulation of gross locomotor activity in adult and larval zebrafish [48, 49]. These authors automated the continuous recording of swimming movements in individual fish by using either infrared motion detectors for adult fish or video image analysis for larval fish. Locomotor activity rhythms in adult zebrafish are somewhat variable with about 70 % of fish showing strong activity patterns, with a characteristic diurnal activity profile; adult fish are more active during the light phase of the cycle than the dark cycle. Under constant conditions, such as continuous dark or light, circadian rhythmicity in locomotor activity continues for up to 10 days. Together these data suggest that zebrafish locomotor activity is regulated by a circadian clock [48].

The locomotor activity of larval zebrafish (5–18 days) is even more consistent, with over 95 % of larvae exhibiting a clear circadian rhythm. Similar to adult fish, larval zebrafish are more active during the subjective day and display a periodicity of about 25 h under dark-adapted conditions [49]. Cahill and colleagues used these behavioral rhythms as an assay to screen for mutants that display altered gross locomotor activity patterns. One such mutant is the *lager and lime* (*lag*^{dg2}) mutant, and its circadian periodicity is shortened by 0.7 h in heterozygotes and by 1.3 h in homozygotes [50]. The *clock1* (*clk*^{dg3}) mutant, which has a mutation in the PAS domain of the Clock1 protein, also displays a shortened biological clock [51].

Behavioral assays designed to screen for deficits in the visual system of larval zebrafish include the optokinetic reflex (OKR) and the visual-motor response (VMR) assays [42, 52]. The OKR is a stereotyped eye movement, exhibited by all vertebrates in response to objects in motion when the head remains stationary. When zebrafish larvae are placed inside a rotating drum with alternating black and white bars moving in either direction, the animals move their eyes with a smooth pursuit until the bar moves out of the field of vision at which point their eyes rapidly move back to the starting position. The latter eye movement is called a saccade. In humans this reflex can be tested in newborns but in larval zebrafish as early as 4-day postfertilization (dpf). The use of this behavioral assay in large-scale genetic screens has led to the isolation of many mutant strains with specific defects in the visual pathway [52]. An especially interesting example is the belladonna mutation, where an axonal misrouting effect at the optic chiasm leads to a reversed OKR [53].

The VMR is a behavioral assay that is based on the ON and OFF responses of larval zebrafish in response to changes in illumination. Similar to the locomotor behavioral assay used to study circadian rhythms, zebrafish larvae are placed individually in 96 well plates and the motion of each fish recorded. However, instead of measuring gross locomotor activity, this assay measures the rate at which larval fish moves in each well with the use of a high-speed camera. The ON responses, i.e., the change in motor activity in response to lights on, consist of a sharp and robust increase of motor activity for the first 20–30 s after the start of the light stimulus. After the initial ON response, the larvae reduce their activity levels dramatically and show a partial freeze response, slowly returning to baseline activity levels within a few minutes. The OFF response, which is elicited by switching the lights off, is also characterized by a sharp and robust increase of motor activity that follows immediately after the illumination change. After the initial sharp OFF response, the animals reduce their activity to an intermediate level, and it typically takes 10–15 min to return to baseline levels [42]. Thus, the VMR ON and OFF responses consist of sharp increases in motor activity immediately after light increments or decrements. There are also small, slow changes in activity that occur, especially after light extinction, that do not appear retinal related. As expected baseline activity levels are higher during the subjective day than during the subjective night [54].

9.3.2 Visual Function and ERG Light Sensitivity in Adult Zebrafish

Do circadian rhythms influence visual function in adult zebrafish? To address this question, a behavioral assay based on the escape response of fish to avoid predators was developed [55]. Normally adult zebrafish swim around slowly when confined in a container that contains a post in its middle. However, when a threatening object

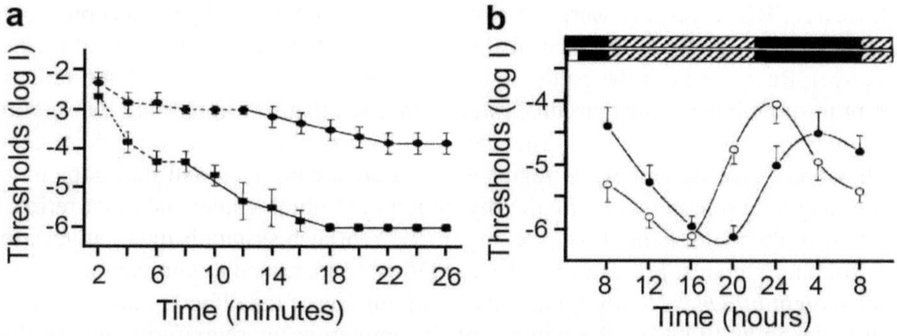

Fig. 9.3 Visual sensitivity of adult zebrafish. (**a**) Dark adaptation curves determined at 4 a.m. (*circles*) and 6 p.m. (*squares*) of cone and rod systems in adult zebrafish maintained in normal LD cycle. *Dashed lines* represent cone function; *continuous lines* represent rod function. Note that both cone and rod thresholds were significantly higher at 4 a.m. as compared to 6 p.m. (**b**) The variation of visual sensitivity during a 24-h period of constant darkness (DD) in control (*closed circles*) and test animals (*open circles*). In the control animals, thresholds were highest in the early morning hours (~4 a.m.) and lowest in the late afternoon hours (~6 p.m.). In the test animals, the rhythm was shifted forward by about 4 h by an early morning 1-h pulse; thresholds were highest at midnight and lowest at about 2 p.m. The *black bars* at the top of the panel represent the subjective night, and the *hatched bars* represent the subjective day. The small *white bar indicates* the 1-h light pulse given between 4 and 5 a.m. to the test animals. Data represents the means ± S.D. Modified from [55] Li and Dowling, 1998. *Visual Neuroscience*, 15(5):851–7. Copyright © 1998 Cambridge University Press

comes into view (i.e., a black segment on a rotating drum), the fish immediately turn and rapidly swim away to hide behind the post. Using this vision-mediated reflex of zebrafish, cone and rod light sensitivity was determined under both cyclic and constant lighting conditions [55]. Under a normal light–dark cycle, the visual system is strongly influenced by the circadian clock such that it is most sensitive in the late afternoon and least sensitive in the early morning hours. Over the course of 24 h, the threshold of light intensity shone on the drum required to elicit a positive response to the visual stimulus fluctuated by about 1.4 log units for cones and about 2.2 log units for rods (see Fig. 9.3a). This rhythm persisted in animals maintained in constant darkness for several days and could be phase shifted by light pulses (see Fig. 9.3b). Also, when fish are kept under constant darkness, the electroretinogram (ERG), a field potential generated primarily in the outer retina by photoreceptor and bipolar cells, shows similar changes in visual sensitivity as a function of time. However, the ERG visual sensitivity changes are about one log unit smaller than the visual changes measured behaviorally. Hence, not all aspects of visual sensitivity rhythmicity are reflected by changes in the outer retina, suggesting circadian regulation of the inner retina and/or higher visual centers is involved as well [55]. To study how dopamine affects visual sensitivity, Li and Dowling depleted the zebrafish retina of dopaminergic neurons (i.e., the interplexiform cells) by 6-OHDA treatment. Under dark-adapted conditions, only cone system function was observed, indicating that dopamine depletion leads to a loss of rod function as measured behaviorally.

However, ERG recordings were not affected, suggesting that rod photoreceptors are fully functional and transmit electrical signals to ON bipolar cells (as judged by the b-wave). Recordings of the ganglion cell discharges in dopamine-depleted retinas point to some inner retinal circuit dysfunction, such that rod signals are blocked in the inner plexiform layer [56]. Together, the results described above indicate that behavioral visual sensitivity is regulated by a circadian clock and that dopamine plays a crucial role in regulating the rhythmicity in both the inner and outer retina.

One of the major benefits of working on the zebrafish system is the relative ease of isolating mutants. For example, Li and Dowling identified a dominant mutation called *night blindness b* (*nbb*) that causes a disruption of the olfactoretinal centrifugal pathway and reduces the number of dopaminergic interplexiform cells in the retina [57]. After prolonged dark adaptation, this mutant displays fluctuations in its visual sensitivity by 2–3 log units from day to day. In contrast, wild-type fish maintain a constant level of visual sensitivity when measured at the same time on different days. ERG recordings indicate that outer retinal function is normal, whereas ganglion cell discharges suggest that the visual deficit relates to abnormalities in the inner retina. Also, ablation of the olfactory epithelium and olfactory bulb in wild-type fish phenocopy the visual abnormalities seen in the nbb mutant [57]. Thus, it is fair to conclude that dopaminergic interplexiform cells and the olfactoretinal efferents that innervate them are crucial in modulating circadian rhythmicity of the visual system.

9.3.3 Visual Responses and Light Sensitivity in Larval Zebrafish

To determine whether vision in larval zebrafish is regulated by a circadian clock, Emran et al. measured the OKR and VMR responses of 5–6-day-old zebrafish larvae at several time points during the day and at night [42]. During the day, fish respond normally in these behavioral tests, whereas during the night, fish display neither OKR nor VMR responses (see Fig. 9.4a). ERG responses recorded during the day and at night substantiated these results; responses were normal in amplitude throughout the day but were almost absent at night, suggesting that vision is severely reduced at night (see Fig. 9.4b). This phenomenon is largely circadian driven as fish show similar dramatic changes in visual responsiveness when raised in constant darkness from 4 to 7 dpf. However, light exposure at night partially restores the responses. Two mechanisms by which visual responsiveness might be decreased were examined: (1) photoreceptor outer segment activity was reduced at night as judged by a decrease in the ERG a-wave, and (2) synaptic ribbons in cone pedicles disassembled at night [42]. It should be pointed out that in this study mainly cone-mediated vision was examined during the day and at night, as rod responses are minimal in 5–6-day-old larvae. It is possible that rod vision is still present at night

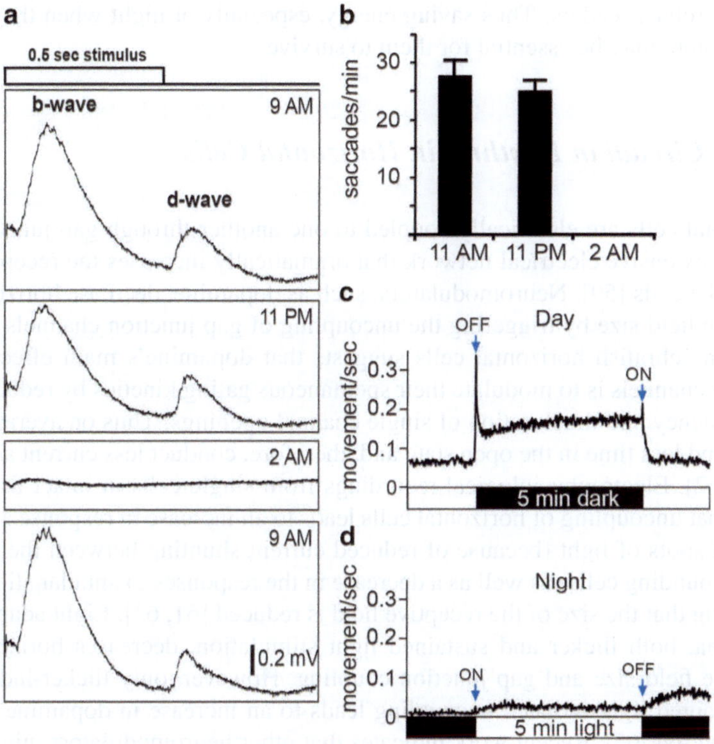

Fig. 9.4 Larval zebrafish exhibit a day/night cycle of visual responsiveness sensitivity. (**a**) Representative ERG recordings to a 0.5 s white light stimulus are shown at four different time points: at 9:00 a.m., when the ambient lights are turned ON; 11:00 p.m., shortly before the lights are turned OFF; 2:00 a.m., after the lights have been OFF for 3 h; and at 9:00 a.m. the next morning, after the lights are turned ON again. Note the large reduction of ERG amplitudes at 2:00 a.m. (**b**) Larval zebrafish exhibit a robust OKR at 11:00 a.m. and 11:00 p.m. (shortly before the lights are turned OFF at night) but not at 2 a.m. (3 h after the lights were turned OFF). (**c**) Larval zebrafish display ON and OFF visual-motor responses (VMR) during the day but fail to exhibit transient startle responses at night (each trace represents an average of 240 responses from 80 individual fish). During the daytime, motor output increased in response to a 5-min light off stimulus with prominent OFF and ON responses (first stimulus was introduced at 2:00 p.m.). (**d**) At night, the VMR responses to a 5-min light stimulus were abolished (first stimulus was introduced at 2:00 a.m.). Modified from [42] Emran et al., 2010. *PNAS*, 107(13):6034–9

and is not under the influence of the circadian clock, but rod vision does not become prominent until zebrafish are 2–3 weeks old.

Why do larval zebrafish turn off their retinas at night? It may be to save energy. The retina is extraordinarily active metabolically at night, because the photoreceptors are maximally active at night. A single dark-adapted mouse rod consumes 10^8 ATP molecules per second, and retinas contain thousands to millions of photoreceptors [58]. At 5–6 days of age, zebrafish have used up much of their yolk and are not yet very robust feeders. Thus saving energy, especially at night when they are not using vision, may be essential for them to survive.

9.3.4 Circadian Rhythms in Horizontal Cells

Horizontal cells are electrically coupled to one another through gap junctions and form an extensive electrical network that dramatically increases the receptive field size of the cells [59]. Neuromodulators, such as dopamine, decrease horizontal cell receptive field size by triggering the uncoupling of gap junction channels [60, 61]. Work on zebrafish horizontal cells suggests that dopamine's main effect on gap junction channels is to modulate their spontaneous gating kinetics by reducing both the frequency and the duration of single channel openings. Thus on average, channels spend less time in the open state and, therefore, conduct less current as a population [62]. Electrophysiological recordings from single cells in intact fish retinas reveal that uncoupling of horizontal cells leads to an increase in response amplitude to small spots of light (because of reduced current shunting between the recorded and surrounding cells) as well as a decrease in the responses to annular stimulation, indicating that the size of the receptive field is reduced [61, 63]. Light adaptation of the retina, both flicker and sustained light stimulation, decreases horizontal cell receptive field size and gap junction coupling. However, only flicker-induced but not sustained light-induced uncoupling leads to an increase in dopamine levels in the retina [64, 65]. Recent work indicates that other neuromodulators, nitric oxide, retinoic acid, and zinc are also involved in the regulation of horizontal cell receptive field size [66–68].

The receptive field size of fish horizontal cells is altered by activation of dopaminergic receptors. As mentioned earlier, D_1 and D_2 are G-protein-coupled receptors that are classified based on their effects on cAMP production. Recordings from pairs of perch and catfish horizontal cells in culture showed that the electrical conductance across these cells is decreased by exogenous application of dopamine or by a D_1 agonist. This decrease in cell to cell communication is also observed when cAMP or agents such as forskolin, which increase the activity of adenylyl cyclase, are applied. Application of protein kinase A (PKA) blockers reverses the action of forskolin suggesting that dopamine uncouples horizontal cells by raising cyclic AMP levels and thereby activating PKA [61, 69]. Thus, dopamine activates D_1 receptors on fish horizontal cells and increases cAMP production and PKA activity, which subsequently leads to a decrease in gap junction coupling and receptive field size. Further studies in goldfish showed that specific activation of D_2 receptors by an agonist, quinpirole, has the opposite effect and increases the electrical coupling between horizontal cells, as indicated by a decrease in responses to centered spots or slits of light [70]. However, D_2 receptors are not found on horizontal cells but on dopaminergic interplexiform cells. Activation of these D_2 receptors decreases dopamine release from the interplexiform cells; hence, the effect of D_2 receptor activation on horizontal cell coupling is indirect, mediated by a decrease of dopamine levels in the retina [70].

The effects of light and dark on horizontal cell coupling are murky and even disputed. Several reports have claimed that when a retina has been in the dark for a prolonged period, the horizontal cells are physiologically uncoupled [63, 71–73], and impressive histological evidence has been provided in support of this result. These data showed that Lucifer yellow, a gap junctional permeable dye, is confined to the cell in which the dye is injected and a few surrounding cells when retinas are thoroughly dark-adapted. As dim light illuminates the retina, however, the cells become coupled, shown both physiologically and histologically, but then with bright illumination of the retina, the cells uncouple again [73]. This has been termed triphasic adaptation and has been observed also in horizontal and amacrine cells of the rabbit retina [74, 75]. Others, on the other hand, claim that in dark-adapted retinas, the horizontal cells are extensively coupled and when light is presented to the retina, the cells uncouple [76, 77]. In other words, these later authors do not observe an uncoupling of the horizontal cells when the retinas are dark-adapted for prolonged periods of time.

There is general agreement that dopamine uncouples horizontal cells, and so one might suppose dopamine levels are high in retinas dark-adapted for a prolonged period, but that their levels decrease in dim to moderate light, but then increase again in bright light. Most (but not all) dopamine release studies [65] have claimed that dopamine release is low in the dark (although there is some ongoing dark release) and that dopamine levels increase substantially in the light. It is important to remember however that uncoupling of horizontal cells in fish retinas can be caused by other neuroactive substances including nitric oxide and retinoic acid, and so which agent is responsible for the uncoupling in prolonged darkness (assuming it does happen) is not known. Clearly more work is needed to sort out the effects of prolonged darkness, dim and bright light on horizontal cell coupling, and determining the role of modulators other than dopamine in regulating the coupling.

How does dopamine affect horizontal cell coupling? Studies in goldfish revealed that retinal dopamine release is under the control of the circadian clock, leading to increased dopamine production during the subjective day [78]. However, no effect on horizontal cell coupling was observed during subjective day when animals were kept continuously in the dark [76, 77]. Mangel and colleagues propose that in the goldfish retina two different dopamine systems are present. One system is controlled by an endogenous clock that activates low-threshold D_2-like receptors during the subjective day—i.e., circadian-dependent release. The second is controlled by light and involves the activation of higher-threshold D_1-like receptors. As mentioned earlier, light increases dopamine release to a greater extent than does the circadian clock. Hence, sufficient dopamine is released to activate D_2-like receptors during the subjective day, but not D_1-like receptors that are involved in horizontal cell uncoupling in the retina [76, 77]. Thus, horizontal cell coupling is not circadian regulated, at least by dopamine.

9.3.5 Circadian Control of Photoreceptor Electrical Coupling

Cone horizontal cells display a substantial increase in light sensitivity, as well as a spectral sensitivity shift from red sensitivity (i.e., cones) to green sensitivity (i.e., rods) under dark-adapted conditions [78–80]. These results suggest that cone horizontal cells receive input from rods when sufficiently dark-adapted since cone horizontal cells receive direct synaptic input only from cones. Consequently, rod input to cone horizontal cells must arrive via rod–cone synapses. Further, the light responses of dark-adapted cone horizontal cells vary in waveform, light sensitivity, and spectral sensitivity depending on the time of day [79, 81], suggesting circadian regulation of these junctions. Cone input to cone horizontal cells predominates in the day and rod input predominates at night, as indicated by the increase in light sensitivity as well as by the spectral sensitivity shift from red to green. Application of exogenous dopamine also induces a shift from rod to cone domination under constant darkness [78, 80, 82]. During the subjective day in animals maintained in the dark, endogenous dopamine levels increase and presumably activate D_2 receptors on both rods and cones. Activation of the D_2 mechanism decreases intracellular cAMP and PKA activity in photoreceptor cells, resulting in the decoupling of rods and cones, thus favoring cone input to cone horizontal cells [78].

Similar to horizontal cells, rods and cones are coupled to each other via gap junctions [83]. Using tracer coupling and recording in the goldfish retina, Ribelayga et al. measured the extent and strength of rod–cone electrical coupling during the day and night [82]. They report that the conductance of rod–cone gap junctions is regulated by a circadian clock and is increased at night. During the subjective day, activation of D_2 receptors decreases rod–cone coupling, whereas at night, rod–cone coupling is extensive and allows cones to receive light signals from rods [82]. Thus, in the goldfish retina, activation of D_2 receptors in the light decreases rod–cone coupling [78, 80, 82] and the same holds true in the rabbit [84]. It is worth pointing out, however, that in other animals, such as in the salamander retina, D_2 receptor activation or light adaptation increases rod–cone coupling [85, 86]. Similarly, rod–cone coupling is greater in the turtle retina when rods are slightly light-adapted as compared to fully dark-adapted rods [87, 88].

How are cells in the retina electrically coupled to each other? The building blocks for gap junctions are specialized proteins called connexins. Six connexins assemble into a connexon, which forms a cytoplasmic bridge by joining with another connexon on an adjacent cell. A gap junction typically consists of many connexons gathered in close proximity. There are many subtypes of connexins in the retina [89], all of which have phosphorylation sites. It is believed that the phosphorylated state of connexins modulates the electrical conductance across a junction, such that phosphorylation decreases conductance between horizontal cells, but increases it between photoreceptor cells [27, 90]. In fish, connexin 35 (homologous to connexin 36 of mammals) is found at both cone–cone and rod–cone gap junctions [91, 92]. A study in zebrafish found that phosphorylation of connexin 35 (Cx35) at serine residues 110 and 276 directly correlate with changes in photoreceptor coupling. Tracer coupling of Neurobiotin is reduced during the day, when phosphorylation is low, and is extensive (especially in the cone network) at night when phosphorylation is high. Phosphorylation of Cx35 is regulated by PKA, as pharmacological manipulation of PKA activity mimics both the daytime and nighttime phosphorylation state and coupling of photoreceptors. Somewhat surprising is the observation that not only is rod–cone coupling enhanced at night, but the cone network is extensively coupled as well [92]. What function enhanced cone–cone coupling serves at night, when the cones are outside their operating range, is not clear.

9.4 Clock Genetics in Zebrafish

9.4.1 Light-Sensitive Peripheral Oscillators

Following the isolation of the first clock gene in mouse, CLOCK [93], the zebrafish homolog of this gene was discovered [6]. As expected, Clock is expressed in the brain, the eye, and the pineal gland in zebrafish. However, unlike its mammalian counterpart, zebrafish Clock displays robust circadian oscillation in gene expression in these defined pacemaker structures as well as in other tissues such as heart and kidney [6]. Moreover, cultured tissues from the heart and kidney maintain rhythmic expression of the *clock* gene under both continuous dark and light–dark conditions. Perhaps the most striking observation is that these cultured organs are themselves light sensitive and can be entrained by light/dark cycles. When the light–dark cycle is reversed, the cultured organs show a reverse cycle of Clock rhythm from day 2 onwards [94]. Cell lines derived from zebrafish embryos, PAC-2 and Z3, also contain an independent, light-entrainable circadian oscillator, suggesting that many cell types in zebrafish contain their own endogenous peripheral clocks [95]. The discovery of light-sensitive peripheral tissue circadian oscillators was first demonstrated in insects [96, 97]. For example, in Drosophila explanted tissues, including head, thorax, and abdominal tissues, show rhythmic expression of another clock gene *period* (*per*), are directly responsive to light [97], and can be entrained independently of the central pacemakers in the brain [98, 99]. This phenomenon has also been validated in mammals. Using a PER2::LUCIFERASE fusion protein reporter in mice, Takahashi and colleagues demonstrated that peripheral tissues express self-sustained circadian oscillations for more than 20 cycles in culture [100]. In mice, however, light cannot reset peripheral oscillators directly as is the case for Drosophila and zebrafish.

9.4.2 Clock Genes

The elucidation of the core circadian clock mechanism occurred first in the fruit fly, Drosophila, whereas Takahashi and colleagues, building upon the invertebrate work, showed that a similar mechanism works in vertebrates [101–103].

The prevailing model for the core molecular mechanism is a transcription–translation feedback loop that cycles with a period of about 24 h. In this feedback loop, a transcription factor, formed by heterodimerization of CLOCK and BMAL1, drives the expression of three Period genes (in mice, *mPer1*, *mPer2*, and *mPer3*) and two Cryptochrome genes (*mCry1* and *mCry2*). The mPER and mCRY proteins heterodimerize to inhibit their own transcription by translocating back to the nucleus, where they bind to the Clock:Bmal1 complex [104]. Due to a late whole-genome duplication event in the teleost lineage [105], extra copies of clock genes have been discovered in the zebrafish as compared to the mouse. For example, zebrafish have six rhythmically expressed cryptochrome genes (*cry1a*, *1b*, *2a*, *2b*, *3*, and *4*) [106]. Phylogenetic analysis revealed that Cry1a, 1b, and 2b share most sequence homology to mCRY1 and can repress transcriptional activation by Clock:Bmal1 dimers. However, Cry3 and Cry4 do not inhibit Clock:Bmal1-mediated transcription. Cry3 shares some sequence similarities with *mCrys*, whereas Cry4 is most divergent from all other vertebrate CRYs and resembles Drosophila CRY. As mentioned earlier (Sect. 4.1), many tissues throughout the body of both zebrafish and Drosophila are photosensitive. But what molecules are responsible for detection of light in these peripheral tissues? Two independent studies have identified the blue-sensitive pigment Cryptochrome (CRY) to be the major contributor for entrainment of peripheral oscillators in Drosophila [107, 108]. Similar results have been reported in zebrafish. Using the light-inducible embryonic zebrafish cell line Z3, Cermakian et al. reported that blue-sensitive pigments, probably Crys, are coupled to the MAPK (mitogen-activated protein kinase) pathway to elicit light-induced transcriptional activation of clock genes [109]. This was later confirmed by Tamai et al., demonstrating that overexpression of Cry1a protein in a Per1:Luciferase zebrafish cell line acts as a strong repressor of clock function and mimics the effects of sustained light [110]. Indeed, *cry1a* shows a light-driven pattern of expression and is not rhythmic in constant conditions, whereas the other *cry* genes are predominantly clock regulated [106].

Zebrafish homologs of the three mammalian per genes have also been identified: two *per1* homologs (*per1a* and *1b*) as well as a single copy of *per2* and *per3* genes. *Per1* and *per3* genes display a similar rhythm in expression pattern. However, *per2* is stimulated by light and is not rhythmic in constant conditions [95, 111]. Similar to *cry1a*, *per2* has also been suggested to act as a photoreceptor and be involved in entrainment pathways [110, 112]. Finally, three *clock* genes (*clock1a*, *1b*, and 2) and three *bmal* genes (*bmal1a*, *1b*, and 2) have been isolated and shown to interact as heterodimers [113–115]. Generally, the expression pattern for *bmal* and *clock* genes parallels each other with a peak just after the LD transition [116].

9.4.3 Clock-Regulated Genes in the Retina

In zebrafish retinas, the early circadian genes (*clock*, *bmal*, *per*, and *cry*) are expressed in several cell types, including photoreceptor cells. Opsin expression of both rhodopsin and cone pigment has been shown to fluctuate between day and night in several vertebrates [117, 118]. Circadian rhythmicity in *long-wavelength cone (LC) opsin* mRNA levels has also been reported in zebrafish [119]. Expression levels reach a peak in the afternoon and a low in the early morning, and these fluctuations persist in constant darkness. To demonstrate that Clock is involved in regulating the circadian rhythms in photoreceptor cells, translation of clock mRNA was inhibited in zebrafish embryos [119]. This resulted in diminished circadian rhythms of *opsin* mRNA expression and also abolished the fluctuations in cAMP concentrations observed in photoreceptor cells. The authors proposed that the regulatory effect of Clock on *LC opsin* mRNA expression is mediated by cAMP-dependent cascades. That is, Clock regulates rhythmic synthesis of cAMP, which in turn regulates circadian expression of *LC opsin* mRNA. To examine whether rhodopsin gene expression also exhibits circadian fluctuation, Li and his team used a transgenic zebrafish line that expresses a short-life GFP under the transcriptional control of the rhodopsin promoter [120]. Similar to *LC opsin* expression, *rhodopsin* promoter expression in rod photoreceptor cells fluctuates rhythmically in constant darkness with an expression peak in the subjective late afternoon. However, not all rod photoreceptor cells are synchronized with each other, and in some cells *rhodopsin* promoter expression peaks in the subjective early morning, whereas in other cells it peaks in the afternoon or at night. Light synchronizes rod photoreceptor oscillations by transiently decreasing *rhodopsin* promoter expression. Application of dopamine or a D_2 receptor agonist also triggers synchronization of *rhodopsin* promoter expression in rod photoreceptor cells. In addition to light and dopamine, the circadian rhythms of *rhodopsin* gene expression may also be synchronized by central mechanisms, including the rhythmic production of melatonin by the pineal gland [120].

Two other retinal rhythms that have been studied in zebrafish include melatonin synthesis and mRNA for interphotoreceptor retinoid binding protein (Irbp). As discussed in more detail above (see Introduction), melatonin synthesis is circadian regulated and persists in retinal organ culture for a few cycles in constant darkness [4]. Irbp is synthesized and secreted by photoreceptors and mediates retinoid trafficking between photoreceptors and pigment epithelium. Expression of *irbp* mRNA is circadian regulated and is higher during the day than during the night. This rhythm persists for 2 days under constant conditions [121]. In addition, mRNA expression of this protein is regulated differentially in various photoreceptor subtypes. For example, *irbp* mRNA is expressed by all cone types and to a lesser extent by rods during the day. At night, *irbp* mRNA expression is restricted to the ultraviolet short single cones [121]. Despite the fluctuation in expression pattern of this protein at the level of photoreceptors, the total protein content in the retina remains constant throughout the day–night cycle [122]. It appears that the turnover rate of Irbp is faster during the day than during the night. Thus, the increased expression of mRNA and protein synthesis of this protein is to offset the increased turnover rate during the day to maintain constant amounts of the protein.

Acknowledgements Special thanks to Manija Emran for creating the retinomotor movement diagram and to the editors for helpful criticisms and suggestions on the chapter.

References

1. Burrill JD, Easter Jr SS. The first retinal axons and their microenvironment in zebrafish: cryptic pioneers and the pretract. J Neurosci. 1995;15(4):2935–47.
2. Larison KD, Bremiller R. Early onset of phenotype and cell patterning in the embryonic zebrafish retina. Development. 1990;109(3):567–76.
3. Easter Jr SS, Nicola GN. The development of vision in the zebrafish (Danio rerio). Dev Biol. 1996;180(2):646–63.
4. Cahill GM. Circadian regulation of melatonin production in cultured zebrafish pineal and retina. Brain Res. 1996;708(1–2):177–81.
5. Kazimi N, Cahill GM. Development of a circadian melatonin rhythm in embryonic zebrafish. Brain Res Dev Brain Res. 1999;117(1):47–52.
6. Whitmore D, Foulkes NS, Strahle U, Sassone-Corsi P. Zebrafish Clock rhythmic expression reveals independent peripheral circadian oscillators. Nat Neurosci. 1998;1(8):701–7.
7. Noche RR, Lu PN, Goldstein-Kral L, Glasgow E, Liang JO. Circadian rhythms in the pineal organ persist in zebrafish larvae that lack ventral brain. BMC Neurosci. 2011;12:7.
8. Li X, Montgomery J, Cheng W, Noh JH, Hyde DR, Li L. Pineal photoreceptor cells are required for maintaining the circadian rhythms of behavioral visual sensitivity in zebrafish. PLoS One. 2012;7(7):e40508.
9. Young RW, Bok D. Participation of the retinal pigment epithelium in the rod outer segment renewal process. J Cell Biol. 1969;42(2):392–403.
10. Young RW. Visual cells and the concept of renewal. Invest Ophthalmol Vis Sci. 1976; 15(9):700–25.
11. Besharse JC. The daily light-dark cycles and rhythmic metabolism in the photoreceptor-pigment epithelial complex. In: Osborne NN, Chader GJ, editors. Progress in retinal research. Oxford: Pergamon Press; 1982.
12. LaVail MM, Ward PA. Studies on the hormonal control of circadian outer segment disc shedding in the rat retina. Invest Ophthalmol Vis Sci. 1978;17(12):1189–93.
13. Besharse JC, Iuvone PM, Pierce ME. Regulation of rhythmic photoreceptor metabolism: a role for post-receptoral neurons. In: Osborne N, Chadar GJ, editors. Progress retinal research. Oxford: Pergamon Press; 1988. p. 21–61.
14. O'Day WT, Young RW. Rhythmic daily shedding of outer-segment membranes by visual cells in the goldfish. J Cell Biol. 1978;76(3):593–604.
15. Ali MA, Wagner HJ. Distribution and development of retinomotor responses. In: Ali MA, editor. Vision in fishes. New York: Plenum Press; 1975. p. 369–96.
16. Levinson G, Burnside B. Circadian rhythms in teleost retinomotor movement. A comparison of the effects of circadian rhythm and light condition on cone length. Invest Ophthalmol Vis Sci. 1981;20(3):294–303.
17. Douglas RH, Wagner HJ, Zaunreiter M, Behrens UD, Djamgoz MB. The effect of dopamine depletion on light-evoked and circadian retinomotor movements in the teleost retina. Vis Neurosci. 1992;9(3–4):335–43.
18. McCormack CA, Burnside B. Effects of circadian phase on cone retinomotor movements in the Midas cichlid. Exp Eye Res. 1991;52(4):431–8.
19. Ali MA. [Retinomotor response: characteristics and mechanisms]. Vision Res. 1971; 11(11): 1225–88. Les reponses retinomotrices: caracteres et mecanismes.
20. Dearry A, Barlow Jr RB. Circadian rhythms in the green sunfish retina. J Gen Physiol. 1987;89(5):745–70.

21. Menger GJ, Koke JR, Cahill GM. Diurnal and circadian retinomotor movements in zebrafish. Vis Neurosci. 2005;22(2):203–9.
22. Hodel C, Neuhauss SC, Biehlmaier O. Time course and development of light adaptation processes in the outer zebrafish retina. Anat Rec A Discov Mol Cell Evol Biol. 2006;288(6): 653–62.
23. Cahill GM, Besharse JC. Resetting the circadian clock in cultured Xenopus eyecups: regulation of retinal melatonin rhythms by light and D2 dopamine receptors. J Neurosci. 1991; 11(10):2959–71.
24. Wirz-Justice A, Da Prada M, Reme C. Circadian rhythm in rat retinal dopamine. Neurosci Lett. 1984;45(1):21–5.
25. Witkovsky P, Schutte M. The organization of dopaminergic neurons in vertebrate retinas. Vis Neurosci. 1991;7(1–2):113–24.
26. Adachi A, Nogi T, Ebihara S. Phase-relationship and mutual effects between circadian rhythms of ocular melatonin and dopamine in the pigeon. Brain Res. 1998;792(2):361–9.
27. Witkovsky P. Dopamine and retinal function. Doc Ophthalmol. 2004;108(1):17–40.
28. Dowling JE, Ehinger B. The interplexiform cell system. I. Synapses of the dopaminergic neurons of the goldfish retina. Proc R Soc Lond B Biol Sci. 1978;201(1142):7–26.
29. Yazulla S, Zucker CL. Synaptic organization of dopaminergic interplexiform cells in the goldfish retina. Vis Neurosci. 1988;1(1):13–29.
30. Burnside B, Nagle B. Retinomotor movements of photoreceptors and retinal pigment epithelium: mechanisms and regulation. In: Burnside B, editor. Progress in retinal research. Oxford: Pergamon Press; 1983.
31. Burnside B, Ackland N. Effects of circadian rhythm and cAMP on retinomotor movements in the green sunfish, Lepomis cyanellus. Invest Ophthalmol Vis Sci. 1984;25(5):539–45.
32. Hillman DW, Lin D, Burnside B. Evidence for D4 receptor regulation of retinomotor movement in isolated teleost cone inner-outer segments. J Neurochem. 1995;64(3):1326–35.
33. Dearry A, Burnside B. Dopamine inhibits forskolin- and 3-isobutyl-1-methylxanthine-induced dark-adaptive retinomotor movements in isolated teleost retinas. J Neurochem. 1985;44(6):1753–63.
34. McCormack CA, McDonnell MT. Circadian regulation of teleost retinal cone movements in vitro. J Gen Physiol. 1994;103(3):487–99.
35. Kolbinger W, Wagner D, Wagner HJ. Control of rod retinomotor movements in teleost retinae: the role of dopamine in mediating light-dependent and circadian signals. Cell Tissue Res. 1996;285(3):445–51.
36. Pierce ME, Besharse JC. Circadian regulation of retinomotor movements. I. Interaction of melatonin and dopamine in the control of cone length. J Gen Physiol. 1985;86(5):671–89.
37. Wagner HJ. Darkness-induced reduction of the number of synaptic ribbons in fish retina. Nat New Biol. 1973;246(150):53–5.
38. Ladman AJ. The fine structure of the rod-bipolar cell synapse in the retina of the albino rat. J Biophys Biochem Cytol. 1958;4(4):459–66.
39. Mercer AJ, Thoreson WB. The dynamic architecture of photoreceptor ribbon synapses: cytoskeletal, extracellular matrix, and intramembrane proteins. Vis Neurosci. 2011;28(6): 453–71.
40. Wagner HJ. Light-dependent plasticity of the morphology of horizontal cell terminals in cone pedicles of fish retinas. J Neurocytol. 1980;9(5):573–90.
41. Allwardt BA, Lall AB, Brockerhoff SE, Dowling JE. Synapse formation is arrested in retinal photoreceptors of the zebrafish nrc mutant. J Neurosci. 2001;21(7):2330–42.
42. Emran F, Rihel J, Adolph AR, Dowling JE. Zebrafish larvae lose vision at night. Proc Natl Acad Sci U S A. 2010;107(13):6034–9.
43. Kaneko A, Yamada M. S-potentials in the dark-adapted retina of the carp. J Physiol. 1972;227(1):261–73.
44. Stell WK, Lightfoot DO. Color-specific interconnections of cones and horizontal cells in the retina of the goldfish. J Comp Neurol. 1975;159(4):473–502.

45. Downing JE, Djamgoz MB. Quantitative analysis of cone photoreceptor-horizontal cell connectivity patterns in the retina of a cyprinid fish: electron microscopy of functionally identified and HRP-labelled horizontal cells. J Comp Neurol. 1989;289(4):537–53.
46. Li YN, Matsui JI, Dowling JE. Specificity of the horizontal cell-photoreceptor connections in the zebrafish (Danio rerio) retina. J Comp Neurol. 2009;516(5):442–53.
47. Wagner HJ, Luo BG, Ariano MA, Sibley DR, Stell WK. Localization of D2 dopamine receptors in vertebrate retinae with anti-peptide antibodies. J Comp Neurol. 1993;331(4):469–81.
48. Hurd MW, Debruyne J, Straume M, Cahill GM. Circadian rhythms of locomotor activity in zebrafish. Physiol Behav. 1998;65(3):465–72.
49. Cahill GM, Hurd MW, Batchelor MM. Circadian rhythmicity in the locomotor activity of larval zebrafish. Neuroreport. 1998;9(15):3445–9.
50. DeBruyne J, Hurd MW, Gutierrez L, Kaneko M, Tan Y, Wells DE, et al. Isolation and phenogenetics of a novel circadian rhythm mutant in zebrafish. J Neurogenet. 2004;18(2):403–28.
51. Tan Y, DeBruyne J, Cahill GM, Wells DE. Identification of a mutation in the Clock1 gene affecting zebrafish circadian rhythms. J Neurogenet. 2008;22(2):149–66.
52. Brockerhoff SE, Hurley JB, Janssen-Bienhold U, Neuhauss SC, Driever W, Dowling JE. A behavioral screen for isolating zebrafish mutants with visual system defects. Proc Natl Acad Sci U S A. 1995;92(23):10545–9.
53. Huang YY, Neuhauss SC. The optokinetic response in zebrafish and its applications. Front Biosci. 2008;13:1899–916.
54. Prober DA, Rihel J, Onah AA, Sung RJ, Schier AF. Hypocretin/orexin overexpression induces an insomnia-like phenotype in zebrafish. J Neurosci. 2006;26(51):13400–10.
55. Li L, Dowling JE. Zebrafish visual sensitivity is regulated by a circadian clock. Vis Neurosci. 1998;15(5):851–7.
56. Li L, Dowling JE. Effects of dopamine depletion on visual sensitivity of zebrafish. J Neurosci. 2000;20(5):1893–903.
57. Li L, Dowling JE. Disruption of the olfactoretinal centrifugal pathway may relate to the visual system defect in night blindness b mutant zebrafish. J Neurosci. 2000;20(5):1883–92.
58. Okawa H, Sampath AP, Laughlin SB, Fain GL. ATP consumption by mammalian rod photoreceptors in darkness and in light. Curr Biol. 2008;18(24):1917–21.
59. Naka KI, Rushton WA. The generation and spread of S-potentials in fish (Cyprinidae). J Physiol. 1967;192(2):437–61.
60. Negishi K, Drujan BD. Similarities in effects of acetylcholine and dopamine on horizontal cells in the fish retina. J Neurosci Res. 1979;4(5–6):335–49.
61. Lasater EM, Dowling JE. Dopamine decreases conductance of the electrical junctions between cultured retinal horizontal cells. Proc Natl Acad Sci U S A. 1985;82(9):3025–9.
62. McMahon DG, Brown DR. Modulation of gap-junction channel gating at zebrafish retinal electrical synapses. J Neurophysiol. 1994;72(5):2257–68.
63. Mangel SC, Dowling JE. The interplexiform-horizontal cell system of the fish retina: effects of dopamine, light stimulation and time in the dark. Proc R Soc Lond B Biol Sci. 1987;231(1262):91–121.
64. Baldridge WH, Ball AK. Background illumination reduces horizontal cell receptive-field size in both normal and 6-hydroxydopamine-lesioned goldfish retinas. Vis Neurosci. 1991;7(5):441–50.
65. Weiler R, Baldridge WH, Mangel SC, Dowling JE. Modulation of endogenous dopamine release in the fish retina by light and prolonged darkness. Vis Neurosci. 1997;14(2):351–6.
66. Weiler R, Schultz K, Pottek M, Tieding S, Janssen-Bienhold U. Retinoic acid has light-adaptive effects on horizontal cells in the retina. Proc Natl Acad Sci U S A. 1998;95(12):7139–44.
67. Sun Z, Zhang DQ, McMahon DG. Zinc modulation of hemi-gap-junction channel currents in retinal horizontal cells. J Neurophysiol. 2009;101(4):1774–80.
68. Daniels BA, Baldridge WH. The light-induced reduction of horizontal cell receptive field size in the goldfish retina involves nitric oxide. Vis Neurosci. 2011;28(2):137–44.

69. DeVries SH, Schwartz EA. Modulation of an electrical synapse between solitary pairs of catfish horizontal cells by dopamine and second messengers. J Physiol. 1989;414:351–75.
70. Harsanyi K, Mangel SC. Activation of a D2 receptor increases electrical coupling between retinal horizontal cells by inhibiting dopamine release. Proc Natl Acad Sci U S A. 1992; 89(19):9220–4.
71. Mangel SC, Dowling JE. Responsiveness and receptive field size of carp horizontal cells are reduced by prolonged darkness and dopamine. Science. 1985;229(4718):1107–9.
72. Tornqvist K, Yang XL, Dowling JE. Modulation of cone horizontal cell activity in the teleost fish retina. III. Effects of prolonged darkness and dopamine on electrical coupling between horizontal cells. J Neurosci. 1988;8(7):2279–88.
73. Baldridge WH. Triphasic adaptation of teleost horizontal cells. Prog Brain Res. 2001;131: 437–49.
74. Bloomfield SA, Xin D, Osborne T. Light-induced modulation of coupling between AII amacrine cells in the rabbit retina. Vis Neurosci. 1997;14(3):565–76.
75. Xin D, Bloomfield SA. Dark- and light-induced changes in coupling between horizontal cells in mammalian retina. J Comp Neurol. 1999;405(1):75–87.
76. Ribelayga C, Mangel SC. Absence of circadian clock regulation of horizontal cell gap junctional coupling reveals two dopamine systems in the goldfish retina. J Comp Neurol. 2003;467(2):243–53.
77. Ribelayga C, Mangel SC. Tracer coupling between fish rod horizontal cells: modulation by light and dopamine but not the retinal circadian clock. Vis Neurosci. 2007;24(3):333–44.
78. Ribelayga C, Wang Y, Mangel SC. Dopamine mediates circadian clock regulation of rod and cone input to fish retinal horizontal cells. J Physiol. 2002;544(Pt 3):801–16.
79. Mangel SC, Baldridge WH, Weiler R, Dowling JE. Threshold and chromatic sensitivity changes in fish cone horizontal cells following prolonged darkness. Brain Res. 1994;659 (1–2):55–61.
80. Wang Y, Mangel SC. A circadian clock regulates rod and cone input to fish retinal cone horizontal cells. Proc Natl Acad Sci U S A. 1996;93(10):4655–60.
81. Baldridge WH, Weiler R, Dowling JE. Dark-suppression and light-sensitization of horizontal cell responses in the hybrid bass retina. Vis Neurosci. 1995;12(4):611–20.
82. Ribelayga C, Cao Y, Mangel SC. The circadian clock in the retina controls rod-cone coupling. Neuron. 2008;59(5):790–801.
83. Raviola E, Gilula NB. Gap junctions between photoreceptor cells in the vertebrate retina. Proc Natl Acad Sci U S A. 1973;70(6):1677–81.
84. Ribelayga C, Mangel SC. Identification of a circadian clock-controlled neural pathway in the rabbit retina. PLoS One. 2010;5(6):e11020.
85. Krizaj D, Gabriel R, Owen WG, Witkovsky P. Dopamine D2 receptor-mediated modulation of rod-cone coupling in the Xenopus retina. J Comp Neurol. 1998;398(4):529–38.
86. Yang XL, Wu SM. Effects of background illumination on the horizontal cell responses in the tiger salamander retina. J Neurosci. 1989;9(3):815–27.
87. Schwartz EA. Rod-rod interaction in the retina of the turtle. J Physiol. 1975;246(3):617–38.
88. Copenhagen DR, Owen WG. Coupling between rod photoreceptors in a vertebrate retina. Nature. 1976;260(5546):57–9.
89. Dermietzel R, Kremer M, Paputsoglu G, Stang A, Skerrett IM, Gomes D, et al. Molecular and functional diversity of neural connexins in the retina. J Neurosci. 2000;20(22):8331–43.
90. Kothmann WW, Massey SC, O'Brien J. Dopamine-stimulated dephosphorylation of connexin 36 mediates AII amacrine cell uncoupling. J Neurosci. 2009;29(47):14903–11.
91. O'Brien J, Nguyen HB, Mills SL. Cone photoreceptors in bass retina use two connexins to mediate electrical coupling. J Neurosci. 2004;24(24):5632–42.
92. Li H, Chuang AZ, O'Brien J. Photoreceptor coupling is controlled by connexin 35 phosphorylation in zebrafish retina. J Neurosci. 2009;29(48):15178–86.
93. Vitaterna MH, King DP, Chang AM, Kornhauser JM, Lowrey PL, McDonald JD, et al. Mutagenesis and mapping of a mouse gene, Clock, essential for circadian behavior. Science. 1994;264(5159):719–25.

94. Whitmore D, Foulkes NS, Sassone-Corsi P. Light acts directly on organs and cells in culture to set the vertebrate circadian clock. Nature. 2000;404(6773):87–91.
95. Pando MP, Pinchak AB, Cermakian N, Sassone-Corsi P. A cell-based system that recapitulates the dynamic light-dependent regulation of the vertebrate clock. Proc Natl Acad Sci U S A. 2001;98(18):10178–83.
96. Giebultowicz JM, Riemann JG, Raina AK, Ridgway RL. Circadian system controlling release of sperm in the insect testes. Science. 1989;245(4922):1098–100.
97. Plautz JD, Kaneko M, Hall JC, Kay SA. Independent photoreceptive circadian clocks throughout Drosophila. Science. 1997;278(5343):1632–5.
98. Hege DM, Stanewsky R, Hall JC, Giebultowicz JM. Rhythmic expression of a PER-reporter in the Malpighian tubules of decapitated Drosophila: evidence for a brain-independent circadian clock. J Biol Rhythms. 1997;12(4):300–8.
99. Tanoue S, Krishnan P, Krishnan B, Dryer SE, Hardin PE. Circadian clocks in antennal neurons are necessary and sufficient for olfaction rhythms in Drosophila. Curr Biol. 2004; 14(8):638–49.
100. Yoo SH, Yamazaki S, Lowrey PL, Shimomura K, Ko CH, Buhr ED, et al. PERIOD2::LUCIFERASE real-time reporting of circadian dynamics reveals persistent circadian oscillations in mouse peripheral tissues. Proc Natl Acad Sci U S A. 2004;101(15):5339–46.
101. King DP, Zhao Y, Sangoram AM, Wilsbacher LD, Tanaka M, Antoch MP, et al. Positional cloning of the mouse circadian clock gene. Cell. 1997;89(4):641–53.
102. Gekakis N, Staknis D, Nguyen HB, Davis FC, Wilsbacher LD, King DP, et al. Role of the CLOCK protein in the mammalian circadian mechanism. Science. 1998;280(5369):1564–9.
103. Darlington TK, Wager-Smith K, Ceriani MF, Staknis D, Gekakis N, Steeves TD, et al. Closing the circadian loop: CLOCK-induced transcription of its own inhibitors per and tim. Science. 1998;280(5369):1599–603.
104. Ko CH, Takahashi JS. Molecular components of the mammalian circadian clock. Hum Mol Genet. 2006;15 Spec No 2:R271–7.
105. Postlethwait JH, Yan YL, Gates MA, Horne S, Amores A, Brownlie A, et al. Vertebrate genome evolution and the zebrafish gene map. Nat Genet. 1998;18(4):345–9.
106. Kobayashi Y, Ishikawa T, Hirayama J, Daiyasu H, Kanai S, Toh H, et al. Molecular analysis of zebrafish photolyase/cryptochrome family: two types of cryptochromes present in zebrafish. Genes Cells. 2000;5(9):725–38.
107. Emery P, So WV, Kaneko M, Hall JC, Rosbash M. CRY, a Drosophila clock and light-regulated cryptochrome, is a major contributor to circadian rhythm resetting and photosensitivity. Cell. 1998;95(5):669–79.
108. Stanewsky R, Kaneko M, Emery P, Beretta B, Wager-Smith K, Kay SA, et al. The cryb mutation identifies cryptochrome as a circadian photoreceptor in Drosophila. Cell. 1998; 95(5):681–92.
109. Cermakian N, Pando MP, Thompson CL, Pinchak AB, Selby CP, Gutierrez L, et al. Light induction of a vertebrate clock gene involves signaling through blue-light receptors and MAP kinases. Curr Biol. 2002;12(10):844–8.
110. Tamai TK, Young LC, Whitmore D. Light signaling to the zebrafish circadian clock by Cryptochrome 1a. Proc Natl Acad Sci U S A. 2007;104(37):14712–7.
111. Delaunay F, Thisse C, Marchand O, Laudet V, Thisse B. An inherited functional circadian clock in zebrafish embryos. Science. 2000;289(5477):297–300.
112. Ziv L, Levkovitz S, Toyama R, Falcon J, Gothilf Y. Functional development of the zebrafish pineal gland: light-induced expression of period2 is required for onset of the circadian clock. J Neuroendocrinol. 2005;17(5):314–20.
113. Hirayama J, Fukuda I, Ishikawa T, Kobayashi Y, Todo T. New role of zCRY and zPER2 as regulators of sub-cellular distributions of zCLOCK and zBMAL proteins. Nucleic Acids Res. 2003;31(3):935–43.
114. Ishikawa T, Hirayama J, Kobayashi Y, Todo T. Zebrafish CRY represses transcription mediated by CLOCK-BMAL heterodimer without inhibiting its binding to DNA. Genes Cells. 2002;7(10):1073–86.

115. Wang H. Comparative analysis of teleost fish genomes reveals preservation of different ancient clock duplicates in different fishes. Mar Genomics. 2008;1(2):69–78.
116. Cermakian N, Whitmore D, Foulkes NS, Sassone-Corsi P. Asynchronous oscillations of two zebrafish CLOCK partners reveal differential clock control and function. Proc Natl Acad Sci U S A. 2000;97(8):4339–44.
117. Korenbrot JI, Fernald RD. Circadian rhythm and light regulate opsin mRNA in rod photoreceptors. Nature. 1989;337(6206):454–7.
118. Pierce ME, Sheshberadaran H, Zhang Z, Fox LE, Applebury ML, Takahashi JS. Circadian regulation of iodopsin gene expression in embryonic photoreceptors in retinal cell culture. Neuron. 1993;10(4):579–84.
119. Li P, Chaurasia SS, Gao Y, Carr AL, Iuvone PM, Li L. CLOCK is required for maintaining the circadian rhythms of opsin mRNA expression in photoreceptor cells. J Biol Chem. 2008;283(46):31673–8.
120. Yu CJ, Gao Y, Li P, Li L. Synchronizing multiphasic circadian rhythms of rhodopsin promoter expression in rod photoreceptor cells. J Exp Biol. 2007;210(Pt 4):676–84.
121. Rajendran RR, Van Niel EE, Stenkamp DL, Cunningham LL, Raymond PA, Gonzalez-Fernandez F. Zebrafish interphotoreceptor retinoid-binding protein: differential circadian expression among cone subtypes. J Exp Biol. 1996;199(Pt 12):2775–87.
122. Cunningham LL, Gonzalez-Fernandez F. Coordination between production and turnover of interphotoreceptor retinoid-binding protein in zebrafish. Invest Ophthalmol Vis Sci. 2000; 41(11):3590–9.
123. Emran F, Rihel J, Adolph AR, Wong KY, Kraves S, Dowling JE. OFF ganglion cells cannot drive the optokinetic reflex in zebrafish. Proc Natl Acad Sci U S A. 2007;104(48):19126–31.

Chapter 10
Circadian Modulation of the *Limulus* Eye for Day and Night Vision

Christopher L. Passaglia and Erik D. Herzog

Abstract The horseshoe crab has been an outstanding model for vision and circadian research. Its lateral eyes are probably the best understood neural structure in the animal kingdom. Fundamental principles of image processing were first gleaned from the animal, including phototransduction, light adaptation, and lateral inhibition. A circadian clock in the brain modulates anatomical and physiological properties of the eyes to increase visual sensitivity each night. As a result, horseshoe crabs can use their eyes to locate mates equally well day and night. This chapter summarizes studies that have localized the clock output neurons that regulate vision, characterized their patterns of electrical activity, and identified octopamine as their primary circadian neurotransmitter. We are at a stage now where we can begin to model the response of each photoreceptor and each optic nerve fiber to natural visual scenes day and night.

Keywords Horseshoe crab • Invertebrate • Retinular cell • Eccentric cell • Cheliceral ganglion • Circadian rhythm • Visual sensitivity • Rhodopsin

---

Dedicated to the memory of Robert B. Barlow, Jr., for his seminal work on the Limulus circadian system.

C.L. Passaglia, Ph.D.
Department of Chemical & Biomedical Engineering, University of South Florida, Tampa, FL, USA
e-mail: passaglia@usf.edu

E.D. Herzog, Ph.D. (✉)
Department of Biology, Washington University in St. Louis, St. Louis, MO, USA
e-mail: Herzog@wustl.edu

G. Tosini et al. (eds.), *The Retina and Circadian Rhythms*, Springer Series in Vision Research 1, DOI 10.1007/978-1-4614-9613-7_10,

10.1 Introduction

The American horseshoe crab (*Limulus polyphemus*) spends much of its life scouring the floor of shallow bays and estuaries along the Atlantic Coast of North America in search of food or a potential mate. Its movements are fairly slow by human standards (10–20 cm/s) but nevertheless graceful, with five pairs of walking legs lifting its helmet-shaped carapace off the sandy bottom and in rhythmic motion propelling the animal forward, while its long spikelike tail whisks behind like the rudder of a ship, lending balance and turning support in the face of underwater currents. Protruding from the sides of its carapace are two compound eyes that have captured the imagination and interest of neuroscientists for over a century now.

The initial source of fascination was the approximately 1,000 ommatidial receptors that comprise each lateral eye. The receptors are so big that they can be seen by the naked eye. Moreover, the receptors are connected to the brain by a nerve that is unusually long for an invertebrate due to the large size of the carapace and the location of the brain deep inside. In adult crabs, the optic nerves can be over 15 cm in length. Optic nerve fibers must therefore fire trains of action potentials in order to communicate visual signals rapidly and reliably to target neurons in the brain, the same mode of information transmission used by vertebrate eyes. The big receptors and their ability to fire spikes that are readily recorded with an extracellular electrode made it possible to study for the first time the light responses of individual visual cells of an animal [1, 2]. The research led to the groundbreaking discoveries that (1) the receptors transduce light into discrete bumps in membrane potential which adapt in size to steady illumination and which summate in time to produce optic nerve spike trains [3, 4] and that (2) the receptors transmit the light signal not only to optic ganglia in the brain but also to neighboring receptors in the eye through lateral inhibitory connections [5]. Adaptation and lateral inhibition have since been found in every vertebrate retina examined and throughout the nervous system, attesting to their fundamental importance to vision and neural information processing in general. The latter discovery gave birth to the first quantitative neural network models of an eye [6, 7], from which our current understanding of lateral inhibition as a biological mechanism of contrast enhancement is derived. This principle has been applied in engineered devices and software applications, such as the common "sharpen edges" command of photo-editing programs.

The finding of a retinal network that performs visual computations implied that the lateral eyes served a purpose that was greater than merely light sensing. For many decades that purpose was uncertain until the important roles of vision and circadian rhythms in the mating behavior of horseshoe crabs were revealed [8]. This chapter aims to synthesize what is known about the role of the circadian clock in retinal processing of visual scenes and the impact on visually guided behaviors. It begins with a description of the anatomical and physiological organization of the circadian system. It follows with a survey of neural network mechanisms of the eyes in their daytime and clock-driven nighttime states. It concludes with a discussion of the visual abilities of the animal day and night. Highlighted are similarities and differences from other organisms that make horseshoe crabs a special model for vision and circadian research.

10.2 Anatomical Organization of the Circadian System

One of the most attractive features of the *Limulus* circadian system from a scientific perspective is its relatively simple organization. In mammals and other vertebrates, cells that express 24-h activity rhythms have been found throughout the nervous system, including the retina and olfactory bulb, in addition to the suprachiasmatic nucleus (SCN) where the master timekeeper is known to reside [9–13]. Circadian oscillations have also been demonstrated in nonneural tissues like the heart, liver, adrenal gland, and lung [14–16]. The peripheral oscillator cells do not have the ability to synchronize on their own, so the master SCN clock employs a complex web of neuronal and hormonal processes that provide tissue-specific timing cues to coordinate the multitude of activity rhythms [17–19]. In contrast, the circadian system of horseshoe crabs is thought to consist of just the master clock and a cluster of clock output neurons. This simplicity has proven valuable for understanding how the circadian system can modulate the anatomical and physiological properties of tissues and organs and how that modulation can benefit an animal behaviorally.

Location of the Clock. The existence of a circadian clock was first evidenced by a daily rhythm in visual sensitivity of the lateral eyes [20]. The sensitivity rhythm went long unnoticed because standard practice was to isolate the eye in a recording chamber, but the retina loses circadian rhythmicity when excised from the animal and reverts to its daytime state within a few hours [21]. Horseshoe crabs, along with other chelicerates (scorpions and spiders) and crustaceans, thus represent an exception to the rule that all retinas contain circadian oscillators. The source of the circadian signal was pinpointed to efferent fibers that project from the brain to the eye, as severing the optic nerve in vivo abolished the daily rhythm in visual sensitivity [20, 21]. This also indicated that hormonal pathways are not sufficient to drive the sensitivity rhythm since the blood supply to the eye remained intact. Figure 10.1 shows a diagram of the horseshoe crab brain, which is defined as the tissue anterior to the circumesophageal ring known as the protocerebrum [22]. The protocerebrum connects directly to the eyes via the ventral, median, and lateral optic nerves and indirectly to the walking legs and tail via pedal and abdominal nerve bundles of the circumesophageal ganglia. The precise location of the circadian clock within the brain has not been determined. Hemisection experiments have revealed that each side of the protocerebrum contains a circadian oscillator and that the two oscillators are coupled together by crossing fibers to generate the master signal that rhythmically modulates visual sensitivity [23, 24].

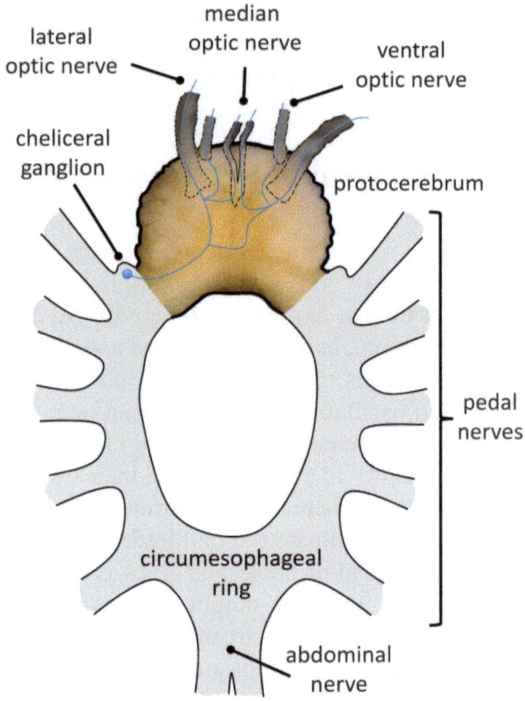

Fig. 10.1 A photograph of the *Limulus* brain (protocerebrum), with the lateral, median, and ventral optic nerves that connect to the animal's eyes outlined with *dashed lines*. The protocerebrum extends anteriorly from the circumesophageal ring, which is illustrated in schematic form along with the abdominal nerve and the five pairs of pedal nerves. Efferent projections of the circadian clock to the eyes proposed by [25] are shown for an individual cheliceral ganglion cell

Pathways from the Clock. The efferent nerve fibers that transmit circadian signals to the eyes originate from a bilateral cluster of cells in the cheliceral ganglion [25], which lies at the anterolateral junction of the protocerebrum and circumesophageal ring (Fig. 10.1). There are approximately 20 of these "clock cells" based on backfilling the optic nerves with tracer molecules. Their intrinsic physiological properties have not yet been described, so cheliceral ganglion cells could be the master oscillators as well as clock messengers. The axons of the clock cells project bilaterally down all optic nerves and to multiple optic and nonoptic neuropils in the brain [25, 26]. The clock efferents are among the smallest fibers in the optic nerves (0.5–2 μm) and terminate on all cell types within the lateral eye, both visual and nonvisual [27, 28]. Their terminal endings are thickened, bulbous, and filled with both vesicles and dense granules, which are characteristic features of neurosecretory synapses in arthropods [29]. Octopamine serves as a principal neurotransmitter output from the circadian clock to the eyes. The clock cells synthesize, store, and release octopamine [26, 30–33], and circadian changes in retinal properties can be mimicked by exogenous octopamine administration and blocked by octopamine antagonists [34]. It is not yet clear whether octopamine is the only circadian neurotransmitter in the *Limulus* visual system [35–39].

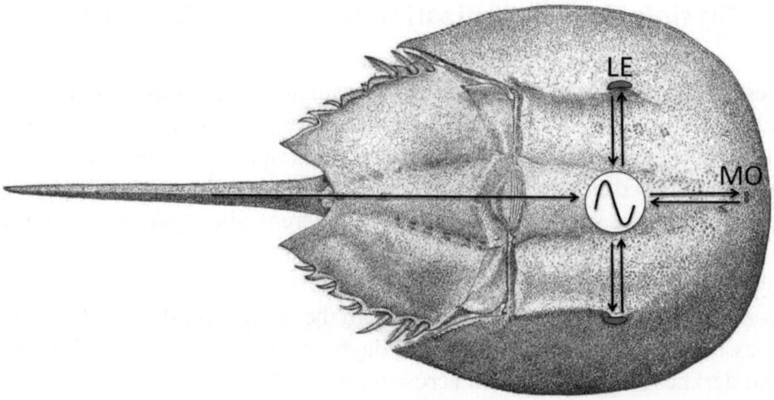

Fig. 10.2 A diagram of the *Limulus* circadian system and its known input and output pathways. A circadian clock located in the brain entrains to photic cues provided by the lateral eyes (LE), median ocelli (MO), ventral photoreceptors (not shown), four rudimentary eyes (not shown), and the tail. The clock also sends circadian signals back to the eyes, which alter retinal structure and function to increase photic sensitivity at night

Pathways to the Clock. Horseshoe crabs have over ten light-sensing organs: two lateral eyes, two median eyes, two rudimental lateral eyes, two rudimentary median eyes, a ventral eye, and ectopic photoreceptors scattered along the tail and over the brain (Fig. 10.2). All of these organs provide the master clock with photic input, and each is individually capable of entraining the clock to an imposed light cycle [24, 40, 41]. The photic signals are communicated to clock cells indirectly through the optic neuropils of the brain since afferent optic nerve fibers do not terminate in the cheliceral ganglion [42]. The circadian system integrates signals from all the input pathways and advances or delays the phase of the master oscillator so as to maintain photic entrainment. Signals from the tail are weighted heaviest by the central time-keeping mechanism, producing phase shifts equal to that of whole animal illumination and greater than the lateral, median, and ventral eye signals combined [40, 41]. The clock thus synchronizes more quickly to a light cycle delivered to just its tail than to all eight of its eyes. The special connection between the circadian clock and tail photoreceptors may be related to the burrowing nature of the animal. Horseshoe crabs can bury themselves for long stretches of time in the act of digging for food, building a nest, or hibernating. Often their telson can be found sticking out the ocean floor presumably to keep their endogenous rhythms in synch with the outside world. This may explain why photoreceptors are also present in the tails of scorpions, another chelicerate that likes to burrow in the sand [43]. The phase-setting process is one aspect of the circadian system that appears more complicated than that of mammals. It combines photic input not just from two eyes, but from several eyes in an unequal manner.

10.3 Physiological Organization of the Circadian System

Another attractive feature of the *Limulus* circadian system is that clock output and its effects on retinal physiology can be concurrently studied in living animals. Clock output can be accessed by opening a hole in the carapace in front of the lateral eye, guiding the exposed nerve into a recording chamber, and teasing away individual nerve bundles into a suction electrode [44]. Efferent fiber spike trains are usually found in dorsolateral bundles of the optic nerve [45]. Since efferent fibers project to both eyes, changes in visual sensitivity induced by clock activity can be monitored by recording electroretinograms (ERGs) from the contralateral eye. ERG amplitude oscillates between a nighttime high and daytime low with a period that ranges in constant darkness from 22 to 25 h across animals [23].

Efferent Spike Trains of the Clock. Circadian rhythms in the eyes are driven by clock output neurons that are active at night and quiescent during the day [20]. Their firing rate ramps up at dusk to a nighttime average of ~1 spike/s per neuron and then ramps down at dawn [45]. This activity rhythm is opposite in phase to most SCN neurons, which fire more during the daytime [13]. Efferent spike trains are transmitted synchronously down the lateral, median, and ventral optic nerves, and they are surprisingly rich in fine temporal structure. Figure 10.3 shows a representative trace of clock efferent activity recorded in constant darkness. A striking feature of the activity pattern is the semi-regular sequence of spike bursts that occur at intervals of around 0.5–2 s [23, 24]. Each spike in a burst represents the discharge of one clock output neuron, and the near simultaneity of their discharge in a burst indicates that these neurons are strongly coupled. Detailed inspection of the activity records suggests there is a hierarchy to the coupling as clock output neurons exhibit a preferred firing order. Spike bursts are typically produced in clusters of 10–30 burst events that are separated by silent periods lasting tens to hundreds of seconds [45]. The functional significance of the multilayered structure is not entirely clear. It has been suggested that these ultradian oscillations could underlie the generation of circadian rhythms [46]. The fine structure of efferent firing patterns could also be important for communicating the circadian message to target organs. The latter possibility can be evaluated by stimulating clock efferent fibers with current pulses applied through a suction electrode to the optic nerve. Figure 10.4 shows that stimulation rates corresponding to the intrinsic burst period of clock output neurons produce ERG changes comparable in amplitude to that of the circadian clock; whereas, stimulation at much lower or higher rates induces little or no ERG amplitude change. This suggests that octopamine release at efferent terminals in the eye might be tuned for specific depolarization patterns or that the postsynaptic machinery of retinal neurons might be geared to operate at certain octopamine release levels. Much like delivering melatonin in a circadian profile can impose rhythms and photoperiodic responses to some birds and mammals (see Chaps. 4 and 5), this provides a rare example of how circadian clock output can be decoded.

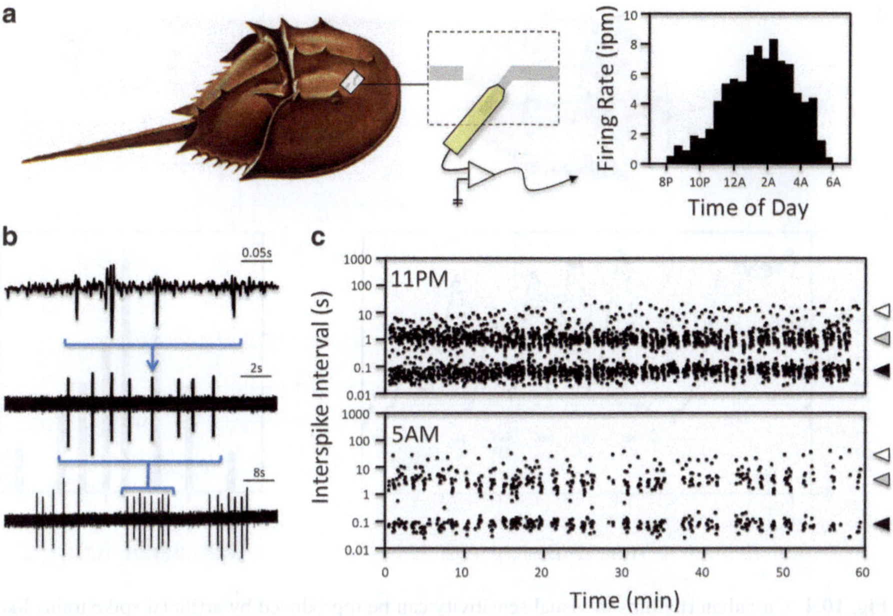

Fig. 10.3 Representative output of efferent nerve fibers of the circadian clock. (**a**) Efferent fiber activity is recorded by surgically exposing the optic nerve, cutting it, and inserting the end connected to the brain in a suction electrode. Clock efferent fibers are silent during the day and active throughout the night. (**b**) Clock spike trains exhibit fine structure at multiple timescales. Efferent fibers fire synchronous bursts of spikes with each fiber contributing one spike per burst (*top*). In this experiment the electrode recorded from six efferent fibers. The spike bursts repeat every few seconds, forming a cluster of burst events (*middle*). The burst clusters then repeat after tens of seconds (*bottom*). (**c**) The temporal patterning of efferent spikes varies over the course of the night. Time intervals between spikes within a burst (*black triangles*) remain constant from dusk to dawn, but interburst intervals (*grey triangles*) and intercluster intervals (*white triangles*) gradually lengthen as dawn approaches when activity ceases altogether

Light Response of the Clock. The circadian clock responds to light in a physical and functional manner. The physical response is an immediate stimulus-evoked change in efferent spike rate that happens only at night when clock output neurons are active [24, 45]. The response is central in origin because it can be elicited by light shined on other eyes. The response delay with respect to stimulus onset was ~0.8 s measured at the lateral eye, most of which can be attributed to nerve conduction time to and from the brain [47]. The form of the response depends on the photic organ stimulated. Lateral eye illumination by a 50-ms flash in the middle of the night evoked a barrage of spikes for tens of seconds that was followed by a quiet period lasting hundreds of seconds during which the ongoing spike activity rhythm was completely suppressed [45]. Median eye stimulation also elicits efferent spikes, but ventral eye stimulation does not, and light shined on the excised brain inhibits efferent activity [24]. The connection of ectopic brain receptors to the clock is presumably a vestige of the juvenile stage of the animal when its carapace is translucent.

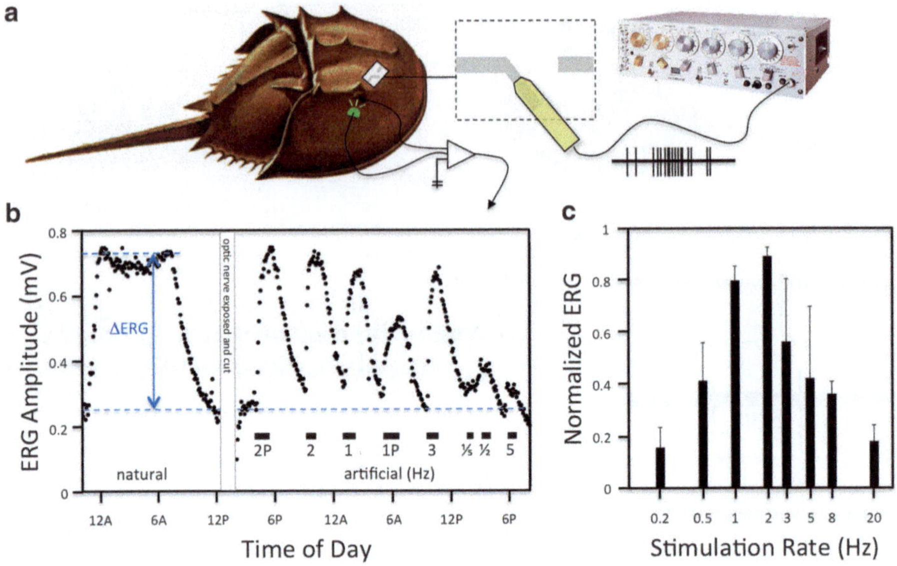

Fig. 10.4 Circadian rhythms in visual sensitivity can be reproduced by artificial spike trains like those of clock efferent fibers. (**a**) Circadian messages are decoded by inserting the end of a cut optic nerve connected to the eye into a suction electrode and stimulating the efferent fibers electrically with trains of current pulses. Effects of pulse trains are monitored by flashing an LED and recording the evoked ERG. (**b**) ERG amplitude depends on the pattern of artificial stimulation, with some pulse trains generating ERG amplitude changes comparable to the circadian clock. The natural ERG rhythm was recorded in constant darkness, and then the optic nerve was cut for purpose of stimulation, eliminating endogenous clock input to the eye. *Bars* indicate episodes of current stimulation, with pulse trains delivered at the specified rate either continuously or intermittently (*P*=pulse train repeatedly presented for 10s on and 10s off). (**c**) ERG amplitude changes are greatest for pulse trains of approximately 1–2 Hz. Current-evoked responses were normalized by dividing the peak ERG amplitude by the maximum endogenous amplitude

The functional response of the circadian clock to light is a shift in phase of the master rhythm. The phase shift might be the functional manifestation of the prolonged physical suppression of clock output neurons that follows a light flash. Figure 10.5 shows the phase response curve (PRC) of the lateral eye ERG rhythm for pulses of illumination delivered to one or more photic organs at various times during the clock cycle. Light pulses at circadian times (CT) 8–18, which run from 4 h before subjective dusk to 6 h into the night, cause phase delays up to 4 h; whereas, light pulses at CT 18–24, which spans the 6 h before dusk, cause phase advances up to 2 h. The magnitude of phase shifts increases with light intensity and duration. This photic PRC resembles that of mammalian circadian systems [48–50].

Neuromodulatory Processes of the Clock. The spike-triggered release of octopamine from efferent nerve terminals of the circadian clock initiates a cascade of biochemical events [reviewed in [51, 52]]. Figure 10.6 shows the steps in the

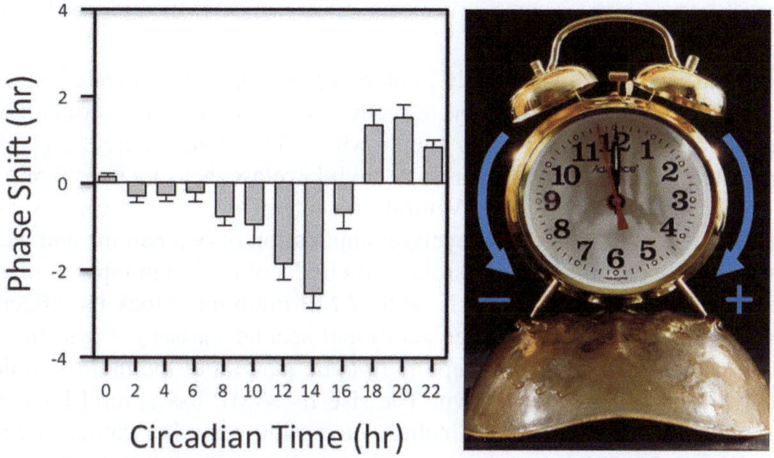

Fig. 10.5 The Phase response curve (PRC) of the lateral eye ERG rhythm for pulses of illumination at different times during the clock cycle. Light during the late subjective night phase advances and light during the early subjective night phase delays the ERG rhythm. The PRC was constructed by pooling data from published studies [23, 40, 41] in 2-h bins. The PRC data had the same general form across these studies though different animals, photic organs, and light pulses were involved. Each bin contained at least three data points. The image illustrates that backwards shifts (negative values) result in a phase delay, whereas forward shifts (positive values) advance circadian rhythms

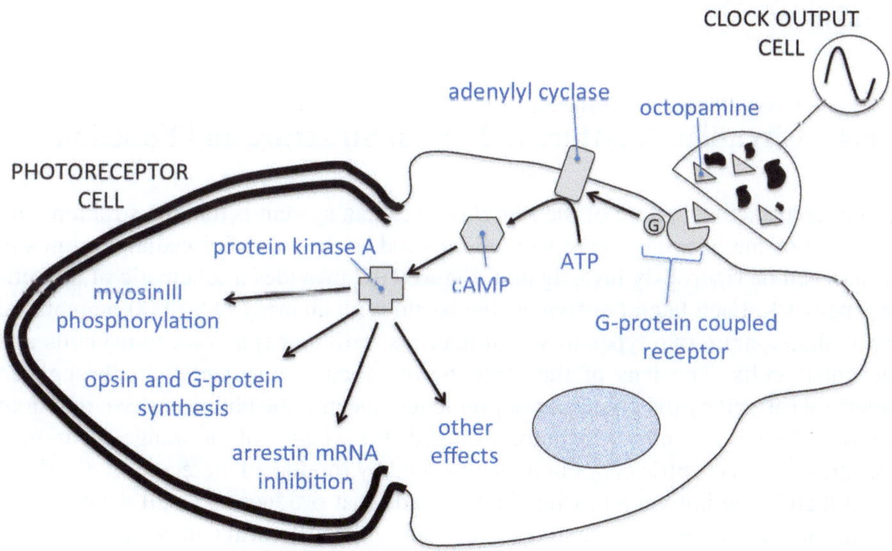

Fig. 10.6 A model of the signaling cascade underlying circadian clock control of photoreceptor sensitivity. At night, efferent terminals of clock output neurons release octopamine in the eye, which binds G-coupled receptors in the membrane to activate the adenylyl cyclase second messenger system. This leads to an increase in intracellular cAMP levels, which stimulates protein kinase A to induce a variety of biochemical changes that change photoreceptor structure and function at night. *Double lines* represent rhabdomeric membrane of photoreceptors. Modified from [59]

cascade that have been identified in photoreceptor cells. The cascade begins with octopaminergic activation of G-protein-coupled receptors. In invertebrates, this receptor family has been shown to couple with at least four different second messenger systems that activate or repress adenylyl cyclase, activate phospholipase C, or open chloride channels [53]. Ventral and lateral eye photoreceptors express receptors that couple to adenylyl cyclase; application of octopamine and adenylyl cyclase activators mimics the physiological effects of circadian input at night and increases intracellular cAMP levels, and cAMP inhibitors block the effects [54, 55]. Whether octopamine activates additional second messenger systems in the eyes and whether octopamine receptors in optic ganglia of the brain also depend on adenylyl cyclase are not known. The rise in cAMP has several biochemical effects on photoreceptors, which include increasing phosphorylation of *Limulus* myosin III [32, 56–58], increasing the concentration of phototransduction proteins opsin and $G_q\alpha$ [52], and decreasing the concentration of arrestin mRNA [52, 59]. All three effects were shown to involve activation of cAMP-dependent protein kinase A (PKA), indicating that PKA plays a major downstream role in circadian neuromodulation of the eyes. There might also be other neuromodulatory processes of the clock besides the octopamine cascade. The efferent terminals contain both synaptic vesicles and dense-core granules generally associated with peptidergic cells [28, 60], suggesting they may co-release neuropeptides and other bioactive molecules [34]. Co-release in neurosecretory cells is likely driven by spike bursts [61, 62], which are a distinctive feature of clock efferent spike trains (Fig. 10.3).

10.4 Circadian Rhythms in Retinal Structure and Function

Another attractive feature of the *Limulus* circadian system is that the structure and function of the lateral eyes are well understood so the role of circadian rhythms in vision can be rigorously investigated. Figure 10.7 provides a schematic of the retinal network. Each lateral eye views the world with an array of ~1,000 ommatidial units that contain two types of visual neurons: retinular (photoreceptor) cells and eccentric cells. The lens of the ommatidium focuses incident light through an aperture formed by the processes of pigment cells onto the photosensitive rhabdom of 10–12 retinular cells, which are arranged like wedges of an orange around the dendrite of an eccentric cell. Photons absorbed by rhodopsin molecules in the rhabdom membrane initiate a biochemical cascade that produces a small depolarizing bump in the membrane potential of retinular cells. At sufficient photon rate, the bumps summate in time to form the photoresponse [3, 4]. The bumps also adapt dynamically in amplitude and duration in response to changes in light intensity [63, 64], allowing the animal to see over a wide range of ambient illumination levels. The excitatory photocurrents of retinular cells collectively propagate to the

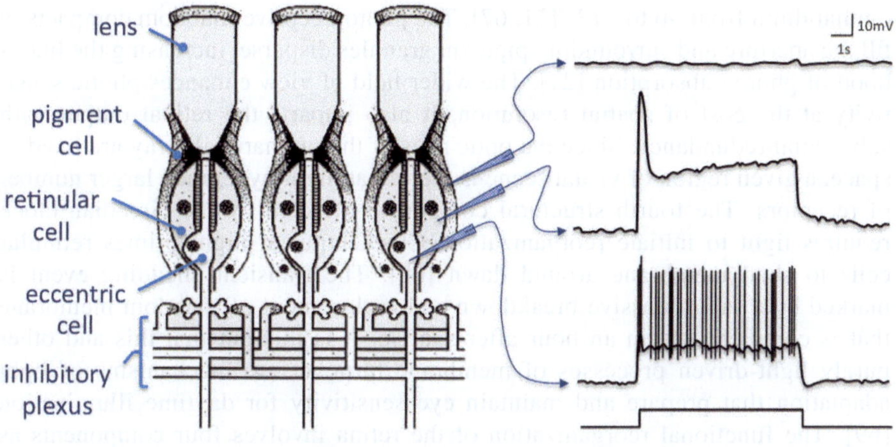

Fig. 10.7 Schematic of the lateral eye and the light responses of its major cell types. Each ommatidium contains two types of visual neurons: retinular cells and eccentric cells. The retinular cells transduce incident light into an excitatory signal. The eccentric cells integrate that signal with spike-driven self- and lateral inhibitory signals from a plexus of synaptic connection with other cells and encode the result with trains of spikes that are sent to the brain and to neighboring ommatidia. Illustrated are intracellular voltage records from pigment cells, retinular cells, and eccentric cells for a 7 s flash of light. Modified from [44]

eccentric cell through gap junctions in its dendrite, where they summate with inhibitory synaptic currents driven by action potentials fired by the eccentric cell itself and by neighboring eccentric cells [5–7]. The eccentric cell then encodes the resultant signal with a train of action potentials that are transmitted across the retinal array and to optic neuropils in the brain. The optical transformation, phototransduction process, self- and lateral inhibitory networks, and spike-firing mechanism have been described in quantitative detail to the point that computational models can accurately simulate the response of the lateral eye to videos acquired with an underwater camera mounted to horseshoe crabs moving in the ocean [65, 66]. The model simulations also demonstrate that the spatiotemporal properties of the daytime eye are tuned to detect objects resembling an adult crab in size and speed.

Circadian input to the lateral eyes from the brain causes a daily reorganization of retinal structure and function. Many of the changes are mimicked by the application of octopamine or cAMP analogs to the eye in its daytime state. The structural reorganization of the retina has four components: pigment cell retraction, rhabdom compression, pigment cell dispersion, and photoreceptor membrane turnover. The first three are driven solely by the circadian clock. At night when clock output neurons are active, the distal pigment cells that line each lens move laterally to increase the aperture, doubling the acceptance angle of each ommatidium from ~6 to ~12° [21, 67]. The photoreceptive rhabdom compacts to fill the aperture and surrounding pigment granules disperse, increasing the likelihood of photon absorption [21]. The wider field of view enhances photic sensitivity at the cost of spatial resolution. It also imparts the retinal output with substantial redundancy. Since the optic axes of the ommatidial array are fixed in space, a given region of visual scene is viewed at night by a much larger number of receptors. The fourth structural component affected by the circadian clock requires light to initiate reorganization. Clock input at night primes retinular cells to shed membrane around dawn [68]. The transient shedding event is marked by a rapid massive breakdown and replacement of rhabdom membrane that is complete within an hour after sunrise. It is thought that this and other purely light-driven processes of membrane turnover are mechanisms of light adaptation that prepare and maintain eye sensitivity for daytime illumination [69]. The functional reorganization of the retina involves four components as well: photoreceptor gain, bump noise, temporal filtering, and lateral inhibition. At low light levels, small bumps in photoreceptor voltage can trigger large regenerative fluctuations in membrane potential, known as LPFs, which eccentric cells encode with high fidelity as bursts of spikes. Efferent activity at night increases photoreceptor gain so that each absorbed photon yields a larger and longer quantum bump [20, 70–73], elevating the likelihood of bump summation and LPF generation. The longer bumps also lower the temporal cutoff frequency of the eye [74], resulting in slower light responses. Remarkably, the spontaneous bump rate also decreases in photoreceptors even though photon catch and gain are high. The reduction in retinal noise is mediated, at least in part, by a clock-driven pH shift that stabilizes rhodopsin molecules [75]. In these three ways, clock input leads to a coordinated increase in signal-to-noise ratio of excitatory photoreceptor signals at night. In addition, there is evidence that clock input weakens inhibitory interactions within the retinal network [74]. Apparently the cost of inhibition for photic sensitivity outweighs its benefits for contrast enhancement when light is scarce. As dawn approaches, the circadian clock anticipates the impending rise in illumination level as ceases efferent signal transmission, allowing the eye to revert back to its default daytime state of low luminance sensitivity and high spatial acuity.

10.5 Clock Influence on Visual Behavior

Each spring along the east coast of North America, horseshoe crabs migrate to the water's edge at high tide to locate mates and bury their eggs [76, 77]. Chemical signals and pheromones are important cues the animals use to identify areas to explore [78–80], and, once at a mating beach, vision plays a role in the detection of potential mates by male crabs and the selection of nesting site by female crabs.

Crabs whose lateral eyes were occluded either by human-applied acrylic paint [8] or by naturally occurring epibionts like barnacles [81] did not locate mates as well as sighted animals. From videotapes of crab movements in the shallows, researchers found that males will turn and investigate objects the size of an adult female viewed at distances of up to 1 m [82, 83]. Reducing the contrast or size of the object required males to pass even closer to see the target. This means that males use signals from at least ten photoreceptors to detect females during the day [84]. Horseshoe crabs can also see remarkably well at night. Although light levels drop over a million fold, the combination of light and circadian adaptation in the retina allows males to detect crab-size objects in moonlight almost as well as sunlight. The probability of a male turning at a given distance to hit the target did not significantly differ from day to night [82]. For example, both day and night, 63 % of males approached a black target once it subtended 0.1 sr. Because the acceptance angle of each photoreceptor nearly doubles each night, the animals may rely on compromising spatial resolution for visual sensitivity and increasing the number of photoreceptors mediating mate detection.

To further probe whether the circadian increase in visual sensitivity was relevant to the animal, Powers and Barlow [85] measured behavioral responses to light flashes day and night. When kept in constant darkness in the lab, *Limulus* lift their tail, reduce their respiration rate, and increase their heart rate in response to short flashes (~5 s) of light. During the subjective night, the probability of a tail movement following a dim light flash increased from 5 to 80 %, and the threshold light intensity needed to evoke a response decreased approximately tenfold. The authors concluded that the daily rhythms in retinal sensitivity regulate circadian rhythms in behavioral threshold.

Why have circadian rhythms in retinal function? Two basic functions have been identified: to anticipate the daily increase in light intensity so that the retina does not saturate and to anticipate the nightly decrease in light intensity so that retina remains sensitive enough to support spatial vision. Retinal, and consequently behavioral, sensitivity to light thus anticipates reliable, daily changes in light intensity and adjusts to transient changes in ambient light levels. These two processes come together in a striking fashion as horseshoe crabs approach the beach at night. As animals move from deep water into the shallows, incident levels of ultraviolet light (UV) levels increase dramatically (Fig. 10.8). This UV is detected by the median ocelli of Limulus. During the night, the ocelli can detect UV from moonlight at depths up to 10 m [86, 87]. Strikingly, the UV signal from the ocelli increases sensitivity of the lateral compound eyes [88]. Thus, the circadian increase in sensitivity of the lateral eyes is augmented further as the ocelli increase their firing when the animals approach the beach. UV through the ocelli thus improves the chances that a male will find a female by helping the male find the shallows and increasing the sensitivity of his image-forming eyes.

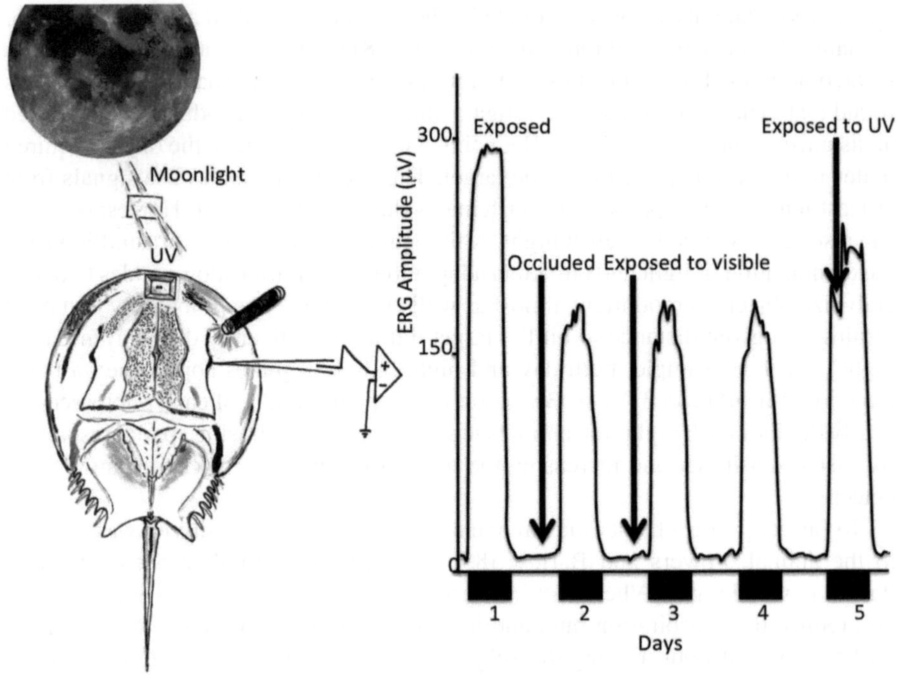

Fig. 10.8 Ambient ultraviolet (UV) light increases the amplitude of circadian changes in retinal sensitivity. When the median ocelli were blocked from UV illumination from the nighttime sky, the lateral eye ERG was approximately halved. When only UV light impacted the median ocelli, the lateral eye ERG increased. The filter placed on the median ocelli passed approximately 50 % of the available UV light so that the response of the lateral eyes increased by the predicted amount. Artist's illustration by Virginia Lee. Graph modified from [88]

10.6 Conclusions

Limulus stands as an exception in the history of retinal chronobiology. Its superbly well-described visual system can be modeled from photons to photoreceptor potentials to spike trains in the 1,000 optic nerve fibers. That is, we have a remarkably quantitative understanding of how the retina converts visual scenes to a neural code for the brain. These signals are used for simple behaviors including attraction to mate-sized objects and avoidance of larger, predator-like shadows. Along with other chelicerates (scorpions and spiders) and crustaceans, horseshoe crabs represent an exception to the rule that all retinas are intrinsically circadian. The retinas of these primitive organisms are driven to change by circadian inputs from the brain. The centralized control of visual structures indicates that retinal rhythmicity is behaviorally important, but does not require a pacemaker in the eye. Circadian changes that enhance nocturnal retinal sensitivity at the cost of spatial resolution include increasing the aperture of each photoreceptor cell and augmenting the levels of and stabilization of rhodopsin molecules. This remarkable coordination of anatomical and physiological adjustments has allowed the animal to find mates about equally well day and night over the past 250 Ma.

Acknowledgements Unpublished data on clock firing patterns were collected by Jiahui Liu and supported by NSF CAREER Award BES-0547457 (C.L.P.). Research in the Herzog lab is supported by grants NIMH 63104, NIGMS 96873 and NIGMS 104991.

References

1. Hartline HK, Graham CH. Nerve impulses from single receptors in the eye. J Cell Comp Physiol. 1932;1(2):277–95.
2. Hartline HK. The dark adaptation of the eye of Limulus, as manifested by its electric response to illumination. J Gen Physiol. 1930;13(3):379–86.
3. Fuortes MGF, Hodgkin AL. Changes in time scale + sensitivity in ommatidia of Limulus. J Physiol. 1964;172(2):239–63.
4. Fuortes MGF. Initiation of impulses in visual cells of Limulus. J Physiol. 1959;148(1):14–28.
5. Hartline HK. Inhibition of activity of visual receptors by illuminating nearby retinal areas in the Limulus eye. Fed Proc. 1949;8(1):69.
6. Hartline HK, Ratliff F. Spatial summation of inhibitory influences in the eye of Limulus, and the mutual interaction of receptor units. J Gen Physiol. 1958;41(5):1049–66.
7. Hartline HK, Ratliff F. Inhibitory interaction of receptor units in the eye of Limulus. J Gen Physiol. 1957;40(3):357–76.
8. Barlow RB, Ireland LC, Kass L. Vision has a role in Limulus mating-behavior. Nature. 1982;296(5852):65–6.
9. Granados-Fuentes D, Herzog ED. The clock shop: coupled circadian oscillators. Exp Neurol. 2013;243:21–7.
10. Piggins HD, Guilding C. The neural circadian system of mammals. Essays Biochem. 2011;49:1–17.
11. Abe M, Herzog ED, Yamazaki S, Straume M, Tei H, Sakaki Y, et al. Circadian rhythms in isolated brain regions. J Neurosci. 2002;22(1):350–6.
12. Tosini G, Menaker M. Circadian rhythms in cultured mammalian retina. Science. 1996;272(5260):419–21.
13. Inouye ST, Kawamura H. Persistence of circadian rhythmicity in a mammalian hypothalamic island containing the suprachiasmatic nucleus. Proc Natl Acad Sci U S A. 1979;76(11):5962–6.
14. Andrews RV, Folk GE. Circadian metabolic patterns in cultured hamster adrenal glands. Comp Biochem Physiol. 1964;11(4):393–409.
15. Tharp GD, Folk GE. Rhythmic changes in rate of mammalian heart and heart cells during prolonged isolation. Comp Biochem Physiol. 1965;14(2):255–73.
16. Yamazaki S, Numano R, Abe M, Hida A, Takahashi R, Ueda M, et al. Resetting central and peripheral circadian oscillators in transgenic rats. Science. 2000;288(5466):682–5.
17. Kornmann B, Schaad O, Bujard H, Takahashi JS, Schibler U. System-driven and oscillator-dependent circadian transcription in mice with a conditionally active liver clock. PLoS Biol. 2007;5(2):179–89.
18. Welsh DK, Takahashi JS, Kay SA. Suprachiasmatic nucleus: cell autonomy and network properties. Annu Rev Physiol. 2010;72:551–77.

19. Yoo SH, Yamazaki S, Lowrey PL, Shimomura K, Ko CH, Buhr ED, et al. PERIOD2: LUCIFERASE real-time reporting of circadian dynamics reveals persistent circadian oscillations in mouse peripheral tissues. Proc Natl Acad Sci U S A. 2004;101(15):5339–46.
20. Barlow RB, Bolanowski SJ, Brachman ML. Efferent optic-nerve fibers mediate circadian-rhythms in Limulus eye. Science. 1977;197(4298):86–9.
21. Barlow RB, Chamberlain SC, Levinson JZ. Limulus brain modulates the structure and function of the lateral eyes. Science. 1980;210(4473):1037–9.
22. Chamberlain SC, Wyse GA. An atlas of the brain of the horseshoe-crab Limulus-polyphemus. J Morphol. 1986;187(3):363–86.
23. Barlow RB. Circadian-rhythms in the Limulus visual-system. J Neurosci. 1983;3(4):856–70.
24. Kass L, Barlow RB. A circadian clock in the Limulus brain transmits synchronous efferent signals to all eyes. Vis Neurosci. 1992;9(5):493–504.
25. Calman BG, Battelle BA. Central origin of the efferent neurons projecting to the eyes of Limulus-polyphemus. Invest Ophthalmol Vis Sci. 1991;32(4):1151.
26. Lee HM, Wyse GA. Immunocytochemical localization of octopamine in the central-nervous-system of Limulus-polyphemus—a light and electron-microscopic study. J Comp Neurol. 1991;307(4):683–94.
27. Fahrenbach WH. Morphology of Limulus visual system. 5. Protocerebral neurosecretion and ocular innervation. Z Mikrosk Anat Forsch. 1973;144(2):153–66.
28. Fahrenbach WH. The morphology of the horseshoe-crab (Limulus-polyphemus) visual-system. 7. Innervation of photoreceptor neurons by neurosecretory efferents. Cell Tissue Res. 1981;216(3):655–9.
29. Fleissner G. Efferent neurosecretory fibers as pathways for circadian clock signals in the scorpion. Naturwissenschaften. 1983;70(7):366–8.
30. Battelle BA. Neurotransmitter candidates in the visual-system of Limulus-polyphemus—synthesis and distribution of octopamine. Vision Res. 1980;20(11):911–22.
31. Battelle BA, Evans JA, Chamberlain SC. Efferent fibers to Limulus eyes synthesize and release octopamine. Science. 1982;216(4551):1250–2.
32. Edwards SC, Battelle BA. Octopamine-stimulated and cyclic amp-stimulated phosphorylation of a protein in Limulus ventral and lateral eyes. J Neurosci. 1987;7(9):2811–20.
33. Evans JA, Chamberlain SC, Battelle BA. Autoradiographic localization of newly synthesized octopamine to retinal efferents in the Limulus visual-system. J Comp Neurol. 1983;219(4): 369–83.
34. Kass L, Barlow RB. Efferent neurotransmission of circadian-rhythms in Limulus lateral eye. 1. Octopamine-induced increases in retinal sensitivity. J Neurosci. 1984;4(4):908–17.
35. Chamberlain SC, Engbretson GA. Neuropeptide immunoreactivity in Limulus. 1. Substance P-like immunoreactivity in the lateral eye and protocerebrum. J Comp Neurol. 1982;208(3): 304–15.
36. Bolbecker AR, Lim-Kessler CC, Li J, Swan A, Lewis A, Fleets J, et al. Visual efference neuromodulates retinal timing: in vivo roles of octopamine, substance P, circadian phase, and efferent activation in Limulus. J Neurophysiol. 2009;102(2):1132–8.
37. Lewandowski TJ, Lehman HK, Chamberlain SC. Immunoreactivity in Limulus. 3. Morphological and biochemical-studies of fmrfamide-like immunoreactivity and colocalized substance-P-like immunoreactivity in the brain and lateral eye. J Comp Neurol. 1989;288(1): 136–53.
38. Mancillas JR, Brown MR. Neuropeptide modulation of photosensitivity. I. Presence, distribution, and characterization of a substance P-like peptide in the lateral eye of Limulus. J Neurosci. 1984; 4(3):832–46.
39. Mancillas JR, Selverston AI. Neuropeptide modulation of photosensitivity. II. Physiological and anatomical effects of substance P on the lateral eye of Limulus. J Neurosci. 1984;4(3): 847–59.
40. Hanna WJB, Horne JA, Renninger GH. Circadian photoreceptor organs in Limulus. 2. The telson. J Comp Physiol A. 1988;162(1):133–40.

41. Horne JA, Renninger GH. Circadian photoreceptor organs in Limulus. 1. Ventral, median, and lateral eyes. J Comp Physiol A. 1988;162(1):127–32.
42. Chamberlain SC, Barlow RB. Neuroanatomy of the visual afferents in the horseshoe-crab (Limulus-polyphemus). J Comp Neurol. 1980;192(2):387–400.
43. Zwicky KT. A light response in tail of urodacus a scorpion. Life Sci. 1968;7(6p2):257–62.
44. Liu JS, Passaglia CL. Using the horseshoe crab, Limulus polyphemus, in vision research. J Vis Exp. 2009 (29). pii: 1384.
45. Liu JS, Passaglia CL. Spike firing pattern of output neurons of the Limulus circadian clock. J Biol Rhythms. 2011;26(4):335–44.
46. Dowse HB, Ringo JM. Further evidence that the circadian clock in Drosophila is a population of coupled ultradian oscillators. J Biol Rhythms. 1987;2(1):65–76.
47. Snodderly DM. Processing of visual inputs by brain of Limulus. J Neurophysiol. 1971;34(4):588–611.
48. Kripke DF, Elliott JA, Youngstedt SD, Rex KM. Circadian phase response curves to light in older and young women and men. J Circadian Rhythms. 2007;5:4.
49. Ruger M, St Hilaire MA, Brainard GC, Khalsa SB, Kronauer RE, Czeisler CA, et al. Human phase response curve to a single 6.5 h pulse of short-wavelength light. J Physiol. 2013;591 (Pt 1):353–63.
50. Summer TL, Ferraro JS, McCormack CE. Phase-response and Aschoff illuminance curves for locomotor activity rhythm of the rat. Am J Physiol. 1984;246(3 Pt 2):R299–304.
51. Battelle BA. Circadian efferent input to Limulus eyes: anatomy, circuitry, and impact. Microsc Res Tech. 2002;58(4):345–55.
52. Battelle BA. What the clock tells the eye: lessons from an ancient arthropod. Integr Comp Biol. 2013;53:1–10.
53. Roeder T. Octopamine in invertebrates. Progr Neurobiol. 1999;59(5):533–61.
54. Kass L, Pelletier JL, Renninger GH, Barlow RB. Efferent neurotransmission of circadian-rhythms in Limulus lateral eye. J Comp Physiol A. 1988;164(1):95–105.
55. Kaupp UB, Malbon CC, Battelle BA, Brown JE. Octopamine stimulated rise of camp in Limulus ventral photoreceptors. Vision Res. 1982;22(12):1503–6.
56. Cardasis HL, Stevens Jr SM, McClung S, Kempler KE, Powell DH, Eyler JR, et al. The actin-binding interface of a myosin III is phosphorylated in vivo in response to signals from a circadian clock. Biochemistry. 2007;46(48):13907–19.
57. Edwards SC, Andrews AW, Renninger GH, Wiebe EM, Battelle BA. Efferent innervation to Limulus eyes in vivo phosphorylates a 122 Kd protein. Biol Bull. 1990;178(3):267–78.
58. Kempler K, Toth J, Yamashita R, Mapel G, Robinson K, Cardasis H, et al. Loop 2 of Limulus myosin III is phosphorylated by protein kinase A and autophosphorylation. Biochemistry. 2007;46(14):4280–93.
59. Dalal JS, Battelle BA. Circadian regulation of Limulus visual functions: a role for octopamine and cAMP. Curr Zool. 2010;56(5):518–36.
60. Elekes K. Ultrastructural aspects of peptidergic modulation in the peripheral nervous system of Helix pomatia. Microsc Res Tech. 2000;49(6):534–46.
61. Van Swigchem H. Endogenous bursting properties of light yellow neurosecretory-cells in the freshwater snail Lymnaea-stagnalis (L). J Exp Biol. 1979;80:55–67.
62. Whim MD, Church PJ, Lloyd PE. Functional roles of peptide cotransmitters at neuromuscular synapses in Aplysia. Mol Neurobiol. 1993;7(3–4):335–47.
63. Wong F, Knight BW. Adapting-bump model for eccentric cells of Limulus. J Gen Physiol. 1980;76(5):539–57.
64. Wong F, Knight BW, Dodge FA. Dispersion of latencies in photoreceptors of Limulus and the adapting-bump model. J Gen Physiol. 1980;76(5):517–37.
65. Passaglia C, Dodge F, Herzog E, Jackson S, Barlow R. Deciphering a neural code for vision. Proc Natl Acad Sci U S A. 1997;94(23):12649–54.
66. Passaglia CL, Dodge FA, Barlow RB. Cell-based model of the Limulus lateral eye. J Neurophysiol. 1998;80(4):1800–15.

67. Chamberlain SC, Barlow RB. Control of structural rhythms in the lateral eye of Limulus—interactions of natural lighting and circadian efferent activity. J Neurosci. 1987;7(7):2135–44.
68. Chamberlain SC, Barlow RB. Light and efferent activity control rhabdom turnover in Limulus photoreceptors. Science. 1979;206(4416):361–3.
69. Pieprzyk AR, Weiner WW, Chamberlain SC. Mechanisms controlling the sensitivity of the Limulus lateral eye in natural lighting. J Comp Physiol A Neuroethol Sens Neural Behav Physiol. 2003;189(8):643–53.
70. Barlow RB, Kaplan E, Renninger GH, Saito T. Circadian-rhythms in Limulus photoreceptors. 1. Intracellular studies. J Gen Physiol. 1987;89(3):353–78.
71. Kaplan E, Barlow RB. Circadian clock in Limulus brain increases response and decreases noise of retinal photoreceptors. Nature. 1980;286(5771):393–5.
72. Kass L, Renninger GH. Circadian change in function of Limulus ventral photoreceptors. Vis Neurosci. 1988;1(1):3–11.
73. Renninger GH, Schimmel R, Farrell CA. Octopamine modulates photoreceptor function in the Limulus lateral eye. Vis Neurosci. 1989;3(2):83–94.
74. Batra R, Barlow RB. Efferent control of temporal response properties of the Limulus lateral eye. J Gen Physiol. 1990;95(2):229–44.
75. Barlow RB, Birge RR, Kaplan E, Tallent JR. On the molecular-origin of photoreceptor noise. Nature. 1993;366(6450):64–6.
76. Barlow RB, Powers MK, Kass L, Fiordalice RW, Camara MD, Howard HA. Vision in Limulus mating-behavior during the day and at night. Biol Bull. 1984;167(2):522–3.
77. Rudloe AE, Herrnkind WF. Orientation by horseshoe crabs, Limulus-polyphemus, in a wave tank. Mar Behav Physiol. 1980;7(3):199–211.
78. Hassler C, Brockmann HJ. Evidence for use of chemical cues by male horseshoe crabs when locating nesting females (Limulus polyphemus). J Chem Ecol. 2001;27(11):2319–35.
79. Saunders KM, Brockmann HJ, Watson WH, Jury SH. Male horseshoe crabs Limulus polyphemus use multiple sensory cues to locate mates. Curr Zool. 2010;56(5):485–98.
80. Schwab RL, Brockmann HJ. The role of visual and chemical cues in the mating decisions of satellite male horseshoe crabs, Limulus polyphemus. Anim Behav. 2007;74:837–46.
81. Brockmann HJ, Penn D. Male mating tactics in the horseshoe-crab, Limulus-polyphemus. Anim Behav. 1992;44(4):653–65.
82. Herzog ED, Powers MK, Barlow RB. Limulus vision in the ocean day and night: effects of image size and contrast. Vis Neurosci. 1996;13(1):31–41.
83. Powers MK, Barlow Jr RB, Kass L. Visual performance of horseshoe crabs day and night. Vis Neurosci. 1991;7(3):179–89.
84. Herzog ED, Barlow RB. The Limulus-eye view of the world. Vis Neurosci. 1992;9(6):571–80.
85. Powers MK, Barlow RB. Behavioral-correlates of circadian-rhythms in the Limulus visual-system. Biol Bull. 1985;169(3):578–91.
86. Jerlov NG. Ultra-violet radiation in the sea. Nature. 1950;166(4211):111–2.
87. Stair R, Johnston R. The ultraviolet spectral radiant energy from the moon. J Opt Soc Am. 1953;43(4):328.
88. Herzog ED, Barlow RB. Ultraviolet-light from the nighttime sky enhances retinal sensitivity of Limulus. Biol Bull. 1991;181(2):321–2.

Chapter 11
Molluskan Ocular Pacemakers: Lessons Learned

Gene D. Block and Christopher S. Colwell

Abstract Invertebrate models have proved invaluable in understanding of fundamental properties of the nervous system including the ionic basis of the resting potential, impulse production, synaptic transmission, and the cellular basis of learning and memory. Invertebrate models have also provided important insights into the cellular mechanisms underlying circadian rhythms. Study of the retinae of several opisthobranch mollusks, which contain circadian clocks, has provided lasting insights into the cellular/molecular basis of biological timing. Key "lessons" from the molluskan retina include the cell-autonomous nature of neuronal clocks, the mechanisms underlying coupling within multi-oscillator ensembles, and the biochemical/ionic mechanisms involved in synchronization of the clock by environmental timing cycles. Much of what was learned from molluskan retinae remains highly relevant to study of mammalian circadian system.

Keywords *Aplysia* • *Bulla* • Circadian • Clock • Eye • Mollusks

11.1 Introduction

During the past 2 decades, we have witnessed a profound growth in knowledge about mammalian circadian rhythms. This has occurred at many levels. There is now an impressive amount of molecular detail about the various negative and positive feedback loops that give rise to circadian rhythms [1]. There are also relatively

---

G.D. Block, Ph.D. (✉) • C.S. Colwell, Ph.D.
Laboratory of Circadian and Sleep Medicine, Department of Psychiatry and Biobehavioral Sciences, David Geffen School of Medicine, University of California, Los Angeles, CA 90024, USA
e-mail: gblock@conet.ucla.edu; ccolwell@mednet.ucla.edu

G. Tosini et al. (eds.), *The Retina and Circadian Rhythms*, Springer Series in Vision Research 1, DOI 10.1007/978-1-4614-9613-7_11,

new insights into the role of neuronal networks in modulating and stabilizing the rhythms generated by individual oscillator neurons [2]. We have an increased appreciation of the role of circadian clock genes in regulating neural structures as well as nonneural organs and tissues located throughout the organism. Increasingly, we are confronted about issues regarding the regulation of a clock-shop that go beyond the understanding of individual clocks. Such discovery has increased our biological understanding significantly and, perhaps not surprisingly, has revealed a significantly more pervasive role for circadian timing within the bodily economy.

It is now difficult to appreciate that there was a time, not long ago, when knowledge about the electrical properties of mammalian clock cells was very limited, there was no information about the molecular mechanisms underlying rhythm generation in mammals, and there was only fragmentary knowledge about what processes might underlay synchronization of the SCN by light cycles. It was during this period, starting about a half century ago, when invertebrate models were proving invaluable to our understanding about the neurophysiological mechanisms controlling behavior. The most celebrated examples of invertebrate models were in the fields of motor control, where crustaceans afforded great opportunities for study [3], and in the field of "learning and memory," where the Kandel laboratory and others exploited the large neurons in the abdominal ganglion of *Aplysia*, to study the biophysical changes that occur during the sensitization and operant learning process [4]. Also, in this time, the first accessible neural models to study circadian rhythmicity were discovered and exploited. The most productive of these invertebrate models for neurophysiological investigation were the marine opisthobranchs, *Aplysia californica* and *Bulla gouldiana*. With the benefit of hindsight, it is satisfying now to observe that these organisms informed us, in important ways, about the mechanisms underlying mammalian circadian systems. First, and perhaps most importantly, is that rhythm generation was primarily the property of individual neurons but also that these central ocular pacemakers are modulated by other parts of the nervous system, that the canonical photoreceptors were not the only light receptors within the retina, that circadian rhythms in neuronal rhythmicity involve circadian regulation of potassium currents, and that synchronization by light involves pacemaker membrane depolarization while modulation by some other factors involves hyperpolarization. Finally, these molluskan models provided insights into how internal oscillator synchrony is maintained.

In the past 20 years, there have been few publications reporting research on circadian rhythms in molluskan eyes. Consequently, much of the literature in this area has been comprehensively reviewed and we refer you to these reviews for more complete coverage [5–7]. In the chapter that follows, we focus on valuable insights or "lessons" that we believe have been learned from the molluskan studies—how this work helped inform the questions that we now ask in mammals. Remarkably, much of what was learned from invertebrate models remains relevant today.

11.2 Lesson One: Central Clock Neural Rhythms Are "Diurnal" Whereas Circadian Behaviors Are Nocturnal or Diurnal

The retinae of a number of opisthobranch mollusks contain circadian pacemakers [8–11]. The rhythm can be easily demonstrated by removing the eye with optic nerve and recording, in vitro, with a suction electrode attached to the optic nerve stump. In darkness, the opisthobranch retina produces low-frequency compound action potentials as revealed by the variable amplitude of the impulses. When summed over time, a very clear circadian rhythm is present in the frequency of spontaneous impulses. Typically, for an animal synchronized by a 24 h light cycle consisting of 12 h of light and darkness, spontaneous impulse activity begins just before dawn and peaks shortly thereafter. Interestingly, although some opisthobranchs that have been studied appear nocturnal (e.g., Bulla, Haemonea) and some diurnal (Aplysia, Bursatella, Navanax), in all cases the ocular rhythms are diurnally expressed. Similarly, neurons in the central clock (suprachiasmatic nucleus, SCN) of both diurnal and nocturnal mammals exhibit circadian rhythms in firing action potentials which peak in the middle of the day [12, 13]. To date, in every case that the neural activity of a central clock in animals has been measured, the neurons were spontaneously active with peak activity during the day (Fig. 11.1).

11.3 Lesson Two: Circadian Entrainment Occurs via Nontraditional Photoreceptors

Soon after discovery of a circadian rhythm in the retina of opisthobranchs [6], efforts were directed towards determining where and how the rhythm is generated. Early studies seemed to suggest that the circadian rhythm was an emergent property of a population of interacting higher frequency oscillators [14]. With regard to cell types involved in rhythm generation, positive identification of the neurons involved in rhythm generation was complicated in Aplysia due to the fact that there are at least five morphologically distinct receptor-like cells surrounding the lens and another two neuronal-type cells at the base of the retina, all in rather close proximity [15]. Surgical reduction experiments suggested that the oscillator cells were located in the basal retina. Experiments with focused irradiation confirmed that only cells in the basal retinal region were critical for the expression of the circadian rhythm [16].

The localization and identification of circadian clock cells were more successful in the Bulla retina, primarily due to comparative simplicity and better spatial separation between cell layers in the retina. The Bulla eye consists of a large solid spherical lens partially surrounded by a layer of photoreceptors and cells in a basal region, relatively distant from the photoreceptor layer and forming a ring around a central neuropil (Fig. 11.2). There are approximately 1,000 photoreceptor-like cells that surround the lens [17, 18]. The neurons at the base of the retina have been referred

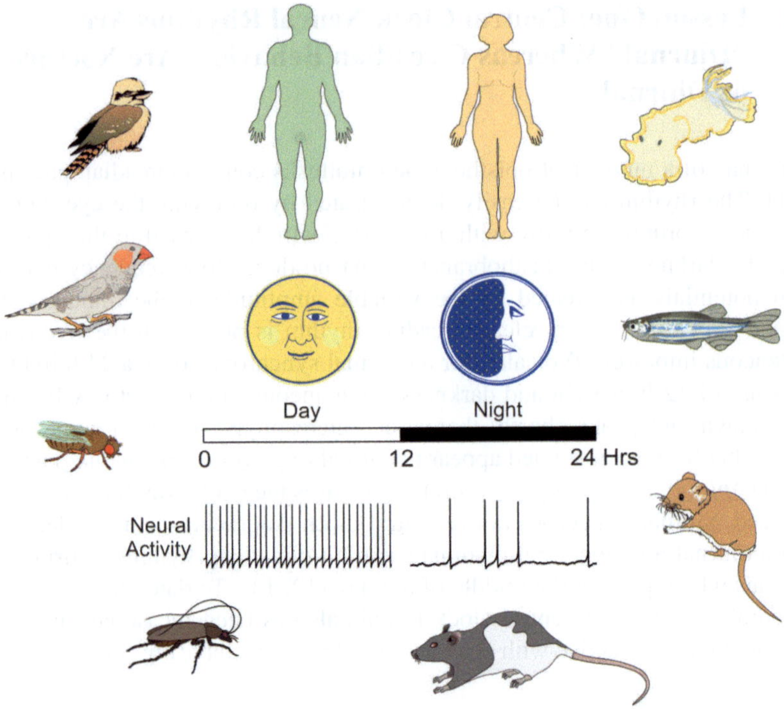

Fig. 11.1 Neural activity rhythms hallmark feature of circadian pacemaker neurons in a variety of species. Circadian rhythms in behavior are synchronized by central pacemaker neurons located in a small subset of cells in the central nervous system. In all of the animal species examined, these pacemaker neurons exhibit circadian rhythms in spontaneous neural activity. One of the conserved features of these neural activity rhythms is that the spontaneous activity is highest during the day whether the species is diurnal (*Aplysia*) or nocturnal (*Bulla*). The impact of light on the system is to increase neural activity during the night. The mechanism by which the molecular clockwork within single cells drives rhythms in neural activity is largely unknown

to as Basal Retinal Neurons (BRN) based on their location. There are approximately 130 BRNs surrounding a central neuropil region. The BRNs are electrically coupled to one another and fire in synchrony with the optic nerve recorded compound action potentials that display a circadian rhythm. They appear to be a relatively homogenous group (15–25 μM) in diameter with axons that extend out of the optic nerve [17].

The entire ocular circadian pacemaker system appears to reside in the BRNs. Intracellular recordings from the most common cell type, the "R-type," indicated that these cells do not generate impulses or exhibit a circadian rhythm in resting potential [10]. In addition, surgical removal of the receptor layer did not impact generation of the circadian rhythm, nor did it impact the ability of the circadian rhythm to be phase shifted by light pulses [10, 19]. Thus, circadian photoreception occurs outside the layer of classical photoreception, among what might be thought of as "ganglion cells" in vertebrate retinae. These cells exhibit sustained impulse activity in response to long light pulses, most likely a requirement for large stable

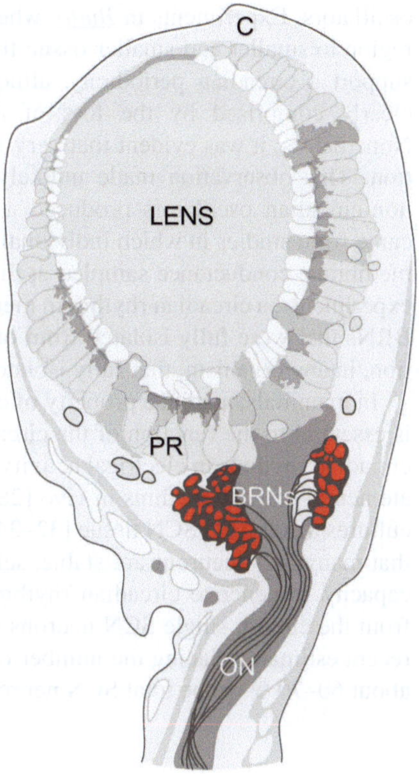

Fig. 11.2 The *Bulla* eye consists of a large solid spherical lens partially surrounded by a layer of photoreceptors and cells in a basal region, relatively distant from the photoreceptor layer and forming a ring around a central neuropil. There are approximately 1,000 photoreceptor-like cells that surround the lens. The neurons at the base of the retina have been referred to as Basal Retinal Neurons (BRN) based on their location (shown here in *red*). The BRNs are electrically coupled to one another and fire in synchrony with the optic nerve recorded compound action potentials that display a circadian rhythm. After [17]

phase shifts. This is in contrast to the light-induced responses of the receptor layer, which are transient in nature.

The mammalian circadian system also relies on nontraditional, specialized photoreceptors (see Chap. 2). In brief, the intrinsically photosensitive retinal ganglion cells (ipRGCs) express the photopigment melanopsin and receive information from rods and cones [20–22]. These ipRGCs thus encode ambient lighting [23] and generate action potentials that travel down the RHT and innervate the SCN. The RHT terminals release glutamate and, under certain conditions, the neuropeptide PACAP [24]. Thus in both the SCN and BRN neurons, the net result of photic stimulation is a slow and persistent increase in firing rate.

11.4 Lesson Three: Single Neurons Can Act as Circadian Clocks

An early study in *Aplysia* had suggested that the circadian rhythm might emerge from the interactions of higher frequency, non-circadian, oscillators. Consequently, a fundamental question was whether individual neurons are competent circadian oscillators. Experiments in Bulla, where it was possible to reduce the basal retinal region to smaller and smaller tissue fragments, revealed that even six BRNs could support a circadian periodicity, although the waveform was more variable and clearly comprised by the loss of many of the contributing oscillators [19]. Nonetheless, it was evident that very few cells were required for circadian generation. This observation made unlikely the possibility that a network of coupled non-circadian oscillators produced a circadian output. More definitive evidence came from studies in which individual BRNs were isolated in microwell dishes and membrane conductance sampled at different times in the circadian cycle. In these experiments, a circadian rhythm in membrane conductance could be measured from BRNs that were fully isolated from other cells [25]. This was the first demonstration, in any organism, that fully isolated neurons can express a circadian rhythm.

In mammals, a unique property of central clock neurons in the SCN and one that is essential to the function of the circadian timing system is the ability to generate circadian rhythms in electrical activity [26, 27]. The SCN has been shown to generate neural activity rhythms in vivo [28–30], in brain slice preparations [31], and in cultures made from SCN tissue [32–34]. These findings are consistent with the idea that many SCN neurons are stable, self-sustained oscillators that have the intrinsic capacity to generate circadian rhythms in electrical activity. Even when isolated from the circuit, single SCN neurons can exhibit a rhythm in firing rate [32], with recent estimates placing the number of SCN neurons that exhibit such a rhythm at about 60–70 % of the total SCN neuron population [33, 35, 36].

11.5 Lesson Four: Clock Neuron Rhythms Are Expressed via Modulation of Potassium Conductances

Another area of commonality is in the regulation of membrane conductance by the molecularly-based circadian clock. In opisthobranch retinas, the circadian rhythm is most easily measured in the frequency of compound action potentials recorded from the optic nerve stump. Long-duration intracellular recordings in *Bulla* from BRNs reveal a rhythm in membrane potential and impulse production that suggests that an underlying rhythm in membrane potential drives the impulse rhythm measured in the optic nerve. BRNs are depolarized during the day by approximately 13 mV when compared to nighttime resting potential levels [37]. The rhythm in membrane potential is associated with a rhythm in membrane conductance, which is elevated during the (subjective) day compared to the (subjective) night [38]. The difference in conductance between the daytime and nighttime is not observed in the presence of tetraethylammonium suggesting that the underlying rhythm is due to a decrease in a potassium (K^+) conductance during the daytime, resulting in membrane depolarization [25]. Similar to Bulla, Aplysia retinal neurons, maintained in cell culture, express a circadian rhythm in specific K^+ currents [39].

SCN neurons generate action potentials in the absence of synaptic drive and can therefore be considered endogenously active neurons [40]. SCN neurons demonstrate a diurnal rhythm in membrane potential, such that cells are depolarized by about 6–10 mV during the day compared to the night [12, 41, 42]. This day-night difference in resting membrane potential is largely mediated by a hyperpolarizing K^+-dependent conductance that is active at night at resting membrane potential and inactive during the day [12]. Resting membrane potentials of neurons are largely set by a class of two-pore-domain K^+ channels (K2P and TASK/TREK) [43, 44]. These K^+ channels are active over the whole voltage range (unlike other K^+ channels) and are referred to as providing "leak" or background currents. The K2P channels are coded for by the *Kcnk* gene family. The transcripts for *Kcnk1* and *Kcnk2* are expressed in the SCN [45], with transcripts for *Kcnk1* exhibiting a robust rhythm in the SCN [46]. Unfortunately, specific pharmacological blockers are not available, but the rhythmic pattern of expression in the SCN makes the K2P channels a promising candidate for driving the nightly hyperpolarization in membrane potential.

BK currents are also involved in the nightly silencing of SCN neurons. The expression of *Kcnma1*, the gene encoding the pore-forming subunit of the BK channel, peaks in the middle of the night [47] and the relative contribution of the BK current to the outward currents is larger in the night than in the day. In addition, deletion of *Kcnma1* increases nighttime firing in SCN neurons, although it does not completely abolish the day-night difference in firing rate [48, 49].

11.6 Lesson Five: Clock Synchronization by Light Occurs via Clock Neuron Membrane Depolarization and a Calcium Flux

To be useful chronometers, biological clocks need to be set to "local time." In the molluskan retina and in the mammalian SCN, the process appears to involve increases in cGMP (demonstrated only in *Aplysia*), membrane depolarization, and calcium (Ca^{2+}) influx. In *Aplysia*, pulses of 8-bromo-cGMP lead to increases in optic nerve impulse activity and stable phase shifts with a phase response curve (PRC) similar to light [50]. These phase shifts are not additive with light increases the levels of cGMP in the eye, suggesting that increases in cGMP in response to light are part of the entrainment cascade. In both *Aplysia* and *Bulla*, light depolarizes pacemaker neurons [51]. Depolarization alone can mimic light as a synchronizing agent. This was first demonstrated in *Aplysia* by providing pulses of bathing solutions with elevated potassium levels, which generated a PRC similar to that of light [52]. A more direct demonstration was possible in *Bulla* where direct current injection into a BRN, which depolarizes the population of BRNs via electrical coupling, yielded phase shifts similar to light pulses. In addition, injecting hyperpolarizing current into BRNs attenuated light-induced phase shifts [53].

Depolarization appears to exert phase control by inducing a Ca^{2+} flux. Lowering extracellular Ca^{2+} concentration into the nanomolar range by the addition of EGTA prevents light-induced phase shifts in both *Bulla* and *Aplysia* [53–55]. Supporting these observations is the fact that nickel, a blocker of voltage-gated Ca^{2+} channels, reduced light-induced phase shifts in both *Bulla* and *Aplysia* eyes [54, 55]. Measurements of intracellular Ca^{2+} concentrations in dispersed BRN neurons reveal a sustained increase in intracellular Ca^{2+} in response to depolarization, which is blocked by nickel [56].

During the night, SCN neurons are normally silent but do respond to photic stimulation transduced by ipRGCs that generate action potentials up to 20 Hz [29, 57–60]. This light-induced increase in neural activity drives synaptic communication with the rest of the cells in the circuit. The retinal recipient cells use GABA and peptide transmitters such as GRP and vasoactive intestinal peptide (VIP) to communicate with the rest of the circuit. The impact of GRP and VIP is to produce long-term increases (lasting hours) in excitability in SCN neurons [61, 62]. Interestingly, these studies indicate that the peptide-driven increases in firing rate are dependent upon CREB pathway and activation of *Per1*. These changes in electrical activity also trigger intracellular signaling cascades in SCN neurons.

The RHT stimulation during the night and resulting increase in the frequency of action potentials produces a robust increase in Ca^{2+} in SCN neurons. In dendrites, the Ca^{2+} influx is likely mediated by NMDA receptors, with a major contribution from the NR2B subunit [41, 63, 64], whereas in the soma, the L-type voltage-sensitive Ca^{2+} channels are the main contributors [60]. There is also evidence indicating that T-type Ca^{2+} currents are required for glutamate-induced phase shifts in the SCN neural activity rhythm in rats [65]. The concentration of intracellular Ca^{2+} in neurons is tightly controlled by various channels, pumps, and buffers. In addition, ryanodine receptors have a role in mediating the effects of light- and glutamate-induced phase delays of the circadian system and in regulation of the electrical activity of SCN neurons [66]. Thus, RHT-evoked Ca^{2+} influx is likely to be a major transducer of light information to the circadian system.

11.7 Lesson Six: Clock Neuron Rhythms Are Modulated by Efferent Signals

In both *Aplysia* and *Bulla*, the waveform of the circadian rhythm and phase is modified by the presence of efferent activity [67–70]. Efferent modulation can also affect the ability of the eye to reset to light pulses. Prichard and Lickey reported that the response of the eye clock to the onset of darkness was dependent upon activity of efferent neurons. The onset of darkness, under some conditions, produces only small resets of the eye clock when efferent activity is absent but produces large resets when efferent activity is present [69]. In *Aplysia*, serotonin (5-HT) appears to be one of the neurotransmitters involved in efferent regulation of the retinal clock by other regions of the nervous system. 5-HT is found in fibers within the optic nerve that terminate in and around the eye [71]. The role of 5-HT in efferent regulation appears to be complex. Neural enhancement of resetting appears to be mediated by 5-HT. Exogenous 5-HT mimics the effects of efferent activity on the neural enhancement of resetting [72]. In other experiments, 5-HT has been shown to modulate the impact of light on the ocular pacemaker [73].

By itself, 5-HT is a potent regulator of phase of the ocular pacemaker in *Aplysia*. When 5-HT is applied in the absence of light, these treatments cause stable phase shifts of the circadian rhythm; however, the PRC for 5-HT is shifted approximately 180° on the time axis with respect to light, with phase delays in the early subjective day and phase advances in the late subjective daytime [74, 75]. 5-HT exerts its actions by increasing cAMP levels [76, 77]. The action of cAMP can be mimicked by forskolin, an adenylate cyclase activity, or by 8-benzylthio-cAMP [76, 77]. In *Bulla*, the efferent transmitter acting similarly to 5-HT appears to be FMRFamide. FMRFamide is present in fibers entering the eye as well as in cell bodies in the cerebral ganglia [78, 79]. Similar to 5-HT in *Aplysia*, FMRFamide generates phase shifts with a similar shape to the 5-HT PRC [79–81]. The effects of FMRFamide can be mimicked by cAMP analogs [82]; the FMRFamide-activated signal transduction cascade has not been well studied in this species.

The mechanism by which 5-HT and FMRFamide influence the ocular clocks appears to be via their effect on membrane potential (Fig. 11.3). Both transmitters hyperpolarize and reduce impulse activity in clock neurons [78, 83]. In *Aplysia*, 5-HT-induced phase shifts can be blocked by application of high potassium in the bathing medium [84]. The K^+ channel blocker barium also blocks 5-HT-induced phase shifts [81]. In *Bulla*, the K^+ channel blocker tetraethylammonium blocks cAMP-mediated phase shifts [82]. Similar to the light-entrainment pathway, efferent regulation may involve modulation of a Ca^{2+} flux. In the case of light, Ca^{2+} influx increases in response to membrane depolarization. With efferent modulatory signals which hyperpolarize clock neurons, it seems likely that the reduction in impulse activity leads to a reduced Ca^{2+} flux. In *Aplysia* and *Bulla*, removing Ca^{2+} from the bathing medium leads to phase shifts that are similar to those induced by hyperpolarizing clock neurons [55, 85].

The mammalian SCN is modulated by a number of transmitters. While this is a much larger topic than we can cover here, it is worth briefly considering the role of 5-HT in the SCN. The SCN receives a dense serotonergic projection from the midbrain raphe nuclei that terminates predominantly in the retinorecipient region of the nucleus. By themselves, 5-HT receptor agonists directly cause phase shifts of the circadian system both in vitro [86, 87] and in vivo [88]. Evidence suggests a role for this pathway in mediating activity-induced phase shifts of the circadian system [88–90]. In addition, there is also evidence that this serotonergic projection can modulate light input to the SCN [90–95]. Thus, it appears that two major transmitter systems (glutamatergic and serotonergic) that convey information to the SCN can

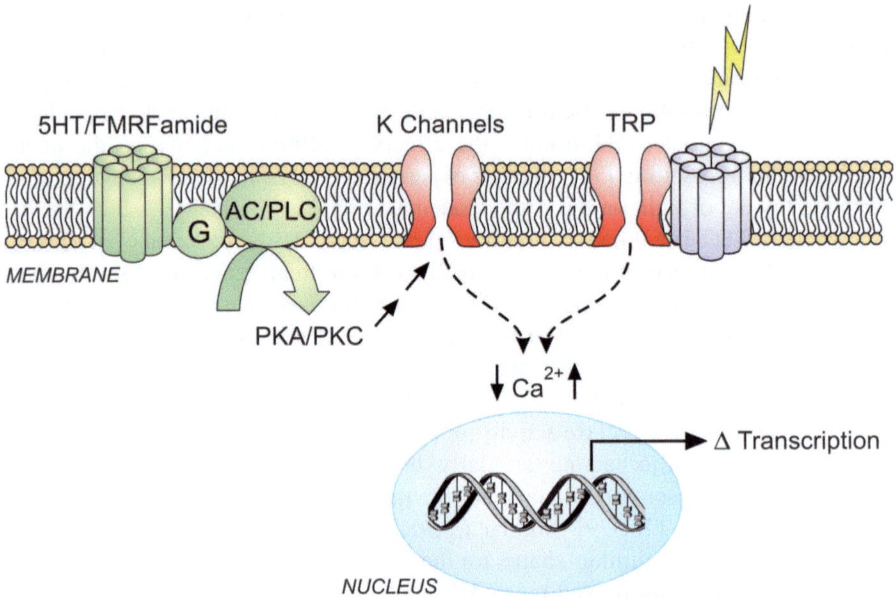

Fig. 11.3 The mechanisms that underlie the phase shifting impact of light and efferent transmitters have been explored in the molluskan system. This body of work suggests that exposure to light depolarizes the membrane. The assumption is that this depolarization is mediated by transient receptor potential channels (TRP channels) but this has not been tested. Light exposure does result in an increase in intracellular Ca^{2+}. This Ca^{2+} signal drives changes in transcription and translations at specific phase of the daily cycle that drive phase shifts in the molecular clock. The efferent transmitters 5-HT and FMRFamide work through G-protein-coupled signaling pathways including protein kinase A and C (PKA/PKC) to hyperpolarize the membrane by increasing K^+ currents. Depending on the phase, this membrane hyperpolarization can modulate the effects of light as well as directly cause phase shifts of the circadian system

interact with each other and we believe that these interactions may determine the ultimate response of the circadian system to environmental stimulation. Again, the role of 5-HT highlights another remarkable parallel between the molluskan and mammalian systems.

11.8 Lesson Seven: Bilateral Coupling of the Central Clocks Helps Maintain Internal Synchrony

The mollusks have two eyes containing circadian clocks and an important issue is how synchrony is maintained in this multi-oscillator, bilaterally distributed clock system. In *Bulla*, it has been possible to investigate clock to clock coupling both in vivo and in vitro. When *Bulla* are placed into darkness for up to 2 weeks, the two eye clocks remain closely in phase with one another. If communication between the two is surgically disrupted, the eye clocks then drift apart [70, 96], suggesting that the two eyes are internally coupled to one another. This hypothesis was experimentally tested, in vitro, by removing the eyes along with the central ganglia. The clock in one eye was delayed by either cooling or bathing in manganese seawater, which resulted in a delay of approximately 6 h. The two eyes were then allowed to interact in darkness for period up to 72 h. By 72 h, the phases of the two eyes had come to within 2 h of one another. In another study, experiments were conducted in vitro with the two eyes attached to the head ganglia [97]. One eye was exposed to bathing solutions that could either lengthen (treated with lithium) or shorten (treated with low chloride) the free-running period. The results confirmed the presence of internal coupling between the two eye clocks. When the free-running period of the eyes was set at about 1 h difference, the two eyes would assume an intermediate period with the eye with the shorter free-running period leading the one with the longer period. An unexpected result was that when the difference in free-running period exceeded the limits of entrainment, eyes would drift out of phase until a phase difference of approximately 10 h was obtained when a large phase shift occurred. The results suggest a novel resetting mechanism mediated by the interocular coupling pathways.

Electrophysiological investigation of the coupling process revealed that optic nerve impulses generated within one eye crossed through the central ganglia and emerged as efferent signals in the contralateral optic nerve [98]. There is anatomical evidence for fibers crossing from one eye to the other [99] and immunohistochemical evidence that glutamate is expressed in these fibers [100]. Recent electrophysiological work also suggests the synaptic input into the BRNs is glutamatergic [101]. The present data is consistent with a model in which FMRFamide represents a modulatory influence of the central ganglia whereas the glutamate pathway couples the two ocular clocks.

The SCN consists of two bilaterally paired nuclei, which contain populations of functionally distinct cells grouped in regions previously described as core and shell of the SCN [101–104]. The activity of these different cell types or regions has to be synchronized with each other to generate a uniform and high-amplitude timing signal. However, gene expression studies suggest that there are conditions in which the left and right SCN can function independently. For example, the left and right SCN of hamsters with desynchronized behavior exhibit antiphase rhythms in clock genes expression suggesting that each SCN can independently drive a behavioral component [105]. In addition, mice exhibiting splitting behavior maintain a coherent rhythm of *Per1* gene expression in neurons within one SCN, but left and right SCN oscillated in antiphase about 12 h apart [106]. There is some information about the pathways or signals that could mediate communication between the left and the right SCN nuclei. Anatomically, several studies have found evidence for neuronal projections between the two SCN [107, 108]. Furthermore, immunohistological studies revealed fiber tracks of neurons expressing VIP and arginine vasopressin

(AVP) crossing the midline between left and right SCN [102]. Recent work investigated the physiological connections between the left and right SCN and, surprisingly, found evidence that glutamate is the transmitter involved [109] in coupling the left/right SCN just as it does in Bulla.

11.9 Lesson Eight: Regulation of a Transmembrane Ca^{2+} Flux Represents a Common Pathway for Clock Phase Control

Numerous studies in Aplysia and Bulla have led to the development of a simple, admittedly incomplete model that provided a unifying explanation for how light, interocular coupling signals and other modulatory influences affect the phase, period, and perhaps amplitude of the ocular clock via a common pathway [5, 83]. The model (Fig. 11.4) is based in part on observation that membrane potential is both an input and output of an intracellularly located clock mechanism. Under constant conditions, during the daytime, the resting membrane potential is depolarized and clock neurons are spiking in comparison to the hyperpolarized and largely silent nighttime level. During the daytime, light leads to depolarization and some Ca^{2+} entry. Light has no effect on phase during this time, either due to the small amount of Ca^{2+} entering when the neurons are already chronically depolarized or because of a concentration at which there is no effect of additional Ca^{2+} entry. Light does have an effect during the nighttime when the membrane is relatively hyperpolarized and

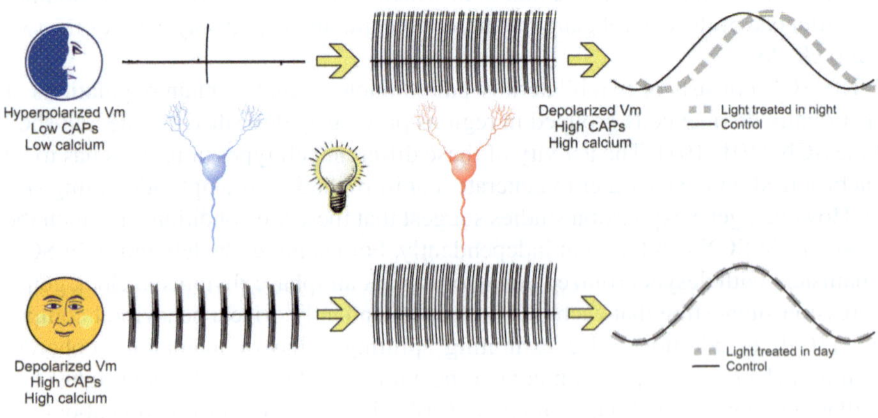

Fig. 11.4 The molluskan model of the daily gating of light input to circadian system. Central to the model is the hypothesis that a circadian rhythm in membrane potential (Vm) drives a rhythm in Ca^{2+} influx through voltage-sensitive Ca^{2+} channels and that light acts to cause a Ca^{2+} influx in these cells. During the day, the Vm of the BRNs is relatively depolarized and Ca^{2+} enters the cell through voltage-sensitive Ca^{2+} channels. Light does not cause phase shifts because intracellular Ca^{2+} is already elevated. During the night, the membrane is relatively hyperpolarized and light-induced depolarization can cause a Ca^{2+} influx, leading to a shift in the phase of the rhythm. Elements of this model are now being successfully applied to the mammalian SCN

silent, and thus, light leads to significant depolarization and Ca^{2+} entry. 5-HT (Aplysia) and FMRFamide (Bulla) are effective phase shifting agents during the daytime. These agents hyperpolarize the membrane during the daytime (see above) and reduce impulse activity and presumably Ca^{2+} entry. These agents would not be expected to have impact during the nighttime when the clock neuron membrane is already hyperpolarized.

Conceptually, elements of this model have begun to be applied to the SCN [60, 63]. The basic idea is that the circadian pacemaker neurons are already electrically active during the day and that additional excitatory input from the RHT does not induce many additional action potentials. In contrast, during the night, when the neurons are electrically silent, RHT stimulation results in a large change in electrical activity and the system responds to the change in action potential frequency with a phase shift.

11.10 Lesson Nine: The Circadian Cycle in Clock Neurons Requires Daily Transcription and Translation as Part of the Timing Loop

While the molluskan system is not suited to molecular genetic analysis, some important lessons about the molecular clockwork were learned from experiments in which transcription and translation were interrupted in very systematic ways. Pulses of the translation inhibitors puromycin, cycloheximide, or anisomycin generate phase delays when applied to Aplysia eyes during the subjective night [110–112]. Pulses of cycloheximide, the most quickly reversing of the inhibitors, revealed that the sensitive phase, when inhibiting translation shifted the clock, occurred during the late night and early morning [113]. Importantly, inhibition at other times in the cycle did not appear to impact the progression of the rhythm. In Bulla, it was possible to demonstrate that the circadian clock cannot complete a full cycle if protein synthesis is blocked [114]. Employing a "wedge experiment," Bulla eyes were exposed to cycloheximide or anisomycin for different amounts of time from 5 to 52 h. It was found that if the inhibitor was present through the time of subjective dawn, the phase of the restored rhythm after the drug pulse was washed out was dependent only upon the time at which the drug was removed. The results suggested that the clock is stopped at dawn by the inhibition of translation. If the drug is applied before dawn, the clock continues to run until dawn and is then stopped for the duration of the treatment. It is likely that in Aplysia, the shorter pulses likewise lead to a brief stopping of the clock near dawn. Thus, the experiments in both Aplysia and Bulla support the idea that there is a critical time near dawn when protein synthesis is required for the cycle to continue. Transcriptional inhibitors also appear to stop the clock but the sensitive period is longer than for translation inhibitors. The reversible transcriptional inhibitor 5,6-dichlorobenzimidazole riboside (DRB) delays the Aplysia rhythm when present during the 12 h period from subjective dawn to dusk [115, 116]. Experiments using a wide range of pulse lengths in

Bulla show that DRB stops the clock with a sensitive period that includes the end of the night and most of the subjective day [117]. The relatively long period of sensitivity of the rhythm to transcriptional inhibitors suggested that either transcription itself is part of the timing loop. Using an ingenious "two-pulse design" originally used by Pittendrigh on *Drosophila*, it was possible to show by probing with light pulses immediately following DRB pulses that phase shifts of the rhythm occur immediately, even though the DRB pulse is given at a time (ZT 6–12) when protein synthesis inhibitors are not effective. This result strongly suggests that the transcription process itself is part of the timing loop [117].

Again, the role of transcriptional/translational feedback loops in the mammalian circadian system is a large topic that is beyond the scope of this chapter. In mammals, the ability to manipulate genes has provided critical insights into the role of identified clock genes and their products in generating the 24 h rhythm. While there may be situations in which rhythmic gene expression is not required to express circadian rhythms [118], in most mammalian cell populations, rhythmic transcription and translation of key genes is a common feature [1, 119].

11.11 Lesson Ten: Age-Driven Changes Occur in the Central Clock

Some of the first studies examining how aging impacts the circadian system were conducted in the molluskan retina. The *Aplysia* eye, from aged animals, exhibits reduced amplitude rhythms, a slightly longer free-running period, and a change in the phase relationship to projected dawn [120]. Although these changes in aged isolated eyes could reflect a change in viability of older tissue maintained in vitro, this possibility was eliminated by also recording from the optic nerves of intact aged and young *Aplysia*. These in vitro experiments confirmed that the reductions in rhythm amplitude actually occurred in vivo, and provided the first evidence, in any model system, that aging leads to reduced amplitude in the central clock and is not simply the result of age-related changes in the pathways by which rhythmic behaviors are expressed.

Disruptions in the circadian system, including decreased amplitude of rhythmic behaviors and fragmentation of the activity/rest episodes, are commonly associated with aging in humans and other mammals [121]. While undoubtedly many factors contribute to these changes, a variety of data [122–126] is emerging that is consistent with the hypothesis that an age-related decline in the output of the central circadian clock in the SCN is a critical factor. For example, in recent work, in vivo multiunit recordings were carried out from the SCN and a brain region that receives robust innervation from the SCN (subparaventricular zone, SPZ) in freely moving animals. The amplitude of the day-night difference in neural activity was substantially reduced in both brain regions of middle-aged mice [127]. Another striking feature was the increase in variation in the levels of the spontaneous activity.

In contrast, the molecular clockwork in the SCN as measured by PERIOD2 levels was not disrupted in middle-aged mice. These results suggest that the age-related disruption in the circadian output occurs before any disruption of the molecular clockwork. The mechanisms underlying the age-related decline in SCN neural activity are unknown but are an important area for future research.

11.12 Concluding Remarks

In the ebb and flow of scientific research, models come in and go out of favor. It is possible that the 2 decades of work on the circadian system of marine opisthobranchs mollusks *Bulla gouldiana* and *Aplysia californica* will become a footnote in the expanding universe of circadian research. Of course, these models may also enjoy resurgence in the future. Regardless, we look backward with pride and satisfaction about a body of work that provided the first comprehensive analysis of the cellular basis of biological timing within higher animals. What is most gratifying, however, is the persisting relevance of experimental approaches, results, and models that emerged from this effort. Indeed, it was a good run for mollusks and we look forward to seeing what the future will bring!

References

1. Mohawk JA, Green CB, Takahashi JS. Central and peripheral circadian clocks in mammals. Annu Rev Neurosci. 2012;35:445–62.
2. Welsh D, Takahashi J, Kay S. Suprachiasmatic nucleus: cell autonomy and network properties. Annu Rev Physiol. 2010;72:551–5.
3. Selverston AI, Kennedy D. Structure and function of identified nerve cells in the crayfish. Endeavour. 1969;28(105):107–13.
4. Kandel ER, Abrams T, Bernier L, Carew TJ, Hawkins RD, Schwartz JH. Classical conditioning and sensitization share aspects of the same molecular cascade in Aplysia. Cold Spring Harb Symp Quant Biol. 1983;48(Pt 2):821–30.
5. Block GD, Khalsa SB, McMahon DG, Michel S, Guesz M. Biological clocks in the retina: cellular mechanisms of biological timekeeping. Int Rev Cytol. 1993;146:83–144.
6. Jacklet JW. Neurobiology of circadian generators. Trends Neurosci. 1995;8:69–73.
7. Blumenthal EM, Block GD, Arnold E. Cellular and molecular analysis of molluscan circadian pacemakers. In: Takahashi JS, editor. Handbook of behavioral neurobiology: circadian clocks. New York: Plenum; 2001. p. 371–400.
8. Jacklet JW. Circadian rhythm of optic nerve impulses recorded in darkness from isolated eye of Aplysia. Science. 1969;164(879):562–3.
9. Eskin A, Harcombe E. Eye of Navanax: optic nerve activity, circadian rhythm and morphology. Comp Biochem Physiol A-Comp Physiol. 1977;57A:443–9.
10. Block GD, Wallace SF. Localization of a circadian pacemaker in the eye of a mollusc, Bulla. Science. 1982;217:155–7.
11. Roberts MH, Block GD, Luska AE. Comparative studies of circadian pacemaker coupling in opisthobranch molluscs. Brain Res. 1987;423(1–2):286–92.

12. Kuhlman S, McMahon D. Encoding the ins and outs of circadian pacemaking. J Biol Rhythms. 2006;21:470–81.
13. Ko G, Shi L, Ko M. Circadian regulation of ion channels and their functions. J Neurochem. 2009;110:1150–69.
14. Jacklet JW, Geronimo J. Circadian rhythm: population of interacting neurons. Science. 1971;174(4006):299–302.
15. Herman KG, Strumwasser F. Regional specializations in the eye of Aplysia, a neuronal circadian oscillator. J Comp Neurol. 1984;230(4):593–613.
16. Woolum JC, Strumwasser F. The differential effects of ionizing radiation on the circadian oscillator and other functions in the eye of Aplysia. Proc Natl Acad Sci U S A. 1980;77(9): 5542–6.
17. Jacklet JW, Colquhoun W. Ultrastructure of photoreceptors and circadian pacemaker neurons in the eye of a gastropod, Bulla. J Neurocytol. 1983;12(4):673–96.
18. Block GD, McMahon DG. Cellular analysis of the Bulla ocular circadian pacemaker system III. Localization of the circadian pacemaker. J Comp Physiol A. 1984;155:387–95.
19. Block GD, McMahon DG. Cellular analysis of the Bulla ocular circadian pacemaker system I. A model for retinal organization. J Comp Physiol A. 1984;155:365–78.
20. Panda S. Multiple photopigments entrain the mammalian circadian oscillator. Neuron. 2007;53(5):619–21.
21. Ecker JL, Dumitrescu ON, Wong KY, Alam NM, Chen SK, LeGates T, Renna JM, Prusky GT, Berson DM, Hattar S. Melanopsin-expressing retinal ganglion-cell photoreceptors: cellular diversity and role in pattern vision. Neuron. 2010;67(1):49–60.
22. Lall GS, Revell VL, Momiji H, Al Enezi J, Altimus CM, Güler AD, Aguilar C, Cameron MA, Allender S, Hankins MW, Lucas RJ. Distinct contributions of rod, cone, and melanopsin photoreceptors to encoding irradiance. Neuron. 2010;66(3):417–28.
23. Berson DM. Strange vision: ganglion cells as circadian photoreceptors. Trends Neurosci. 2003;26(6):314–20.
24. Hannibal J, Hindersson P, Knudsen SM, Georg B, Fahrenkrug J. The photopigment melanopsin is exclusively present in pituitary adenylate cyclase-activating polypeptide-containing retinal ganglion cells of the retinohypothalamic tract. J Neurosci. 2002;22:RC191.
25. Michel S, Geusz ME, Zaritsky JJ, Block GD. Circadian rhythm in membrane conductance expressed in isolated neurons. Science. 1993;259(5092):239–41.
26. Schwartz W, Gross R, Morton M. The suprachiasmatic nuclei contain a tetrodotoxin-resistant circadian pacemaker. Proc Natl Acad Sci U S A. 1987;84:1694–8.
27. Albus H, Bonnefont X, Chaves I, Yasui A, Doczy J, van der Horst GT, Meijer JH. Cryptochrome-deficient mice lack circadian electrical activity in the suprachiasmatic nuclei. Curr Biol. 2002;12(13):1130–3.
28. Inouye ST, Kawamura H. Persistence of circadian rhythmicity in a mammalian hypothalamic "island" containing the suprachiasmatic nucleus. Proc Natl Acad Sci U S A. 1979;76: 5962–6.
29. Meijer J, Watanabe K, Schaap J, Albus H, Détári L. Light responsiveness of the suprachiasmatic nucleus: long-term multiunit and single-unit recordings in freely moving rats. J Neurosci. 1998;18:9078–87.
30. Nakamura W, Yamazaki S, Nakamura TJ, Shirakawa T, Block GD, Takumi T. In vivo monitoring of circadian timing in freely moving mice. Curr Biol. 2008;18(5):381–5.
31. Schaap J, Pennartz C, Meijer J. Electrophysiology of the circadian pacemaker in mammals. Chronobiol Int. 2003;20:171–88.
32. Welsh DK, Logothetis DE, Meister M, Reppert SM. Individual neurons dissociated from rat suprachiasmatic nucleus express independently phased circadian firing rhythms. Neuron. 1995;14(4):697–706.
33. Aton SJ, Colwell CS, Harmar AJ, Waschek J, Herzog ED. Vasoactive intestinal polypeptide mediates circadian rhythmicity and synchrony in mammalian clock neurons. Nat Neurosci. 2005;8(4):476–83.

34. Herzog ED, Geusz ME, Khalsa SB, Straume M, Block GD. Circadian rhythms in mouse suprachiasmatic nucleus explants on multimicroelectrode plates. Brain Res. 1997;757(2):285–90.
35. Webb AB, Angelo N, Huettner JE, Herzog ED. Intrinsic, nondeterministic circadian rhythm generation in identified mammalian neurons. Proc Natl Acad Sci U S A. 2009;106(38):16493–8.
36. Honma S, Shirakawa T, Katsuno Y, Namihira M, Honma K. Circadian periods of single suprachiasmatic neurons in rats. Neurosci Lett. 1998;250(3):157–60.
37. McMahon DG, Wallace SF, Block GD. Cellular analysis of the Bulla ocular circadian pacemaker system: II. Neurophysiological basis of circadian rhythmicity. J Comp Physiol. 1984;155:379–85.
38. Ralph MR, Block GD. Circadian and light-induced conductance changes in putative pacemaker cells of Bulla gouldiana. J Comp Physiol A. 1990;166(5):589–95.
39. Barnes S, Jacklet JW. Ionic currents of isolated retinal pacemaker neurons, projected daily phase differences and selective enhancement by a phase-shifting neurotransmitter. J Neurophysiol. 1997;77(6):3075–84.
40. Colwell CS. Linking neural activity and molecular oscillations in the SCN. Nat Rev Neurosci. 2011;12(10):553–69.
41. Pennartz CM, Bierlaagh MA, Geurtsen AM. Cellular mechanisms underlying spontaneous firing in rat suprachiasmatic nucleus: involvement of a slowly inactivating component of sodium current. J Neurophysiol. 1997;78(4):1811–25.
42. Belle MD, Diekman CO, Forger DB, Piggins HD. Daily electrical silencing in the mammalian circadian clock. Science. 2009;326(5950):281–4.
43. Mathie A. Neural two-pore-domain potassium channels and their regulation by G protein-coupled receptors. J Physiol. 2007;578:377–85.
44. Bayliss DA, Barrett PQ. Emerging roles for two-pore-domain potassium channels and their potential therapeutic impact. Trends Pharmacol Sci. 2008;29(11):566–75.
45. Lein E, et al. Genome-wide atlas of gene expression in the adult mouse brain. Nature. 2007;445:168–76.
46. Panda S, Antoch MP, Miller BH, Su AI, Schook AB, Straume M, Schultz PG, Kay SA, Takahashi JS, Hogenesch JB. Coordinated transcription of key pathways in the mouse by the circadian clock. Cell. 2002;109(3):307–20.
47. Pitts GR, Ohta H, McMahon DG. Daily rhythmicity of large-conductance Ca^{2+}-activated K+ currents in suprachiasmatic nucleus neurons. Brain Res. 2006;1071:54–62.
48. Meredith AL, Wiler SW, Miller BH, Takahashi JS, Fodor AA, Ruby NF, Aldrich RW. BK calcium-activated potassium channels regulate circadian behavioral rhythms and pacemaker output. Nat Neurosci. 2006;9:1041 9.
49. Kent J, Meredith AL. BK channels regulate spontaneous action potential rhythmicity in the suprachiasmatic nucleus. PLoS One. 2008;3(12):e3884.
50. Eskin A, Takahashi JS, Zatz M, Block GD. Cyclic guanosine 3′:5′-monophosphate mimics the effects of light on a circadian pacemaker in the eye of aplysia. J Neurosci. 1984;4(10):2466–71.
51. Block GD, McMahon DG. Localized illumination of the Aplysia and Bulla eye reveals new relationships between retinal layers. Brain Res. 1983;265(1):134–7.
52. Eskin A. Phase shifting a circadian rhythm in the eye of Aplysia by high potassium pulses. J Comp Physiol. 1972;80:353–76.
53. McMahon DG, Block GD. The Bulla ocular circadian pacemaker. I. Pacemaker neuron membrane potential controls phase through a calcium-dependent mechanism. J Comp Physiol A. 1987;161(3):335–46.
54. Khalsa SB, Block GD. Calcium channels mediate phase shifts of the Bulla circadian pacemaker. J Comp Physiol A. 1988;164(2):195–206.
55. Colwell CS, Whitmore D, Michel S, Block GD. Calcium plays a central role in phase shifting the ocular circadian pacemaker of Aplysia. J Comp Physiol A. 1994;175(4):415–23.

56. Geusz ME, Michel S, Block GD. Intracellular calcium responses of circadian pacemaker neurons measured with fura-2. Brain Res. 1994;638(1–2):109–16.
57. Berson D, Dunn F, Takao M. Phototransduction by retinal ganglion cells that set the circadian clock. Science. 2002;295(5557):1070–3.
58. Warren E, Allen C, Brown R, Robinson D. Intrinsic light responses of retinal ganglion cells projecting to the circadian system. Eur J Neurosci. 2003;17:1727–35.
59. Tu D, Zhang D, Demas J, Slutsky E, Provencio I, Holy T, Van Gelder R. Physiologic diversity and development of intrinsically photosensitive retinal ganglion cells. Neuron. 2005;48: 987–99.
60. Irwin R, Allen C. Calcium response to retinohypothalamic tract synaptic transmission in suprachiasmatic nucleus neurons. J Neurosci. 2007;27:11748–57.
61. Gamble KL, Allen GC, Zhou T, McMahon DG. Gastrin-releasing peptide mediates light-like resetting of the suprachiasmatic nucleus circadian pacemaker through cAMP response element-binding protein and Per1 activation. J Neurosci. 2007;27(44):12078–87.
62. Kudo T, Tahara Y, Gamble KL, McMahon DG, Block GD, Colwell CS. Vasoactive intestinal peptide produces long-lasting changes in neural activity in the suprachiasmatic nucleus. J Neurophysiol. 2013;110:1097–106.
63. Colwell CS. NMDA-evoked calcium transients and currents in the suprachiasmatic nucleus: gating by the circadian system. Eur J Neurosci. 2001;13:1420–8.
64. Wang LM, Schroeder A, Loh D, Smith D, Lin K, Han JH, Michel S, Hummer DL, Ehlen JC, Albers HE, Colwell CS. Role for the NR2B subunit of the N-methyl-D-aspartate receptor in mediating light input to the circadian system. Eur J Neurosci. 2008;27(7):1771–9.
65. Kim DY, Choi HJ, Kim JS, Kim YS, Jeong DU, Shin HC, Kim MJ, Han HC, Hong SK, Kim YI. Voltage-gated calcium channels play crucial roles in the glutamate-induced phase shifts of the rat suprachiasmatic circadian clock. Eur J Neurosci. 2005;21(5):1215–22.
66. Aguilar-Roblero R, Mercado C, Alamilla J, Laville A, Díaz-Muñoz M. Ryanodine receptor Ca^{2+}-release channels are an output pathway for the circadian clock in the rat suprachiasmatic nuclei. Eur J Neurosci. 2007;26:575–82.
67. Eskin A. Properties of the Aplysia visual system: in vitro entrainment of the circadian rhythm and centrifugal regulation of the eye. Z Vergl Physiol. 1971;74:353–71.
68. Block GD. In vivo recording of the ocular circadian rhythm in Aplysia. Brain Res. 1981; 222(1):138–43.
69. Prichard RG, Lickey ME. In vitro resetting of the circadian clock in the Aplysia eye. I. Importance of efferent activity in optic nerve. J Neurosci. 1981;1(8):835–9.
70. Roberts MH, Block GD. Analysis of mutual circadian pacemaker coupling between the two eyes of Bulla. J Biol Rhythms. 1985;1(1):55–75.
71. Takahashi JS, Nelson DE, Eskin A. Immunocytochemical localization of serotonergic fibers innervating the ocular circadian system of Aplysia. Neuroscience. 1989;28(1):139–47.
72. Nadakavukaren JJ, Lickey ME, Jordan WP. Regulation of the circadian clock in the Aplysia eye: mimicry of neural action by serotonin. J Neurosci. 1986;6(1):14–21.
73. Colwell CS. Light and serotonin interact in affecting the circadian system of Aplysia. J Comp Physiol A. 1990;167(6):841–5.
74. Corrent G, McAdoo DJ, Eskin A. Serotonin shifts the phase of the circadian rhythm from the Aplysia eye. Science. 1978;202(4371):977–9.
75. Corrent G, Eskin A, Kay I. Entrainment of the circadian rhythm from the eye of Aplysia: role of serotonin. Am J Physiol. 1982;242(3):R326–32.
76. Eskin A, Corrent G, Lin CY, McAdoo DJ. Mechanism for shifting the phase of a circadian rhythm by serotonin: involvement of cAMP. Proc Natl Acad Sci U S A. 1982;79(2):660–4.
77. Eskin A, Takahashi JS. Adenylate cyclase activation shifts the phase of a circadian pacemaker. Science. 1983;220(4592):82–4.
78. Jacklet JW, Klose M, Goldberg M. FMRF-amide-like immunoreactive efferent fibers and FMRF-amide suppression of pacemaker neurons in eyes of Bulla. J Neurobiol. 1987; 18(5):433–49.
79. Roberts MH, Moore RY. Localization of neuropeptides in efferent terminals of the eye in the marine snail, Bulla gouldiana. Cell Tissue Res. 1987;248(1):67–73.

80. Colwell CS, Khalsa SB, Block GD. FMRFamide modulates the action of phase shifting agents on the ocular circadian pacemakers of Aplysia and Bulla. J Comp Physiol A. 1992; 170(2):211–5.
81. Colwell CS, Michael S, Block GD. Evidence that potassium channels mediate the effects of serotonin on the ocular circadian pacemaker of Aplysia. J Comp Physiol A. 1992;171(5): 651–6.
82. Ralph MR, Khalsa SBS, Block GD. Cyclic nucleotides and circadian rhythm generation in Bulla gouldiana. Comp Biochem Physiol. 1992;101A:813–7.
83. McMahon DG, Block GD. The Bulla ocular circadian pacemaker. II. Chronic changes in membrane potential lengthen free running period. J Comp Physiol A. 1987;161(3):347–54.
84. Eskin A. Increasing external K+ blocks phase shifts in a circadian rhythm produced by serotonin or 8-benzylthio-cAMP. J Neurobiol. 1982;13(3):241–9.
85. Khalsa SB, Block GD. Calcium in phase control of the Bulla circadian pacemaker. Brain Res. 1990;506(1):40–5.
86. Medanic M, Gillette MU. Serotonin regulates the phase of the rat suprachiasmatic circadian pacemaker in vitro during the subjective day. J Physiol. 1992;450:629–42.
87. Prosser RA, Dean RR, Edgar DM, Heller HC, Miller JD. Serotonin and the mammalian circadian system: I. In vitro phase shifts by serotonergic agonists and antagonists. J Biol Rhythms. 1993;8:1–16.
88. Edgar DM, Miller JD, Prosser RA, Dean RR, Dement WC. Serotonin and the mammalian circadian system II: phase shifting rat behavioral rhythms with serotonergic agonists. J Biol Rhythms. 1993;8:17–31.
89. Marchant EG, Watson NV, Mistlberger RE. Both neuropeptide Y and serotonin are necessary for entrainment of circadian rhythms in mice by daily treadmill running schedules. J Neurosci. 1997;17(20):7974–87.
90. Morin LP, Blanchard J. Depletion of brain serotonin by 5-7-DHT modifies hamster circadian rhythm response to light. Brain Res. 1991;566:173–85.
91. Dudley TE, DiNardo LA, Glass JD. Endogenous regulation of serotonin release in the hamster suprachiasmatic nucleus. J Neurosci. 1998;18(13):5045–52.
92. Rea MA, Glass JD, Colwell CS. Serotonin modulates the photic response of the circadian oscillator in the hamster suprachiasmatic nucleus. J Neurosci. 1994;14:3635–42.
93. Meyer-Bernstein EL, Blanchard JH, Morin LP. The serotonergic projection from the median raphe nucleus to the suprachiasmatic nucleus modulates activity phase onset, but not other circadian rhythm parameters. Brain Res. 1997;755:112–20.
94. Ying SW, Rusak B. Effects of serotonergic agonists on firing rates of photically responsive cells in the hamster suprachiasmatic nucleus. Brain Res. 1994;651:37–46.
95. Flett J, Colwell CS. Serotonin modulation of calcium transients in cells in the suprachiasmatic nucleus. J Biol Rhythms. 1999;14(5):354–63.
96. Roberts MH, Block GD. Mutual coupling between the ocular circadian pacemakers of Bulla gouldiana. Science. 1983;221(4605):87–9.
97. Page TL, Nalovic KG. Properties of mutual coupling between the two circadian pacemakers in the eyes of the mollusc Bulla gouldiana. J Biol Rhythms. 1992;7(3):213–26.
98. Block GD, Roberts MH, Lusska AE. Cellular analysis of ocular circadian pacemaker coupling in Bulla: role of efferent impulses in phase shifting. J Biol Rhythms. 1986;1(3): 199–217.
99. Lacroix L, Strack S, Olson LM, Jacklet JW. Axons of circadian pacemaker neurons in the eye of Bulla project to the central nervous system and the contralateral eye. Comp Biochem Physiol A. 1991;98:383–91.
100. Michel S, Schoch K, Stevenson PA. Amine and amino acid transmitters in the eye of the mollusc Bulla gouldiana: an immunocytochemical study. J Comp Neurol. 2000;425(2): 244–56.
101. Ehnert C, Stevenson PA, Schildberger K, Michel S. Role of glutamate in coupling between bilaterally paired circadian clocks in Bulla gouldiana. Neuroscience. 2012;202:267–75.
102. Abrahamson EE, Moore RY. Suprachiasmatic nucleus in the mouse: retinal innervation, intrinsic organization and efferent projections. Brain Res. 2001;916(1–2):172–91.

103. Antle MC, Silver R. Orchestrating time: arrangements of the brain circadian clock. Trends Neurosci. 2005;28(3):145–51.
104. Golombek DA, Rosenstein RE. Physiology of circadian entrainment. Physiol Rev. 2010; 90(3):1063–102.
105. De la Iglesia HO, Meyer J, Carpino Jr A, Schwartz WJ. Antiphase oscillation of the left and right suprachiasmatic nuclei. Science. 2000;290:799–801.
106. Ohta H, Yamazaki S, McMahon DG. Constant light desynchronizes mammalian clock neurons. Nat Neurosci. 2005;8:267–9.
107. Pickard GE. The afferent connections of the suprachiasmatic nucleus of the golden hamster with emphasis on the retinohypothalamic projection. J Comp Neurol. 1982;211:65–83.
108. Leak RK, Card JP, Moore RY. Suprachiasmatic pacemaker organization analyzed by viral transsynaptic transport. Brain Res. 1999;819:23–32.
109. Michel S, Marek R, Vanderleest HT, Vansteensel MJ, Schwartz WJ, Colwell CS, Meijer JH. Mechanism of bilateral communication in the suprachiasmatic nucleus. Eur J Neurosci. 2013;37(6):964–71.
110. Rothman BS, Strumwasser F. Phase shifting the circadian rhythm of neuronal activity in the isolated Aplysia eye with puromycin and cycloheximide. Electrophysiological and biochemical studies. J Gen Physiol. 1976;68(4):359–84.
111. Jacklet JW. Neuronal circadian rhythm: phase shifting by a protein synthesis inhibitor. Science. 1977;198(4312):69–71.
112. Lotshaw DP, Jacklet JW. Serotonin induced protein phosphorylation in the Aplysia eye. Comp Biochem Physiol C. 1987;86(1):27–32.
113. Yeung SJ, Eskin A. Involvement of a specific protein in the regulation of a circadian rhythm in Aplysia eye. Proc Natl Acad Sci U S A. 1987;84(1):279–83.
114. Khalsa SB, Whitmore D, Block GD. Stopping the circadian pacemaker with inhibitors of protein synthesis. Proc Natl Acad Sci U S A. 1992;89(22):10862–6.
115. Raju U, Koumenis C, Nunez-Regueiro M, Eskin A. Alteration of the phase and period of a circadian oscillator by a reversible transcription inhibitor. Science. 1991;253(5020):673–5.
116. Koumenis C, Tran Q, Eskin A. The use of a reversible transcription inhibitor, DRB, to investigate the involvement of specific proteins in the ocular circadian system of Aplysia. J Biol Rhythms. 1996;11(1):45–56.
117. Khalsa SB, Whitmore D, Bogart B, Block GD. Evidence for a central role of transcription in the timing mechanism of a circadian clock. Am J Physiol. 1996;271(5 Pt 1):C1646–51.
118. O'Neill JS, Maywood ES, Hastings MH. Cellular mechanisms of circadian pacemaking: beyond transcriptional loops. Handb Exp Pharmacol. 2013;217:67–103.
119. Dibner C, Schibler U, Albrecht U. The mammalian circadian timing system: organization and coordination of central and peripheral clocks. Annu Rev Physiol. 2010;72:517–49.
120. Sloan MA, Levenson J, Tran Q, Kerbeshian M, Block GD, Eskin A. Aging affects the ocular circadian pacemaker of Aplysia californica. J Biol Rhythms. 1999;14(2):151–9.
121. Van Someren EJ. Circadian rhythms and sleep in human aging. Chronobiol Int. 2000; 17:233–43.
122. Satinoff E, Li H, Tcheng TK, Liu C, McArthur AJ, Medanic M, Gillette MU. Do the suprachiasmatic nuclei oscillate in old rats as they do in young ones? Am J Physiol. 1993;265: R1216–22.
123. Turek FW, Penev P, Zhang Y, van Reeth O, Zee P. Effects of age on the circadian system. Neurosci Biobehav Rev. 1995;19:53–8.
124. Watanabe A, Shibata S, Watanabe S. Circadian rhythm of spontaneous neuronal activity in the suprachiasmatic nucleus of old hamster in vitro. Brain Res. 1995;695:237–9.
125. Aujard F, Herzog ED, Block GD. Circadian rhythms in firing rate of individual suprachiasmatic nucleus neurons from adult and middle-aged mice. Neuroscience. 2001;106(2):255–61.
126. Biello SM. Circadian clock resetting in the mouse changes with age. Age. 2009; 31(4):293–303.
127. Nakamura TJ, Nakamura W, Yamazaki S, Kudo T, Cutler T, Colwell CS, Block GD. Age-related decline in circadian output. J Neurosci. 2011;31(28):10201–5.

Index

G. Tosini et al. (eds.), *The Retina and Circadian Rhythms*, Springer Series
in Vision Research 1, DOI 10.1007/978-1-4614-9613-7,

MIX
Papier aus verantwortungsvollen Quellen
Paper from responsible sources
FSC® C105338

If you have any concerns about our products,
you can contact us on
ProductSafety@springernature.com

In case Publisher is established outside the EU,
the EU authorized representative is:
Springer Nature Customer Service Center GmbH
Europaplatz 3, 69115 Heidelberg, Germany

Printed by Libri Plureos GmbH
in Hamburg, Germany